FISHERIES RESEARCH IN THE HUDSON RIVER

FISHERIES RESEARCH IN THE HUDSON RIVER

Hudson River Environmental Society
Edited by C. Lavett Smith

State University of New York Press

Published by
State University of New York Press, Albany

© 1988 State University of New York

All rights reserved

Printed in the United States of America

No part of this book may be used or reproduced
in any manner whatsoever without written permission
except in the case of brief quotations embodied in
critical articles and reviews.

For information, address State University of New York
Press, State University Plaza, Albany, N.Y., 12246

Library of Congress Cataloging-in-Publication Data

Fisheries research in the Hudson River.

   Proceedings of a conference sponsored by the Hudson
River Environmental Society, Sept., 1981.
   Bibliography: p.
   Includes index.
   1. Fisheries—Hudson River (N.Y. and N.J.)—Congresses.
2. Fishes—Hudson River (N.Y. and N.J.)—Congresses.
3. Fishes—Hudson River (N.Y. and N.J.)—Effect of water
pollution on—Congresses.   4. Water—Pollution—Environ-
mental aspects—Hudson River (N.Y. and N.J.)—Congresses.
I. Smith, C. Lavett, 1927–   . II. Hudson River
Environmental Society.
SH221.5.H83F57   1987   639.3′13′097473   86-14547
ISBN 0-88706-455-8
ISBN 0-88706-456-6 (pbk.)

10 9 8 7 6 5 4 3 2 1

Contents

List of Abbreviations ix

List of Tables x

List of Figures xv

Introduction *C. Lavett Smith* 1

Part I Data Sets

1 Fisheries Data Sets Compiled by Utility-Sponsored Research in the Hudson River Estuary
Ronald J. Klauda, Paul H. Muessig, and John A. Matousek 7

Part II Striped Bass and White Perch

2 Commercial Fishery for Striped Bass in the Hudson River, 1931–80
James B. McLaren, Ronald J. Klauda, Thomas B. Hoff, and Marcia Gardinier 89

3 Age-Specific Variation in Reproductive Effort in Female Hudson River Striped Bass
John R. Young and Thomas B. Hoff 124

4 Feeding Selectivity of Larval Striped Bass and White Perch in the Peekskill Region of the Hudson River
Douglas A. Hjorth 134

5 Patterns of Movement of Striped Bass and White Perch Larvae in the Hudson River Estuary
Thomas L. Englert and David Sugarman 148

Part III Sturgeons

6 Contribution to the Biology of Shortnose Sturgeon in the Hudson River Estuary
Thomas B. Hoff, Ronald J. Klauda, and John R. Young 171

Part IV River Herrings

7 Distributions and Movements of the Early Life Stages of Three Species of *Alosa* in the Hudson River, with

Comments on Mechanisms to Reduce Interspecific
Competition
*Robert E. Schmidt, Ronald J. Klauda, and John M.
Bartels* 193

Part V Tomcod

8 Life History of Atlantic Tomcod, *Microgadus tomcod,*
in the Hudson River Estuary, with Emphasis on Spatio-
Temporal Distribution and Movements
*Ronald J. Klauda, Richard E. Moos, and Robert E.
Schmidt* 219

Part VI Food Chains

9 Food Habits of the Amphipod *Gammarus tigrinus* in
the Hudson River and the Effects of Diet Upon Its
Growth and Reproduction
*Gerald V. Poje, Stacey A. Riordan, and Joseph M.
O'Connor* 255

Part VII Pollution

10 Heavy Metals in Finfish and Selected Macroinverte-
brates of the Lower Hudson River Estuary
*Stephen J. Koepp, Edward D. Santoro, and Gerard
DiNardo* 273

11 Recent Dissolved Oxygen Trends in the Hudson River
*Jeffrey A. Leslie, Karim A. Abood, Edward A. Maikish,
and Pamela J. Keeser* 287

12 PCB Patterns in Hudson River Fish: I. Resident Freshwater Species *R. W. Armstrong and R. J. Sloan* 304

13 PCB Patterns in Hudson River Fish: II. Migrant and
Marine Species *R. J. Sloan and R. W. Armstrong* 325

Part VIII Management

14 Management Recommendations for a Hudson River
Atlantic Sturgeon Fishery Based On an Age-Structured
Population Model
*John R. Young, Thomas B. Hoff, William P. Dey, and
James G. Hoff* 353

CONTENTS

Literature Cited   367

List of Contributors   397

Index   399

List of Abbreviations

Con Ed Consolidated Edison Company of New York

CHGE Central Hudson Gas and Electric Corporation

DOT New York State Department of Transportation

EA EA Science and Technology (Formerly Ecological Analysts, Inc.) A division of EA Engineering, Science, and Technology, Inc.

FERC Federal Energy Regulatory Commission

HRES Hudson River Environmental Society

IA Ichthyological Associates, Inc.

LMS Lawler, Matusky and Skelly, Engineers (Formerly QLM)

NMFS U.S. Department of Commerce National Marine Fisheries Service

NMPC Niagara Mohawk Power Company, Inc.

NYDEC New York State Department of Environmental Conservation

NYPA New York Power Authority

ORU Orange and Rockland Utilities, Inc.

PASNY Power Authority of the State of New York, presently New York Power Authority

QLM Quirk, Lawler and Matusky, Inc.

TI Texas Instruments, Inc.

All common and scientific names of fishes follow the recommendations of the American Fisheries Society Committee on Names of Fishes (1980)

List of Tables

Table 1	Sources of data displays, technical reports, computerized data tapes, and documentation packages for fisheries data sets, 1971–1980, p. 13.
Table 2	SAS hierarchical file structure for a typical Hudson River fisheries data set, p. 14.
Table 3	Summary of near-field distribution, abundance, and species composition data sets for Hudson River fishes collected by Texas Instruments Incorporated, p. 16.
Table 4	Summary of far-field distribution, abundance, and species composition data sets for Hudson River fishes collected by Texas Instruments Incorporated, p. 18.
Table 5	Summary of studies on selected Hudson River fishes conducted by Texas Instruments Incorporated that yielded data on distribution, abundance, and movements, p. 23.
Table 6	Summary of fish distribution, abundance, and species composition data sets for Hudson River generating plants collected by Ecological Analysts Incorporated, p. 26.
Table 7	Summary of fish distribution, abundance, and species composition data sets for Hudson River generating plants collected by Lawler, Matusky, and Skelly Engineers, p. 28.
Table 8	Summary of biological characteristics data sets for selected Hudson River fishes collected by Texas Instruments Incorporated, p. 34.
Table 9	Summary of biological characteristics data sets for selected Hudson River fishes collected by Lawler, Matusky, and Skelly Engineers, p. 38.
Table 10	Summary of data sets related to physiology and behavior of Hudson River fishes collected by Texas Instruments Incorporated, p. 40.
Table 11	Summary of data sets related to physiology and behavior of Hudson River fishes collected by Ecological Analysts Incorporated, p. 42.
Table 12	Summary of fish entrainment abundance data sets for Hudson River generating plants collected by Ecological Analysts Incorporated, p. 44.
Table 13	Summary of fish entrainment survival data sets for Hudson River generating plants collected by Ecological Analysts Incorporated, p. 48.
Table 14	Summary of fish entrainment abundance data sets for Hudson River generating plants collected by Lawler, Matusky and Skelly Engineers, p. 52.
Table 15	Summary of fish impingement data sets collected by Texas

LIST OF TABLES

Instruments Incorporated at the Indian Point generating plant, Hudson River, p. 58.
Table 16 Summary of fish impingement survival data sets for Hudson River generating plants collected by Ecological Analysts Incorporated, p. 62.
Table 17 Summary of fish impingement data sets for Hudson River generating plants by Lawler, Matusky and Skelly Engineers, p. 65.
Table 18 Summary of fisheries gear performance data sets collected by Texas Instruments Incorporated in the Hudson River, p. 70.
Table 19 Summary of gear performance and special studies related to fish entrainment at Hudson River generating plants conducted by Ecological Analysts Incorporated, p. 74.
Table 20 Summary of fisheries gear performance data sets collected by Lawler, Matusky and Skelly Engineers in the Hudson River, p. 78.
Table 21 Summary of special studies related to fish distribution and impingement mitigation for Hudson River generating plants conducted by Lawler, Matusky and Skelly Engineers, p. 82.
Table 22 Summary of striped bass culture data sets collected by Texas Instruments Incorporated in the Hudson River, p. 84.
Table 23 Reported landings (kg) of striped bass in Hudson River Commercial Fishery, New York waters, p. 93.
Table 24 Reported landings (kg) of striped bass by county in Hudson River commercial fishery, New York waters, p. 95.
Table 25 Reported dockside value of striped bass catch to commercial fishermen, nominal and adjusted for inflation, New York waters of Hudson River, p. 97.
Table 26 Effort statistics for Hudson River commercial fishery taking striped bass in New York waters (except where noted), p. 100.
Table 27 Total reported landings of striped bass in stake and anchor gill nets and operating stake and gill net effort adjusted to reflect only reported landings and effort data from gill net fishermen interviewed by NMFS personnel during Hudson River statistical survey, p. 103.
Table 28 Indices of relative abundance for striped bass spawning population in Hudson River based upon commercial fishery records obtained during NMFS interviews with New York commercial fishermen, p. 104.

LIST OF TABLES

Table 29 Gill nets used by Hudson River commercial fishermen sampled 1976–1979, p. 105.

Table 30 Spatial-temporal distribution of striped bass catch by five commercial fishermen in the Hudson River 1976–1979, p. 106.

Table 31 Number of Hudson River striped bass > 250 mm total length collected, released, and recaptured annually, 1976–1979, p. 117.

Table 32 Percentage return of tags from Hudson River striped bass fisheries during 1976–1979, p. 118.

Table 33 Return rates for $5 and $10 tags for Hudson River striped bass during 1978 and 1979, p. 119.

Table 34 Results of analysis of variance of egg index and gonad index for female striped bass, ages five to nine, during the 1978 spawning season, p. 132.

Table 35 Prey utilization of striped bass and white perch larvae in the Hudson River near Peekskill, New York, from 30 May to 27 June, 1978, p. 139.

Table 36 Values of the River Mile Index (RMI) for striped bass and white perch larval stages during 1974–1980, p. 160.

Table 37 Location of saltfront in the Hudson River during May, June, and July of 1974–1980 computed from equation (4), p. 161.

Table 38 Number of yearling and older shortnose sturgeon collected in various gear used in the Hudson River ecological study from 1969 through 1980, p. 177.

Table 39 Meristic and morphometric measurements from seven shortnose sturgeon incidentally caught by Texas Instruments in gill nets during mid to late June 1978 (fish died in the nets), p. 182.

Table 40 Comparative morphometric and meristic data for adult *Acipenser brevirostrum*, p. 185.

Table 41 Number of unidentified sturgeon (*Acipenser* spp.) larvae and juveniles collected in the Hudson River ecological study, 1973–1979, p. 186.

Table 42 Annual ratio of yearling and older Atlantic sturgeon to shortnose sturgeon collected during the Hudson River ecological study, 1972 through 1979, p. 187.

Table 43 Criteria used in visual classification of Atlantic tomcod gonads for state of maturity, p. 224.

Table 44 Total number of adult Atlantic tomcod (males and females) collected by box traps from the Hudson River estuary

LIST OF TABLES

	during spawning period and number of females examined for state of maturity, p. 230.
Table 45	Number of tagged adult Atlantic tomcod recaptured more than 1.6 river kilometers from point of release and within four months of release, p. 236.
Table 46	Regional density (number per 1000 m^3) of Atlantic tomcod yolk-sac larvae, Hudson River estuary, p. 241.
Table 47	Data used to evaluate the hypothesis that during years of relatively high early spring freshwater flows Atlantic tomcod postlarvae in the Hudson River are displaced farther downstream than in years of relatively low flows, p. 246.
Table 48	Chemical analysis of Sterling Lake water, p. 258.
Table 49	Mean lengths and weights, their standard deviations, for *Gammarus tigrinus* at sampling dates after release from females, p. 261.
Table 50	Average increases in length, standard deviation, paired t-statistics, and significance of groups of animals before and after placed on diets, p. 261.
Table 51	Mean lengths, standard deviation, one way analysis of variance, F-statistics, and probabilities that animals show the same response to different diets, p. 263.
Table 52	Mean differences, t-statistics for comparisons and probabilities that diet change affected parameters of reproduction, p. 265.
Table 53	Means, standard deviations, one-way analysis of variance tables, F-statistics, and probabilities that animals show the same response feeding on three diets, p. 266.
Table 54	Power curve regression of female length versus young produced, coefficients of correlation, approximate size at onset of maturity, and young production at 10 mm length for several species of *Gammarus*, p. 268.
Table 55	Trace metal residues (ppm wet weight) in finfish and crustaceans, collected in the Hudson River at Alpine, New Jersey (River Mile 19), p. 277.
Table 56	Trace metal residues (ppm wet weight) in finfish and crustaceans collected in the Hudson River at the George Washington Bridge (River Mile 12), p. 278.
Table 57	Trace metal residues (ppm wet weight) in finfish, crustaceans, and bivalves collected in upper New York Bay at Caven Cove (River Mile − 2), p. 279.
Table 58	Trace metal residues (ppm wet weight) in finfish, crustaceans, and bivalves collected in upper New York Bay

LIST OF TABLES

adjacent to the Military Operations Terminal (River Mile −3), p. 280.
Table 59 Freshwater flows at Green Island, 1971–1979, p. 297.
Table 60 Timing of events related to the Hudson River PCB problem, p. 305.
Table 61 Average correlation coefficients relating total PCB concentrations to length and lipid content of resident fish collected at several Hudson River locations, p. 310.
Table 62 Selected parameters related to PCB concentrations in standard fillets of mature Hudson River fish collected 1977–1980, p. 312.
Table 63 Calculated half-lives and annual rates of change of PCB for collections of resident Hudson River fish, p. 317.
Table 64 Selected parameters related to PCB concentrations in yearling pumpkinseed collected at various Hudson River locations in 1979 and 1980, p. 320.
Table 65 Average levels of PCB in Hudson River resident fish (wet basis), p. 323.
Table 66 Hudson River PCB data summary for migrant and marine fish, p. 327.
Table 67 Comparisons of PCB concentrations between organs in a male shortnose sturgeon and seventeen blue crabs (collected in 1980 and 1979, respectively), p. 331.
Table 68 Average PCB concentrations in American shad roe from Hudson River collections, p. 332.
Table 69 PCB-lipid "saturation" as a function of migrant or resident characteristics related to body size, p. 333.
Table 70 Aroclor 1016/1254 ratios, half-lives, and average annual rates of decline for migrant and marine Hudson River fish, p. 335.
Table 71 Annual percent changes of PCB (total, Aroclors 1016 and 1254) in Hudson River migrant and marine fish, p. 340.
Table 72 Spatial gradient in PCB concentrations indicated by upstream versus downstream collections of migrant and marine species in the Hudson River, p. 346.
Table 73 Approximate average total PCB concentrations in Hudson River migrant and marine fish (wet basis) encountered below Troy, p. 347.
Table 74 Life-history data used in population model, p. 356.
Table 75 Model parameters used in determining first-year survival for population simulations, p. 359.

List of Figures

Figure 1 Sampling area for Hudson River Ecological Study showing sites of focal electrical generating plants, p. 11.

Figure 2 Hudson River from Troy Dam to the Atlantic Ocean, showing location of commercial gill nets fishing effort and fishing locations for commercial fishermen A through E, p. 91.

Figure 3 Reported landings (kg) of striped bass in Hudson River commercial fishery in New York waters, expressed as percent deviation from mean catch, 1931–1980., p. 96.

Figure 4 Reported area (m^2) of stake and anchor gill nets operated in Hudson River commercial fishery (New York waters), expressed as percent deviation from mean area, 1931–1978, p. 99.

Figure 5 Size distribution of striped bass caught by four commercial fishermen in the Hudson River estuary during 1979, p. 110.

Figure 6 Size distribution of striped bass caught by commercial fishermen in the Hudson River estuary, 1976 through 1979, p. 111.

Figure 7 Age composition of striped bass commercial catch sampled in Hudson River estuary, 1976–1979, p. 112.

Figure 8 Proportion of females in the striped bass catch of Hudson River commercial fishery, 1976–1979, p. 113.

Figure 9 Age composition of commercial catch (shaded bars) and spring population (dotted line) of striped bass in Hudson River estuary, 1976–1978, p. 115.

Figure 10 Critical ages of male and female striped bass in Hudson River commercial fishery, p. 121.

Figure 11 Seasonal variation in gonad index of female Hudson River striped bass during the 1978 spawning season, p. 127.

Figure 12 Variations in gonad index for female Hudson River striped bass (a) and temporal distribution of spawning (b) during the 1978 spawning season, p. 128.

Figure 13 Variation in gonad index with age for female Hudson River striped bass during the 1978 spawning season, p. 129.

Figure 14 Variation in egg index with age for female Hudson River striped bass during the 1978 spawning season, p. 130.

Figure 15 Relationship between mean egg diameter and length for female Hudson River striped bass during the 1978 spawning season, p. 131.

LIST OF FIGURES

Figure 16 Location of sampling stations, p. 136.
Figure 17 Feeding selectivity of striped bass and white perch larvae for selected prey taxa (pooled daily data with range), p. 141.
Figure 18 Feeding selectivity of white perch larvae for all rotifers, *Keratella* spp. and *Notholca acuminata* in the Hudson River near Peekskill, New York, 30–31 May 1978 (pooled data with range), p. 142.
Figure 19 Feeding selectivity of striped bass larvae based on prey size in the Hudson River near Peekskill, New York, 1978, p. 143.
Figure 20 Feeding selectivity of white perch larvae based on prey size in the Hudson River near Peekskill, New York, 1978, p. 144.
Figure 21 Comparison of Real-Time Life Cycle Model predictions with field measurements of the spatial distribution of striped bass yolk-sac larvae, 1–7 June 1975, p. 151.
Figure 22 Comparison of Real-Time Life Cycle Model predictions with field measurements (bimodal egg distribution) of the spatial distribution of striped bass yolk-sac larvae, 16–22 May 1976, p. 152.
Figure 23 Comparison of Real-Time Life Cycle Model predictions with field measurements of the spatial distribution of striped bass yolk-sac larvae, 29 May–4 June 1977, p. 153.
Figure 24 Comparison of Real-Time Life Cycle Model predictions with field measurements of the spatial distribution of striped bass yolk-sac larvae, 4–10 June 1978, p. 154.
Figure 25 Comparison of Real-Time Life Cycle Model predictions with field measurements of the spatial distribution of striped bass post yolk-sac larvae, 1–7 June 1975, p. 155.
Figure 26 Comparison of Real-Time Life Cycle Model predictions with field measurements (bimodal egg distribution) of the spatial distribution of striped bass post yolk-sac larvae, 13–19 June 1976, p. 156.
Figure 27 Comparison of Real-Time Life Cycle Model predictions with field measurements of the spatial distribution of striped bass post yolk-sac larvae, 5–11 June 1977, p. 157.
Figure 28 Comparison of Real-Time Life Cycle Model predictions with field measurements of the spatial distribution of striped bass post yolk-sac larvae, 4–10 June 1978, p. 158.

LIST OF FIGURES

Figure 29 Influence of salt front location on River Mile Index for striped bass yolk-sac larvae, p. 162.

Figure 30 Influence of salt front location on River Mile Index for striped bass post yolk-sac larvae, p. 163.

Figure 31 Influence of salt front location on River Mile Index for white perch Yolk-sac larvae, p. 164.

Figure 32 Influence of salt front location on River Mile Index for white perch post yolk-sac larvae, p. 165.

Figure 33 Difference between yolk-sac and post yolk-sac stage locations for striped bass and white perch, p. 167.

Figure 34 Major morphometric characteristics of the Hudson River estuary, p. 174.

Figure 35 Monthly totals of yearling and older shortnose sturgeon collected in the Hudson River Ecological Study, 1969–1980, p. 178.

Figure 36 Regional distribution of yearling and older shortnose sturgeon collected in the Hudson River Ecological Study, 1969–1980, p. 179.

Figure 37 Length-frequency distribution of shortnose sturgeon collected in the Hudson River Ecological Study, 1969–1980, p. 180.

Figure 38 Weight-frequency distribution of shortnose sturgeon collected in the Hudson River Ecological Study 1969–1980, p. 181.

Figure 39 Study area in the Hudson River estuary, showing three sampling zones and the political boundaries of the contiguous states, p. 195.

Figure 40 Spatial and temporal distribution of American shad eggs and larvae in the mainstream Hudson River estuary, 1976–1979, p. 198.

Figure 41 Spatial and temporal distribution of river herring eggs in the mainstream Hudson River estuary, 1976–1979, p. 199.

Figure 42 Mean catch per haul of juvenile American shad in seines from the three zones of the mainstream Hudson River estuary, 1976–1979, p. 201.

Figure 43 Differences in modes of 5 mm length classes between the juvenile American shad collected by all surveys in the Hudson River estuary, 1979, p. 203.

Figure 44 Mean catch per haul of juvenile alewives in seines, 1976–1979, p. 204.

LIST OF FIGURES

Figure 45 Differences in modes of 5 mm length classes between zones in the Hudson River estuary, 1979, p. 206.

Figure 46 Length-frequency distribution for juvenile alewife collected in all surveys in the Hudson River estuary, 1979, p. 207.

Figure 47 Mean catch per haul of juvenile blueback herring in seines in the mainstream Hudson River estuary, 1976–1979, p. 208.

Figure 48 Differences in modes of 5 mm length classes of blueback herring between zones in the Hudson River, 1979, p. 209.

Figure 49 Patterns of freshwater flow in the Hudson River estuary, 1976–1979, p. 212.

Figure 50 Location of sampling regions (with River Km boundaries) in the Hudson River estuary, p. 220.

Figure 51 Diagrammatic cross-section of the Hudson River estuary showing strata sampled by Tucker trawl and epibenthic sled, p. 221.

Figure 52 Typical distribution of box trap sampling sites used to collect adult Atlantic tomcod during the winter spawning period in the Hudson River estuary, p. 223.

Figure 53 Catch rates of adult Atlantic tomcod in box traps during five spawning seasons, Hudson River estuary, p. 228.

Figure 54 Percentage of female Atlantic tomcod in prespawning condition, p. 231.

Figure 55 Late fall and winter water temperatures in the West Point region (km 75–88), Hudson River estuary, p. 232.

Figure 56 Percentage of Atlantic tomcod samples that were significantly different from a 1:1 sex ratio, 1975–1979, p. 233.

Figure 57 Catch rates for adult Atlantic tomcod during five spawning seasons, p. 235.

Figure 58 Recapture locations outside of the Hudson River estuary, 1975–1979, p. 237.

Figure 59 Courtship (prespawning) behavior of Atlantic tomcod in aquaria, p. 239.

Figure 60 Density isopleths of Atlantic tomcod postlarvae collected by Tucker Trawl and epibenthic sled, Hudson River estuary, p. 243.

Figure 61 Typical annual cycle of freshwater flow entering the Hudson River estuary at Troy Dam, p. 244.

Figure 62 Extent of downriver displacement of Atlantic tomcod postlarvae versus weighted mean daily flow entering

LIST OF FIGURES

	Hudson River estuary at the Troy Dam, February–April, p. 247.
Figure 63	Density isopleths (no./1000m³) of Atlantic tomcod juveniles collected by Tucker trawl and epibenthic sled, Hudson River estuary, 1975–1979, p. 248.
Figure 64	Catch-per-unit-effort isopleths for Atlantic tomcod juveniles collected by bottom trawl, Hudson River estuary, p. 249.
Figure 65	Map of the lower Hudson estuary depicting the locations of sampling stations used in this study, p. 275.
Figure 66	Trophic level contributions (ppm wet weight) of mercury, cadmium and lead in aquatic fauna collected at four stations in the lower Hudson estuary, p. 283.
Figure 67	Trophic level contributions (in ppm wet weight) of zinc, arsenic, and copper in aquatic fauna collected at four stations in the lower Hudson estuary, p. 284.
Figure 68	Summer Hudson River dissolved oxygen levels, p. 289.
Figure 69	Predicted summer dissolved oxygen profiles in the Hudson River, p. 291.
Figure 70	Spring Hudson River dissolved oxygen data, p. 293.
Figure 71	NYSDEC dissolved oxygen values at Glenmont, New York, 1970–1979, p. 295.
Figure 72	Monthly dissolved oxygen values and percent saturation at Glenmont, New York, p. 298.
Figure 73	Historical dissolved oxygen and temperature data, p. 299.
Figure 74	Mean dissolved oxygen concentration versus time, West Side Highway project, 1979–1980, p. 301.
Figure 75	Percent dissolved oxygen saturation at West Side Highway Project station WHA3 (Channel), 1979–1980, p. 302.
Figure 76	Variation in total PCB concentration (wet basis) with lipid content of yearling pumpkinseed collected in 1979 at three Hudson River locations, p. 308.
Figure 77	Relationships between total PCB concentration (wet basis) and lipid content for largemouth bass collected at Stillwater in 1977 and 1980, p. 309.
Figure 78	Temporal changes for lipid-based total PCB and its component Aroclors for four species of resident fishes taken at Stillwater from 1977 to 1980, p. 314.
Figure 79	Temporal changes for lipid-based total PCB and its component Aroclors for two species of resident fish taken at Albany and Troy from 1977 to 1980, p. 315.

LIST OF FIGURES

Figure 80 Temporal changes for lipid-based total PCB and its component Aroclors for three species of resident fish taken at Catskill, 1977–1980, p. 316.

Figure 81 PCB trends in American shad collected from the Poughkeepsie area of the Hudson River, p. 338.

Figure 82 PCB trends in American shad collected from the Tappan Zee Bridge area of the Hudson River, p. 339.

Figure 83 PCB trends in American eel collected from the Haverstraw Bay area (Indian point to the Tappan Zee Bridge) of the Hudson River, p. 342.

Figure 84 PCB trends is striped bass collected from riverwide locations (Poughkeepsie to the George Washington Bridge) in the Hudson River, p. 343.

Figure 85 Conceptual life history used as basis for population dynamics model, p. 355.

Figure 86 Linear relationships of density dependent survival with number of age 0, 1, and 2 Atlantic sturgeon assumed for model, p. 357.

Figure 87 Relationship between weight and age for Hudson River Atlantic sturgeon, p. 358.

Figure 88 Model behavior under assumptions to produce fivefold and twentyfold increases in population sizes without fishing mortality, p. 360.

Figure 89 Mean harvest and stock size as a function of exploitation and fishing strategy under fivefold and twentyfold growth assumptions, p. 361.

Figure 90 Predicted annual harvest and stock size for early and late fishing strategies under 0.45 fishing exploitation within Hudson River, p. 363.

Introduction
C. Lavett Smith

In the late 1960s, after the publication of Rachel Carson's *Silent Spring*, society suddenly became aware of the desperate importance of environmental issues. New laws required environmental impact statements to be filed, and it soon became apparent that often the basic data required for intelligent choices were lacking. A whole new industry, biological consulting, developed in response to the need for fundamental biological data; and because the data were often subject to intense scrutiny in the courtroom, new standards of documentation and quality control became the rule. At the same time, the development of electronic data processing made it possible to process and store ever greater quantities of information.

Nowhere has the conflict between society and environment been more intense than in the Hudson River. Although the Hudson River did not prove to be the shortcut to the Orient that its discoverers had at first hoped, as the gateway to the fertile farm land of the central part of the North American continent it played an important role in the economy of the United States. Before the development of railroads in the latter half of the nineteenth century, movement of heavy and bulky goods depended on water transportation, and this meant that the Appalachian Mountains effectively isolated the Mississippi and Great Lakes regions from the population centers of the Atlantic coast. Only three routes provided practical access for large numbers of people and their trade goods: The Mississippi River, the St. Lawrence valley, and the Mohawk-Hudson corridor. Of these the Mohawk-Hudson route was the shortest and most practical, and soon it became the main route to the interior. Construction of the Erie Canal enhanced this role and led inevitably to the establishment of the major population centers along its course.

Industrial development in the Hudson Valley was also spurred by the availability of cheap power from waterfalls close to the main river on almost every tributary. With the head start provided by these natural advantages, the Hudson-Mohawk region soon became the focus of industry in New York, and even today the Hudson Valley has the fastest growing population in the state.

Expanding human populations inevitably bring about an array of environmental problems and conflicts. Some of these problems—overfishing, loss of wetland habitats, and domestic pollution—began almost as soon as the first European colonists arrived. Others, including pollution by industrial wastes, were slower in coming; and some pollutants—such as persistent pesticides and the polychlorinated biphenyls (PCBs) whose critical levels are measured in parts per million, billion, or trillion—are strictly the by-products of twentieth-century technology. Often, as in the case of DDT and PCBs, the potential for adverse health effects was not discovered until after large quantities had been discarded into the nearest waterway.

Contrary to a popular misconception, the Hudson River is today a thriving ecosystem with a viable (and valuable) fishery resource. It contains one of the largest remaining populations of the shortnose sturgeon, and its fishery for the anadromous American shad is now at its highest level in many years.

The Hudson also holds great potential for both commercial and recreational fishing for striped bass, although neither of these fisheries is active at present owing to the high levels of PCB contamination. In spite of this, the Hudson River plays an important role as a spawning and nursery ground for striped bass. Recently, there has been an alarming decline in other Atlantic coast stocks, and the contribution of the Hudson stock to the coastal population therefore is becoming increasingly important to the very survival of the species.

Much of the research described in this volume is the result of the conflict between society's need for electric power and the potential harm to the environment resulting from withdrawing large quantities of water from the tidal estuary. This is a widespread problem that will continue to grow as long as our need for power increases. It is true of nuclear, fossil-fuel, and pumped-storage plants. The electronic age has certainly increased the efficiency with which our power is used, but it has not diminished the need for electric power and is not likely to in the foreseeable future. Power plants, by virtue of their

INTRODUCTION

size and obvious environmental impacts, are prime targets for activist attacks. In the Hudson estuary, the power-plant issue centers on the effects on fish eggs and larval stages when they are in the vicinity of the plants. Because the flow in the estuary is tidal, the fish are repeatedly exposed to entrainment and impingement as they are carried upstream and downstream by the tides. Although the problem can be avoided by using closed circuit cooling towers, such towers are expensive and carry their own set of environmental problems.

On 19 December 1980 the signing of the Hudson River Settlement ushered in a new era of working together to attack environmental problems. Formulated after years of research and court action, the Settlement calls for a ten-year trial period during which the utilities have agreed to institute certain mitigative measures and to carry out studies designed to evaluate the effectiveness of those measures, in return for which the plaintiff groups have agreed to drop their suits with prejudice.

Each ecosystem has its own unique parameters, and the success of a technique in one estuary does not automatically mean that it will work in another. The enormous expense of most mitigation efforts precludes a trial and error approach; at present, the decision to invest in any steps aimed at mitigating the effects of landfills, waste disposal or industrial water use can only be made on the basis of a subjective prediction of the possible harm and benefits. These predictions are based on worst-case analyses of the possible actions, using whatever data are available. Obviously, the better the data base, the better the chance of making accurate predictions and choosing successful mitigation procedures. It has been estimated that between forty and fifty million dollars had been spent on basic fisheries research in the Hudson River by 1981, yet the recent controversy over the possible harm to the fishery that would result from filling in approximately 240 acres along the west side of Manhatten Island for the Westway project illustrates that we still do not understand enough of the fundamental workings of estuarine systems to construct predictive models and use them for making crucial environmental predictions.

With the signing of the Settlement and the change of emphasis of research sponsored by the utilities, many of the most active biologists have left the Hudson Valley to seek new challenges elsewhere. Much of the information gathered by the consultants has been filed in reports with limited distribution so that it is not generally available. Some very excellent basic work was judged not to be directly relevant to the problems at hand and has remained in the investigators' files. Only a small fraction of what is known about the Hudson River ecosystem has become generally available through formal publication.

Recognizing that much of this information is in danger of being lost to science, the Hudson River Environmental Society, an independent nonprofit organization of scientists and others interested in environmental issues in the Hudson Valley, planned a conference that was held at the Norrie Point Environmental Center in September 1981. The conference had the double objective of bringing together those scientists interested in the Hudson system and gathering a selected group of papers of general interest for formal publication. The first objective was accomplished at the conference; the second is met by the present volume. Some of the papers presented at the conference dealt directly with the power-plant issue, others were directed at the pollution problems, and a few were academic studies that addressed fundamental issues of ecosystem structure and function.

The present volume is a selection of the papers that deal more or less directly with fisheries issues, including a guide to data bases, life-history data, an experimental study of an important food source, reports on contaminant levels that document changes that occurred during that critical decade when the trend toward ever greater degradation was finally reversed, and a management proposal for the Atlantic sturgeon. We hope that the publication of this series of papers will stimulate and encourage further analysis of the wealth of information that is now available, for there is still a long way to go before this wealth of data can be transmuted into a real understanding of the Hudson River ecosystem.

Partial support for this publication was provided by the Griffis Foundation Inc., New York.

Part I
Data Sets

1. Fisheries Data Sets Compiled by Utility-Sponsored Research in the Hudson River Estuary

Ronald J. Klauda, Paul H. Muessig, and John A. Matousek

INTRODUCTION

The Hudson River power-plant case stands as a landmark event in the technical and legal development of impact assessment in aquatic ecosystems. Utility-sponsored research since the mid-1960s, collectively referred to as the Hudson River Ecological Study, compiled numerous data sets which contain a wealth of information on the estuarine portion of the river. Although the data were collected for the purpose of assessing ecological impacts of present and proposed electric generating plants and of developing mitigation procedures to reduce impacts, the scopes of many studies were broad and yielded basic information on the aquatic biota of this major East Coast estuary.

Concern over the interactions between electricity production and the Hudson River ecosystem was centered around questions of impact on selected fish populations because of their recreational or commercial value. As a result, the major emphasis of utility-sponsored research has been on finfish species. Hence, the fisheries data sets are the most extensive and complete, and they will be the focus of this paper. Several data sets were also collected on the lower trophic levels (phytoplankton, zooplankton, and benthos), but these are not addressed in this paper.

The Hudson River Ecological Study evolved in size and complexity since the mid-1960s. Early fisheries studies (Perlmutter et al. 1966, 1967, 1968; Carlson and McCann 1969; Raytheon 1971; QLM 1971a, 1971b) approached the state-of-the-art by most standards of that time. As the level of environmental awareness heightened in the United States during the early 1970s, the scope of the Hudson River Ecological study expanded concomitantly. By 1974, it had increased from primarily near-field surveys with fixed sampling stations in the vicinity of several power plants to a monumental effort focused on the entire tidal portions of the river. Between 1974 and 1980, scientists from several environmental consulting companies and universities worked concurrently toward three major goals: (1) to measure the ecological consequences of operating steam electric generating plants with once-through cooling systems; (2) to predict the incremental consequences of constructing and operating a pumped-storage generating project at Cornwall, New York; and (3) to develop mitigation measures to minimize any ecological consequences of operating these plants (Limburg, 1986, Barnthouse et al. 1987).

The negotiated agreement signed by the utilities, regulatory agencies, and concerned citizens organizations on 19 December 1980 essentially closed the power-plant case and sharply curtailed the scope and intensity of the Hudson River Ecological Study (Sandler and Schoenbrod 1981). This settlement-agreement ended an almost twenty-year-old aquatic impact assessment study through compromises that all parties believed to be in the public interest. Utility-sponsored studies, which are being continued during the 1980s will focus primarily on evaluating the effectiveness of the mitigation measures required by the settlement-agreement and on developing and implementing a ten-year monitoring program.

Reams of analyses, numerous technical reports and data displays, stacks of hearings testimony and transcripts, and a few published papers have begun to tap these extensive data bases. But only a small part of this information is easily accessible to individuals who were not directly associated with the data, and many of the data have never been analyzed.

Current interest in the estuary and the austere economic climate should make the existing fisheries data extremely attractive to Hudson River researchers. The settlement agreement created the Hudson River Foundation for Science and Environmental Research, Inc., for the purpose of continuing investigations of basic ecological processes in the estuary. The existing fisheries data sets should be useful in planning this organization's future studies. Striped bass is a main export item from the estuary; therefore, the fisheries data sets for this species should also be of interest to the Federal Striped Bass Emergency Study Team and their search for causes of the current decline of the coastal stocks (Anonymous 1980).

Perhaps most importantly, the fisheries data sets in general should be invaluable to New York's Department of Environmental Conservation in the development and implementation of an effective management plan for the Hudson estuary (NYDEC 1980). Other groups, such as community planners, developers, and industry, may also want to use these data sets.

In addition to the fact that they are extensive and available, many of the fisheries data sets possess a unique property developed in the adversary climate of adjudicatory hearings. The Hudson River data sets have undergone a level of scrutiny that is probably unmatched in the annals of fisheries research. Legal debate on the ecological impact questions in the power-plant case consumed countless hours and pages of testimony and transcripts. These public hearings also stimulated a zealous desire among all parties to ensure that the raw data collected in the Hudson River Ecological Study were sound.

The major ecological consulting companies working under contract to the utilities since the early 1970s developed detailed Standard Operating Procedures (SOPs), employee training courses, and state-of-the-art, statistically-based Quality Control programs to standardize and document sample collection methods and minimize errors in the raw data and subsequent analyses. The regulatory agencies and their consultants received copies of the raw data, and they reanalyzed selected portions of these data using various procedures to check the results given in technical reports and hearings testimony submitted by the utility consultants. Agreement was uncommon among the experts involved in the Hudson River power-plant case with regard to analytical procedures, mathematical models, assumptions, results, and inferences (Christensen et al. 1981). However, the fisheries data sets compiled in utility-sponsored research withstood scrutiny and proved to be reliable sources of information for evaluating the aquatic impact of power plants.

OBJECTIVES

We address three major objectives in this paper: (1) to describe the fisheries data sets compiled by utility-sponsored studies conducted between 1971 and 1980 by Texas Instruments Incorporated (TI), Lawler, Matusky & Skelly Engineers (LMS), and Ecological Analysts, Inc. (EA); (2) to evaluate the relative strengths and weaknesses of each data set for general use, based on specific study objectives, focal species, and sampling designs; and (3) to describe the disposition of these data sets and their accessibility to the scientific community.

Our examination of the fisheries data is intended to be neither a comprehensive description of all existing data sets nor an exhaustive discussion of the data sets we present. Rather, we hope to transmit in a comprehensible manner a large and representative subsample of fisheries data that have been compiled for the Hudson estuary. We also hope that future research can benefit from our description of these studies and build upon the existing foundation to fill critical information gaps rather than waste valuable resources in duplicative efforts.

STUDY AREA AND TERMINOLOGY

Utility-sponsored fisheries studies conducted by TI, LMS, and EA between 1971 and 1980 were focused on assessing the ecological impacts of six operating steam electric generating plants (Bowline Point, Lovett, Indian Point, Roseton, Danskammer Point, and Albany), two proposed fossil-fueled plants (at Ossining and Kingston), and one proposed pumped-storage plant (at Cornwall). Sampling programs at each plant were designed to measure entrainment (passage of organisms through the cooling water system), impingement (entrapment of organisms on cooling water intake screens), thermal discharge effects, construction effects, and seasonal changes in relative abundance, distribution, and species composition of fishes in the vicinity (near-field) of each plant (Figure 1). Estuary-wide (far-field) sampling encompassed the major study area between the George Washington Bridge and the Troy Dam, with some sampling in the upper and lower portions of New York Bay, the western end of Long Island, and beyond (Figure 1).

DESCRIPTIONS OF FISHERIES DATA SETS

Data Categories

Fisheries data sets collected by TI, LMS, and EA can be grouped into seven major categories: (1) distribution, abundance, and species composition; (2) biological characteristics (length, weight, age, maturity, sex, fecundity, and stomach contents); (3) physiology and behavior; (4) entrainment; (5) impingement; (6) gear performance and special studies; and (7) culture.

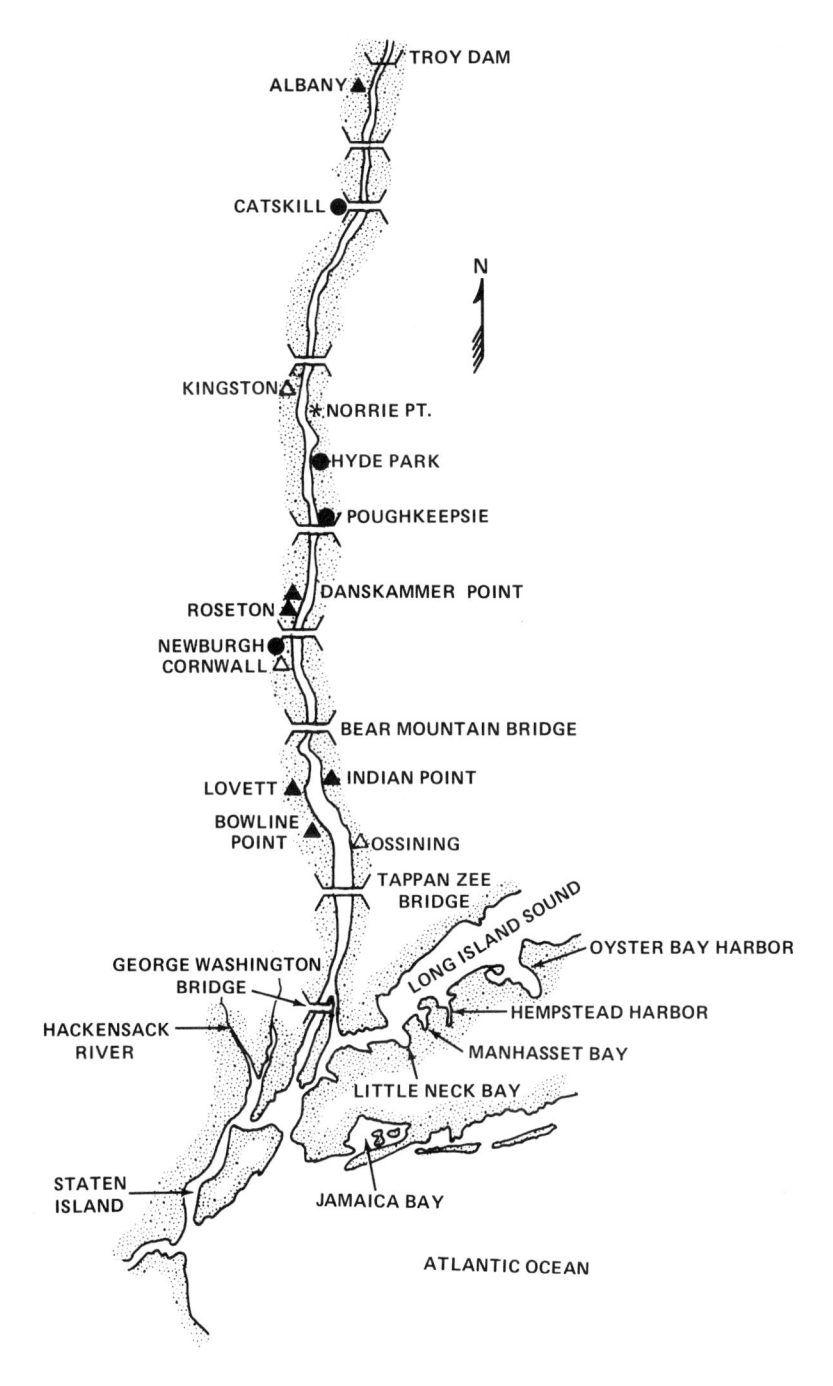

Figure 1. *Sampling area for Hudson River Ecological Study showing sites of focal electrical generating plants (closed triangle for existing or "present" plants, open triangle for proposed plants).*

Data Sets. General descriptions of each data set are presented in the text and tables with discussions of major strengths and weaknesses, as well as disposition and accessibility of the data. TI, LMS, and EA data sets are described separately.

Portions of several data sets, in addition to the descriptions of relevant field and laboratory procedures (including quality control methods), have been published in at least 155 technical reports (cited herein) and several data displays that TI, LMS, and EA submitted to the utilities. These documents should be available from the utility that sponsored the studies (Table 1). Partial collections of these documents are held in the public libraries of Peekskill and Kingston, New York. Summarized versions of portions of the fisheries data sets authored by present and former TI, LMS, and EA scientists that have been published in the journals are cited or listed in the Literature Cited section. However, papers authored by other scientists, even though they were based solely or partly on the fisheries data sets described in this paper, have not been cited or listed here.

Complete collections of most fisheries data sets are located in documented computer files maintained by the utilities. Persons interested in using a particular computer file should contact the appropriate utility (Table 1) for accession procedures and detailed instructions on the format and associated documentation. Access to a computer facility and financial support to make copies of the data tapes and documentation are essential. Some of the TI, LMS, and EA data sets are filed in different formats; hence, the documentation package for each data set is essential for gaining access to the data.

The Hudson River Foundation recently funded the preparation of a user's guide to the fisheries data sets compiled by TI between 1972 and 1980.

The computerized EA data sets were initially entered as card images formatted from the field data sheets and are accessible using most programming languages. Many of the larger data sets were recently reformatted to be compatible with Statistical Analysis System (SAS) programming, documentation and analytical packages.

The computerized TI data sets were also initially entered in card image (eighty characters per line) formats from field and laboratory data sheets. These files are accessible with most programming languages, but the documentation package for each data set must be consulted to access them. A few small data sets were not entered into

FISHERIES DATA SETS

Table 1. *Sources of data displays, technical reports, computerized data tapes, and documentation packages for fisheries data sets collected and compiled by Texas Instruments, Lawler, Matusky & Skelly Engineers, and Ecological Analysts for utility-sponsored studies on the Hudson River estuary, 1971–1980.*

DATA SET	UTILITY SPONSOR	CONTACT INDIVIDUAL
Station-specific and nearfield surveys		
Bowline Point	Orange and Rockland Utilities (ORU)	Mr. Jay B. Hutchison, Jr. Orange and Rockland Utilities, Inc. One Blue Hill Plaza Pearl River, NY 10965
Lovett	Orange and Rockland Utilities	
Indian Point 1 & 2	Consolidated Edison (CE)	Dr. William L. Kirk Consolidated Edison Company of New York, Inc. 4 Irving Place New York, NY 10003
Ossining	Consolidated Edison	
Cornwall	Consolidated Edison	
Indian Point 3	Power Authority of the State of New York (PASNY)	Dr. John W. Blake Power Authority of the State of New York 10 Columbus Circle New York, NY 10019
Roseton	Central Hudson Gas and Electric (CHG&E)	Mr. Thomas G. Huggins Central Hudson Gas and Electric Corporation 284 South Avenue Poughkeepsie, NY 12602
Danskammer Point	Central Hudson Gas and Electric	
Kingston	Central Hudson Gas and Electric	
Albany	Niagra Mohawk Power Corporation (NMPC)	Ms. Cheryl Blum Niagra Mohawk Power Corporation 300 Erie Blvd. West Syracuse, NY 13202
Farfield surveys	ORU, CE, PASNY, CHG&E	Dr. Kirk

computer files, but all data sheets (original and microfilm copies) are stored by Consolidated Edison.

Most of the major TI data sets collected after 1973 have been reformatted into SAS files (Watson et al. 1981). These files reside in Consolidated Edison's computer and can be accessed, retrieved, and manipulated only with SAS commands and procedures (Helwig and

13

Council 1979). Anyone wishing to access these data sets must have use of an IBM 360/370 computer (or plug-compatible machine) under OS or OS/VS that will support SAS. The listing of all variables contained in each data set and a dictionary of variable names, definitions, coding values and a user's guide are key components of the documentation package which should accompany these SAS format data tapes.

Data sets collected by LMS prior to 1974 are maintained as complete laboratory analysis sheets and hand-reduced summary tables. All ecological data collected since 1974 were placed on card-image format data sheets and stored on an HP 3000 computer system in files that are accessible to several computer languages. Fishery data collected for ORU and CHGE have been reformatted into SAS files. These SAS files have been patterned after TI's SAS formats to maximize interdata-base compatibility for computer manipulation.

A brief description of the organization of fisheries data sets in a typical SAS file is useful. The data from each fisheries survey are arranged in hierarchical file structures which can be manipulated with SAS through the use of multiple rectangular data sets and SAS software

Table 2. *SAS hierarchical file structure for a typical Hudson River fisheries data set.*

LEVEL	DATA RESOLUTION	KEY IDENTIFIER	INFORMATION IN LEVEL
0	Project (or Survey)	Project Code	Null data set
1	Year	., Study Year	Null data set
2	Task	.., Task code	Null data set
3	Sample	..., Sample number	Data unique to one sampling attempt
4	Gear	...., Gear code	Data unique to one sampling device
5	Taxon	....., Taxon code	Data unique to each taxon collected
6	Individual	......, ID number	Data unique to each fish collected
7	Sub-individual	......., Analysis code	Data resulting from processing analysis which generate multiple observations on each individual (e.g., length, weight, age, sex)

capabilities. The fisheries data are grouped by survey, and each data set is split into levels of increasing resolution (Table 2). Each level of an SAS file consists of one data subset containing all values pertaining to a particular aspect of the sampling effort and organized in a rectangular matrix of rows and columns. Each observation (row) contains one value for each variable (column). Each observation in a level is uniquely identified by key identifier variables which contain the special identifier for a given level plus the identifiers from all previous levels. The key identifiers make it possible to match each observation in one level with appropriate observations in other levels. Thus, the levels can be recombined in several ways to facilitate a variety of data set manipulations.

DISTRIBUTION, ABUNDANCE, AND SPECIES COMPOSITION

TI Data

NEAR-FIELD SURVEYS. From 1972 to 1980, TI conducted an annual near-field survey in the area of the Indian Point plant (Table 3) and extended the near-field data base at Indian Point begun in 1969 (Raytheon 1971). This fixed-station survey was designed to collect data on annual and seasonal changes in distribution, relative abundance, and species composition primarily of juvenile (young-of-year) and yearling fishes in an 8-km area around the Indian Point plant and to assess the effects of thermal and chemical discharges from the plant on the fish community. Seven shore stations were sampled with a 30.5-m beach seine from early April through December. Seven offshore stations were sampled weekly during the day with a 7.8-m (headrope) otter-type bottom trawl. Each trawl station was also sampled from July through December with a 5.3-m (headrope) surface trawl. Standard physicochemical parameters (water temperature, pH, dissolved oxygen, conductivity, and turbidity) were measured concurrently with each seine haul and trawl tow. This data set contains a continuous series of comparable catch-per-effort estimates of relative abundance for fishes in the Indian Point near-field. Survey methods, results, and summarized portions of this data set were published in a series of data displays and technical reports (Table 3). During 1972–1973, TI also conducted a near-field fisheries survey in an 8-km area of the estuary near Ossining, New York (Table 3). This fixed station survey was designed to obtain baseline data on the aquatic biota adjacent to the site of a proposed fossil-fueled generating plant

Table 3. Summary of near-field distribution, abundance, and species composition data sets for Hudson River fishes collected by Texas Instruments Incorporated.

SURVEY	YEARS	PRIMARY SPECIES	DATA	STRENGTHS	WEAKNESSES	DISPOSITION/ACCESS
Indian Point plant	1972–1980	Striped bass, white perch, Atlantic tomcod, plus other common species (e.g., American shad, blueback herring, alewife, bay anchovy, tesselated darter, spottail shiner, Atlantic silverside, pumpkinseed, banded killifish, hogchoker)	1–Catch per effort 2–Lengths and weights 3–Water chemistry 4–Beach seine, bottom trawl, surface trawl 5–Seven fixed seine stations 6–Seven fixed trawl stations 7–Weekly survey, day, Apr–Dec (surface trawl Jul–Dec)	1–Long-term, comparable data set collected with basically same gear at same stations 2–All fish species counted, measured and weighed 3–Water chemistry measured with each sample	1–Near-field survey 2–No liner in bottom trawl cod end in 1972–1973	Post 1973 data in SAS format at CE; rest in card image format; TI 1972a, b; 1973a, c; 1974c; 1975c; 1976c, f; 1977h; 1979e; 1980b, d
Ossining area	1972–1973	Striped bass, white perch, Atlantic tomcod, plus other common species	1–Catch per effort, no/m³ 2–Lengths and weights 3–Water chemistry 4–Plankton nets, beach seines, bottom trawl, box traps, gill nets 5–Several stations 6–May 1972–Apr 1973	1–Several gear comparable to Indian Point 2–All fish species counted, measured and weighed 3–Water chemistry measured with each sample	1–One-year survey 2–Near-field survey	Data in card image format at CE; TI 1973b

CE = Consolidated Edison Company of New York, Inc.

which was never constructed. The Ossining survey used several types of gear to sample the fish community, including 0.5-m and 1.0-m conical plankton nets, 1.0-m^2 epibenthic sled, 2.0-m^2 Tucker trawl, 23-m and 30.5-m beach seines, 8-m (headrope) semiballoon otter trawl, and various-size box traps and gill nets. Some gear and sampling procedures used in the Ossining survey were similar to the Indian Point survey; hence, comparable portions of these two data sets encompass almost 20 km of the estuary. Standard physicochemical parameters were measured concurrently. Results of this study were published in a technical report (Table 3).

FAR-FIELD SURVEYS. From 1973–1980, TI conducted four or five far-field surveys each year that were designed to examine annual and seasonal changes in distribution, relative abundance, and species composition of eggs through yearling-life stages of fishes in over 200 km of the estuary (Table 4). Early life-history stages (eggs, larvae, and early juveniles) in surface, bottom, and channel areas were sampled by the Ichthyoplankton survey from early April through late July or mid-August. The survey used a 1.0-m^2 epibenthic sled and 1.0-m^2 Tucker trawl. A stratified random design was used with a sample allocation scheme keyed to the distribution of striped bass eggs and larvae. From 1975 through 1980, samples were also collected in February and March, with sample allocations keyed to Atlantic tomcod distribution. Older juveniles and yearlings were sampled from April through December in the Beach Seine survey (30.5-m seine), the Fall Shoals survey (epibenthic sled, August-December), and the Interregional Trawl survey (7.8-m headrope length, 1978–80 only). Stratified random sample designs were used for the Beach Seine and Fall Shoals surveys. The allocation of samples to each stratum was keyed to juvenile striped bass distribution in 1973–1978 and to available habitat accessible to each gear in 1979–1980. The Interregional Trawl survey sampled fixed stations. Samples were randomly collected in the Try Trawl survey, with the allocation to each stratum proportional to available habitat. TI also conducted a survey from 1976–1980 with gill nets and haul seines that was designed primarily to assess the biological characteristics of adult striped bass. The survey also yielded data on the distribution of striped bass during the spawning run (March–June) and incidental catches of other species, especially Atlantic and shortnose sturgeons.

Standard physicochemical parameters were measured concurrently with each far-field survey haul or tow. Although minor changes occurred in each survey between 1973 and 1980, they comprise

Table 4 Summary of far-field distribution, abundance, and species composition data sets for Hudson River fishes collected by Texas Instruments Incorporated.

SURVEY	YEARS	PRIMARY SPECIES	DATA	STRENGTHS	WEAKNESSES	DISPOSITION/ACCESS
Ichthyoplankton	1973–1980	Striped bass, white perch, Atlantic tomcod	1–No./m^3 by life stage 2–Lengths for striped bass larvae in 1976–1980 3–Water chemistry 4–Epibenthic sled, Tucker trawl 5–Stratified random design 6–Weekly runs, day or night, Apr–Jul (Feb–Apr for Atlantic tomcod)	1–Comparable data set for early life stages from 1974–1980 in area from km 22–224. 2–All fish species counted. 3–Water chemistry measured with each sample 4–150 to 210 samples per run	1–1973 sampling design not comparable to 1974–1980. 2–Apr–Dec sample allocations keyed to striped bass 3–Gear relatively ineffective for eggs and early juveniles	Post 1973 data in SAS format at CE, rest in card image format; TI 1974f; 1975a, b; 1976a; 1977e, f; McFadden 1977 TI 1978a; McFadden et al. 1978; TI 1979a; 1980a, e; 1981b
Beach Seine	1973–1980	Striped bass, white perch	1–Catch per effort by age/length group 2–Lengths and weights for juvenile and yearling striped bass and white perch 3–Lengths for juveniles of selected species	1–Comparable data set for juveniles from 1973–1980 in area from km 19–243 2–All fish species counted, juveniles separated 3–Water chemistry measured with each sample	1–Sample allocations in 1973–1978 keyed to striped bass 2–Beach seine ineffective in some near-shore areas (e.g., rip-rap areas)	See Ichthyoplankton Survey

			4–Stratified random design 5–Weekly or biweekly runs, day, Apr–Dec (some night samples in 1973–1974) 6–Water chemistry 7–30.5-m seine	4–Mean area swept per tow known 5–100 samples per run		
Fall Shoals	1973–1980	Striped bass, white perch, Atlantic tomcod	1–No./m^3 by age group length group (except 1973) 2–Lengths and weights for juvenile striped bass and white perch 3–Lengths for juveniles of selected species 4–Biweekly runs, night, Aug–Dec 5–Water chemistry	1–Comparable data set for juveniles from 1974–1980 between km 22–122. 2–All fish species counted, juveniles separate 3–Water chemistry measured with each sample 4–100 or 200 samples per run	1–Sample allocations in 1973–1978 keyed to striped bass 2–No samples near surface or mid-depth and upriver from km 122 prior to 1979 3–Volume per sample not measured in 1973 4–Low (<10%) catch efficiency of gear	See Ichthyoplankton Survey
Interregional Trawl	1973–1980	Striped bass, white perch, Atlantic tomcod	1–Catch per effort by age/length group	1–All fish species counted, juveniles separate	1–Number and location of stations differed among	See Ichthyoplanton Survey

Table 4 *Summary of far-field distribution, abundance, and species composition data sets for Hudson River fishes collected by Texas Instruments Incorporated. (Continued)*

SURVEY	YEARS	PRIMARY SPECIES	DATA	STRENGTHS	WEAKNESSES	DISPOSITION/ACCESS
			2–Lengths and weights for juvenile striped bass and white perch 3–Lengths for juveniles of selected species 4–35–40 fixed station 5–Biweekly runs, day, Apr–Dec 6–Water chemistry	2–Water chemistry measured with each sample 3–One sample per station per run	years 2–Some gear changes among years 3–Sampling area from km 38–122, no effort in upper estuary	
Try Trawl	1978–1980	Striped bass, white perch, Atlantic tomcod	1–Catch per effort by age/length group 2–Lengths and weights for juvenile striped bass and white perch 3–Lengths for juveniles of selected species 4–Stratified random design in depths of 1.5–6 m	1–All fish species counted, juveniles separate 2–100 samples per run	1–Only 3 years of data 2–Conducted only from Sep–Dec in 1978	Data in SAS format at CE; TI 1980e; 1981b

| Southern (Lower) Estuary | 1974–1975 | Striped bass, white perch, Atlantic tomcod | 1–Catch per effort by age/length group
2–Nonrandom design to maximize catch
3–Approx. biweekly runs, day
4–Several gear used
5–Jun 1974–Jul 1975
5–Biweekly runs, day, Apr–Dec
6–Water temperature only | Yielded some information on emigration from study area | 1–Only one year of data
2–Nonrandom sampling design | Data in card image format at CE; TI 1977g |

collectively a long-term data set on the distribution, relative abundance, and species composition of several fish species throughout the estuary. From a general use viewpoint, they have three major weaknesses. (1) The Ichthyoplankton survey was relatively ineffective for eggs and early juveniles. (2) Sample allocations in the Beach Seine and Fall Shoals surveys were keyed to juvenile striped bass distribution from 1973–1978, so effort was weighted more heavily toward the downriver third of the study area. (3) Avoidance of the 1.0-m^2 epibenthic sled used in the Fall Shoals survey by juvenile fishes appeared to be greater than 90%. Descriptions of sampling procedures, results of these far-field surveys for selected species, and summarized portions of this data set have been published elsewhere in journals (Stira and Smith 1976; Bath et al. 1976; Hoff et al. 1977; Tabery et al. 1978; Hoff and Klauda 1979; Klauda et al. 1976, 1980), data displays, and technical reports (Table 4).

In 1974–1975, TI extended the far-field sampling effort to several areas downriver from the George Washington Bridge and adjacent to the western end of Long Island (Figure 1, Table 4). This Southern (Lower) Estuary survey was primarily designed to determine the extent of movement of marked striped bass, white perch, and Atlantic tomcod between the major study area (upriver from the George Washington Bridge) and lower bays such as Little Neck Bay, Manhasset Bay, Hempstead Harbor, Jamaica Bay, Staten Island, Hackensack River, and Oyster Bay Harbor (Figure 1). Since mark-recapture was the primary methodology, the sampling design was nonrandom and several types of gear (61-m haul seine, 7.8-m otter trawl, 1.0-m^2 epibenthic sled, 61-m variable mesh gill net, and 1 × 1 × 2-m box trap) were used. Water temperature and conductivity measurements were taken concurrently with each biological sample. Survey methods, results, and summarized portions of this data set were published in journals (Friedmann and Hamilton 1980) and a technical report (Table 4).

From 1972 to 1980, TI conducted mark-recapture studies designed to estimate the absolute abundance of three selected fish species—striped bass, white perch, and Atlantic tomcod (Table 5). Various combinations and types of finclips and tags were attached to juvenile, yearling, and adult striped bass; juvenile, yearling, and adult white perch; and adult Atlantic tomcod. Recovery of marked individuals also yielded information on fish movements within the study area and migrations to adjacent or distant coastal waters. These mark-recapture studies were ambitious and met with varying degrees of success. For example, in spite of an intensive marking effort, estimates of juvenile

Table 5 *Summary of studies on selected Hudson River fishes conducted by Texas Instruments Incorporated that yielded data on distribution, abundance and movements.*

SPECIES	STUDY	YEARS	DATA	STRENGTHS	WEAKNESSES	DISPOSITION/ACCESS
Striped bass White perch Atlantic tomcod	Mark-Recapture	1972–1980[a]	1–Marked fish releases and recoveries 2–Juvenile and adult striped bass 3–Juvenile and yearling plus white perch 4–Adult Atlantic tomcod	Acceptable numbers of releases and recaptures for adult striped bass, juvenile and yearling plus white perch, adult Atlantic tomcod	1–Insufficient numbers of recaptures for juvenile striped bass in most years 2–Insufficient data on marking-related mortality, tag loss, and non-response rate from fishermen who catch tagged fish	Post 1973 data in SAS format at CE, rest in card image format; TI 1973; 1974c; 1975a, b, c; 1976a, f; 1977e, f, g; McFadden 1977; TI 1978a; McFadden et al. 1978; TI 1979a; 1980a, e; 1981b.
Striped bass	Relative Contribution	1974–1975	1–Stock discrimination 2–Age, length, weight, sex, maturity of adults 3–Biochemical, meristic and morphometric characters	First comprehensive study on contribution of various stocks to coastal fishery	Results for only single year when 1970 Chesapeake Bay year class was predominant	Data in card image format at CE; TI 1975d; 1976d

[a] Most data available on adult striped bass from 1976–1980. Data on adult Atlantic tomcod from 1974–1980.

striped bass population size were generally unreliable because of emigration from the study area and few recaptures. Conversely, estimates of the size of the spawning populations of striped bass (1976–80) and Atlantic tomcod (1973–80) were more reliable.

This data set contains catch data for only striped bass, white perch, Atlantic tomcod, and uncommon (unusual) species. Methods, results, and portions of this data set were published in journals (Klauda et al. 1980, McLaren et al. 1981), data displays, and technical reports (Table 5).

Another TI study also collected data on the migratory patterns of adult striped bass. The Relative Contribution study of 1974–1975 was designed to determine if the populations of striped bass spawning in the Hudson River, Chesapeake Bay, and Roanoke River could be accurately discriminated, and if so, to measure the proportional contribution of each population to the Atlantic coastal fishery from Maine to North Carolina (Table 5). The study was successful, but stock contributions were measured only in 1975 when the 1970 year class from the Chesapeake Bay was predominant. Methods, results and summarized portions of this data set were published in journals (Grove et al. 1976, Berggren and Lieberman 1978) and technical reports (Table 5).

EA Data

During 1976 and 1977, EA conducted surveys to determine the effect of thermal discharges from the Roseton and Danskammer Point plants on the near-field distribution of fish (Table 6) and to examine field data for evidence of thermal preference to be compared with laboratory studies (Table 11). Sampling was conducted at control and discharge stations year-round, except during periods of ice cover. Catch-per-effort, length, body temperature, spawning condition, and surface plume thermal maps were obtained for each sampling effort. Collections were made primarily by electroshocker boat. The major weakness in the data set occurred during periods of very high abundance, when many stunned fish recovered before being collected. Consequently, values for peak abundance periods were underestimates. However, these data were only used for calculations of distribution and relative, not absolute, abundance estimates. All data were hand-tabulated and are available as data summaries in technical reports submitted to CHGE. During 1978 and 1979, EA continued a near-field study conducted by New York University (NYU) between 1971 and 1977 at the Indian Point plant. This study was conducted to evaluate the susceptibility of ichthyoplankton and zooplankton to entrainment and provides temporal and spatial distribution data in the Indian point near-field area. Similar concurrent in-plant entrainment data were collected for comparison with river data. These studies emphasized striped bass ichthyoplankton abundance and include larval length distributions. Sampling occurred at seven standard stations between km 64 and km 70, and along a river transect at the plant intakes. While gear and procedural changes between 1971 and 1979 limit across-year comparability, this data set does provide nine years of similar sampling effort at the same stations in the Indian Point near-field area. The procedures and gear were consistent between 1977 and 1979. Another potential weakness is that the discrete depth samples may be contaminated across depths as a result of deployment and retrieval of the open nets. The data sets are in SAS format files at Consolidated Edison and retrieval requires access to an IBM computer system. The data files may also be accessed with FORTRAN without the versatility of SAS.

LMS Data

NEAR-FIELD SURVEYS. During the period 1971–1980, LMS conducted extensive ecological studies at several locations on the Hudson River estuary. Field programs were generally limited to the period from March to December owing to ice conditions in January and February. The near-field programs were initiated and continuously reviewed and updated to assist in determining: (1) species composition; (2) spatial and temporal distribution patterns; (3) seasonal and annual relative abundance; and (4) morphometric and physiological conditions of the dominant or economically important species. Water quality parameters were monitored during each survey, permitting the evaluation of environmental conditions on the nekton community. Near-field studies conducted by LMS are summarized in Table 7.

Fish sampling programs were initiated at stations in the immediate vicinity of the Danskammer Point and Lovett plants and the Roseton and Bowline Point plants that were under contruction during the summer of 1971. Stations for the near-field programs were located adjacent to the plants as well as away from them in order to address differences in community structure between the two areas. The stations originally inventoried during 1971 were sampled continuously in the Roseton/Danskammer Point vicinity through 1980. The Lovett program was terminated following the 1976 survey.

Table 6. Summary of fish distribution, abundance, and species composition data sets for Hudson River generating plants collected by Ecological Analysts Incorporated.

PLANT	YEARS	PRIMARY SPECIES	DATA	STRENGTHS	WEAKNESSES	DISPOSITION/ACCESS
Roseton, Danskammer Point	1976–1977	Striped bass, white perch, clupeids, white catfish, spottail shiner	1–Distribution and abundance in and around the thermal discharges vs. control sites 2–Electroshocker, seines, traps, body temp., taggings 3–Diversity 4–1976 May–Nov; 15 sample dates 5–1977 Mar–Nov; 24 sample dates	1–Provides data for comparison w/ laboratory thermal preference and avoidance studies 2–Weekly samples during peak abundance	1–Noncomputerized data summaries in report 2–Changes in control area due to differences observed in habitat 3–Very high densities at some times in discharge—not all shocked fish collected 4–Some refinements in methodology between years	CHG&E; hand generated; EA 1978c

Indian Point	1977–1979	Microzooplankton Macrozooplankton Ichthyoplankton	1–7 standard stations 2–Day/night distribution 3–Depth distribution 4–Seasonal succession and abundance 5–Concurrent sampling at 3 stations on transect at plant and in plant intake and discharge, SB only 6–Length data 7–Water chemistry	Continuation of 10-year data base with near-field distribution and abundance	1–Low velocities at intake often outside manufacturer's specs on flowmeter 2–Possible contamination of nets at different levels during deployment and recovery 3–Differences in "tow" speeds; possible difference in avoidance for station/plant comparison 4–Differences in methodology and gear during 10-yr study 5–Emphasis on striped bass, no other ichthyoplankton analyzed	Con Edison; SAS format; EA 1980b, c; 1981c

CHG&E = Central Hudson Gas and Electric Corporation

Table 7. *Summary of fish distribution, abundance, and species composition data sets for Hudson River generating plants collected by Lawler, Matusky & Skelly Engineers.*

PLANT	YEARS	PRIMARY SPECIES	DATA	STRENGTHS	WEAKNESSES	DISPOSITION/ ACCESS
Bowline Point	1971–1980	White perch, striped bass, Atlantic tomcod, hogchoker, alewife, blueback herring, American shad, bay anchovy, plus other common species (e.g., spottail shiner, pumpkinseed, banded killifish)	1–Bottom trawl, surface trawl, beach seine, gill net, and trap net 2–Three fixed bottom trawl stations 3–Eight fixed beach seine stations 4–Three fixed surface trawl stations 5–Water chemistry 6–Species identification, length and weight analysis 7–Seasonally stratified sampling program 8–Diel sampling during several years	1–Same sampling gear used throughout study period 2–Long-term data set from same stations encompassing pre- and postoperational period 3–Sampling covers entire open water period per year 4–Includes station in and out of the influence of the thermal discharge	1–Winter months not represented in data base 2–Bottom trawl gear used as a surface trawl	Data collected prior to 1974 hand reduced, data from 1974 through 1980 in card image format and SAS format/ QLM 1971b, 1973c, 1974a; LMS 1975b, 1976a, 1977b, 1978c, 1979d, 1981a, 1981c
Lovett	1972–1976	White perch, striped bass, bay anchovy, hogchoker, blueback herring, plus	1–Bottom trawl, surface trawl and beach seine 2–Two fixed trawl	1–Same sampling gear used throughout survey	1–Winter months not represented in data base 2–Bottom trawl	Data collected prior to 1974 hand reduced, data from 1974 through 1980

		other common species	stations and two fixed beach seine stations 3—Water chemistry 4—Species identification, length and weight analysis	2—Thermally impacted and control stations sampled	gear used as a surface trawl	in card image format/QLM 1974a; LMS 1975b, 1976a, 1977c
Roseton/Dan-skammer Point	1971–1979	White perch, blueback herring, Atlantic tomcod, hogchoker, alewife, spottail shiner, plus other common species	1—Bottom trawl, surface trawl, beach seine, trap net, gill net 2—Two long-term trawl stations and four long-term seine stations 3—Large study area (15 km) with nine trawl stations, and eight seine stations 4—Species identification, length and weight analysis 5—Sexual differentiation of major species during 1973–1975 6—Diel sampling program	1—Long-term, comparable data set collected with same gear 2—Collection of lower trophic level populations during 1973–1977 3—River sampling corresponds to plant impingement sampling	1—Winter months, not represented in data base 2—Bottom trawl used as a surface trawl	CHGE; 1971–1974 hand generated, 1975–1979 card image format/QLM 1973g; LMS 1975c, 1978e, 1978f, 1979f, 1980d, 1980e, 1981d

Table 7. *Summary of fish distribution, abundance, and species composition data sets for Hudson River generating plants collected by Lawler, Matusky & Skelly Engineers. (Continued)*

PLANT	YEARS	PRIMARY SPECIES	DATA	STRENGTHS	WEAKNESSES	DISPOSITION/ ACCESS
Albany	1975	White perch, blueback herring, American eel, spottail shiner, plus other common species	1–Bottom trawl, surface trawl, beach seine, gill net 2–Two beach seine stations, six trawl and gill net stations 3–River samples collected during four monthly periods: April, June, August, October 4–Species identifications, length and weight analysis 5—Diel sampling program 6—Water chemistry	1–Same gear used on all sampling data 2–Same gear used as at other Hudson River locations 3–Sampling period corresponds to seasonal peak abundance patterns of dominant species	1–Only for sampling dates 2–Bottom trawl gear used as surface trawl	NMPC; card image format/NMPC 1976
Kingston*	1971–1973	Blueback herring, white perch, spottail shiner, plus other common species	1–Bottom trawl, surface trawl, beach seine, gill net, trap net	1–Same sampling gear used throughout study 2–Simultaneous col-	1–Winter months not represented in data base 2–Small study area	CHGE; hand generated/QLM 1973h; LMS 1975d

2—Species identifica-
tion, length and
weight analysis
3—Three fixed trawl
stations and two
fixed seine
stations
4—Diel sampling
program
5—Water chemistry

lection of lower
trophic level
populations

3—Bottom trawl
used as surface
trawl

NMPC = Niagara Mohawk Power Corporation
CHGE = Central Hudson Gas & Electric Corporation
*Proposed plant site

Short duration near-field programs were conducted north of Kingston, New York, during 1971–1974, and in the vicinity of the Albany Steam Station during 1975. The Kingston study was commissioned by CHGE as a pre-siting survey for a fossil-fueled generating station. The Albany study was conducted for Niagara Mohawk Power Corporation (NMPC) during the second summer of a two-year impingement program that had shown a diverse assemblage of nektonic organisms using the upper Hudson River estuary on a seasonal basis.

Ecological programs conducted by LMS for ORU, CHGE, and NMPC have employed the same fishery sampling gear. A 7.9-m headrope otter trawl was used as a surface and bottom sampler at open-water sites. The near-shore nekton community was sampled using a 30.5-m × 2.4-m bag seine. Other sampling gear included trap nets, 91.4-m × 3.0-m (six panel) experimental gill nets, and 91.4-m × 3.7-m mesh drift nets.

The Lovett near-field program consisted of three fixed trawl stations and two beach seine stations. The trawl stations were along the east and west shores and in the channel directly in front of the plant. Beach seine stations were along the western shore, north and south of the plant.

Trawl stations for the Bowline Point near-field program were located along the western shore of Haverstraw Bay near the thermal discharge diffuser, in Bowline Pond (cooling water intake location), and at a channel station south of the plant site. Additional trawl sites, one in the channel and one along shore, were added at three down-river locations during the fall of 1979 and 1980 to help define young-of-the-year distribution patterns and salt-front associations. A total of four Haverstraw Bay west-shore and four Bowline Pond beach seine sites have been sampled on a diurnal schedule since 1973.

Fishery sampling commenced in northern Newburgh Bay during 1971 with the establishment of two trawl stations: one north of the Danskammer Point plant and a second in front of Roseton. The stations established in 1971 were sampled continuously through 1979. During the period 1973–1977, a total of nine trawl and eight beach seine stations were sampled, resulting in one of the most extensive near-field survey data bases collected on the Hudson River. Trawl stations were reduced to three and beach seine stations to four during 1978–1979, all in the immediate vicinity of the two plants. A diel sampling schedule was followed from 1973 through 1979.

Three trawl and two beach seine stations were sampled in the vicinity of the Kingston-Rhinecliff Bridge between 1971 and 1974 as part of a siting study for a proposed fossil-fueled generating station. Fish samples generally were collected on a diel schedule every other week. Four trawl and beach seine surveys were conducted semimonthly in the vicinity of the Albany Steam Station (river km 228) during 1975. Two beach seine stations (located approximately 1.5 km south of the plant) and six trawl stations (two upstream, two in front of the plant, and two downstream of the plant) were sampled on a diurnal schedule. The Albany near-field fishery survey indicated extensive use of the northern portion of the estuary, especially by migratory species such as American eel, blueback herring, alewife, and white perch.

BIOLOGICAL CHARACTERISTICS

TI DATA. TI compiled data on several biological characteristics for three selected species (striped bass, white perch, and Atlantic tomcod) from 1972 through 1980 (Table 8). Studies designed to monitor the spawning populations throughout the estuary were conducted in 1976–1980 for striped bass, 1979–1980 for white perch, and 1974–80 for Atlantic tomcod. Length, weight, age, sex, maturity, fecundity, and stomach contents data were collected. An extensive data set of lengths and weights also was compiled for all fish species collected in the Indian Point Survey from 1972 through 1980. Length measurements on juveniles of several common species, and length and weight measurements on juvenile and yearling striped bass, white perch, and Atlantic tomcod were recorded in the far-field surveys from 1973 through 1980. Relatively complete data sets exist for juvenile and older striped bass, white perch, and Atlantic tomcod. Biological data for other fish species are limited to lengths and weights in the Indian Point near-field, lengths for juveniles of selected species in the far-field surveys, and food habits of bluefish in the lower river. Methods, results, and portions of this data set were published in journals (Holsapple and Foster 1975, Nittel 1976, Klauda et al. 1980, Dey 1981), data displays, and technical reports (Table 8).

EA DATA. EA did not collect data on the biological characteristics of fishes in the Hudson estuary.

LMS DATA. Length and weight information has been obtained on all fish species collected from all Hudson River surveys conducted by LMS. In addition to these two parameters, meristic and physiological data have been compiled on selected populations, including white perch, striped bass, blueback herring, Atlantic tomcod, and alewife.

Table 8. *Summary of biological characteristics data sets for selected Hudson River fishes collected by Texas Instruments Incorporated.*

SPECIES	YEARS	CHARACTERISTICS	COMMENTS	DISPOSITION/ACCESS
Striped bass	1972–1980	1–Length 2–Weight 3–Stomach contents, juveniles	1–Juvenile stomachs 1972–1980 only 2–Most data on adults from 1976–1980	Most post 1973 data in SAS format at CE, rest in card format; TI 1972a, b; 1973a; 1974c; 1975c; 1976f; 1977h; McFadden 1977; TI 1978a; McFadden et al. 1978; TI 1979a, c; 1980a, c; 1981b.
	1973–1980	1–Age 2–Sex 3–Maturity 4–Fecundity		
	1976–1979	1–Stomach contents, adults 2–Size and age in Hudson commercial catch		
White perch	1972–1980	1–Length 2–Weight 3–Stomach contents, juveniles 4–Age 5–Sex 6–Maturity 7–Fecundity	1–Juvenile stomachs 1972–1980 only 2–Data on adults prior to 1979 from Indian Point area only	See Striped Bass

Atlantic tomcod	1972–1980	1–Length 2–Weight	Sex ratio data for juveniles and adults	See Striped Bass
	1973–1980	1–Age 2–Sex 3–Maturity 4–Fecundity		
	1973–1975	Stomach contents		
Other Species	1972–1980	1–Length 2–Weight	1–Lengths and weights by length group, 1972–1980, from Indian Point survey 2–Juvenile lengths, 1973–1980, for selected species from far-field surveys	See Striped Bass
	1973–1980	Length, juveniles		
Bluefish	1974	Stomach contents	Jun-Sep collections from km 12–101	TI 1976e

The principal biological characteristics evaluated in the LMS near-field studies of distribution and species composition were age and growth based on scale analysis for white perch and striped bass. Scale analyses—including determinations of time of annulus formation, body length to scale length relationships, and back-calculations for annual growth patterns—were done on samples collected in Haverstraw Bay (1972–1980), northern Newburgh Bay (1972–1978), near Kingston (1972–1974), and Albany (1975). Results of scale analyses from discrete surveys and a combined Hudson River data base were used to determine annual growth patterns in relation to abiotic (temperature, freshwater flow) and biotic variables (population density), growth differences among year and age classes, duration of the annual growth season, and influence of sex on growth. Scale analysis has been attempted on alewife, blueback herring, and Atlantic tomcod with limited success.

Determination of sex, spawning condition, and fecundity was done during 1973 and 1974 on alewife, blueback herring, striped bass, white perch, and Atlantic tomcod collected from Haverstraw Bay, northern Newburgh Bay, and Kingston. Special sampling programs were conducted during 1973–1975 in lower Hudson River areas to evaluate feeding preferences for Atlantic tomcod, white perch, and striped bass. Data from these surveys have been published in journals (Dew and Hecht 1976; Grabe 1978) and several technical reports (Table 9).

PHYSIOLOGY AND BEHAVIOR

TI DATA. From 1972 through 1974, TI conducted a series of physiological and behavior studies to evaluate the effects of thermal discharges from the Indian Point plant on fishes passing through or attracted to the thermal plume and/or into the effluent canal (Table 10). Specific studies to measure behavioral responses (thermal preference and avoidance) to above-ambient temperatures and physiological responses (temperature tolerance and active respiration) to rapid increases and declines in water temperature were conducted on: juvenile and yearling striped bass; juvenile, yearling, and adult white perch; and adult Atlantic tomcod. The acute and chronic effects of cooling-tower blowdown and other chemical discharges on white perch and striped bass also were studied. The data set contains information on juvenile and older striped bass, white perch, and Atlantic tomcod; but it is limited, from a general use perspective, because neither larvae nor other fish species common to the Indian Point area were studied. Methods, results, and summarized portions of this data set were published in several technical reports (Table 10).

EA DATA. The physiological and behavioral studies conducted by EA examined the effect of temperature on some of the major species of Hudson River fishes from egg through adult stages (Table 11). The parameters evaluated include cold and heat shock, acute response to extended exposure (ultimate incipient lethal temperature), optimal temperature for normal egg hatching and larval growth, thermal preference and avoidance, and prey susceptibility following sublethal thermal shock. This is an extensive data set with emphasis on striped bass, white perch, Atlantic tomcod, alewife, white catfish, and spottail shiner. Limited data were collected for other fish species depending on availability of test specimens and laboratory equipment scheduling, with priority to the above species. The analysis and presentation of these data in technical reports were directed at the specific question of thermal effluent effects. However, the data are being reanalyzed from a life-history perspective for publication in journals. The hand-tabulated data are currently available in the technical reports (Table 11).

LMS DATA. LMS did not collect data on physiology and behavior of fishes in the Hudson River estuary.

ENTRAINMENT

TI Data

TI did not collect data on entrainment in the Hudson River estuary.

EA Data

Entrainment abundance and survival studies have been conducted by EA since 1975 at Hudson River generating plants (Tables 12 and 13). Most sampling occurred during the spring and summer spawning and nursery periods of the common fish species. Abundance sampling was performed at the intakes with nets and in the discharge areas with pumps (Occhiogrosso et al. 1981) or nets to document the magnitude of entrainment. These estimates were compared with river population estimates and survival data in order to assess ecological impact. The nets typically obtain samples with larger volumes than the pump. However, since the discharge is generally more homogeneous owing to mixing, the variability introduced by sampling errors

Table 9. *Summary of biological characteristics data sets for selected Hudson River fishes collected by Lawler, Matusky & Skelly Engineers.*

SPECIES	YEARS	CHARACTERISTICS	COMMENTS	DISPOSITION/ACCESS
Atlantic tomcod	1971–1980	1–Total length 2–Weight	Collected from northern Newburgh Bay, Haverstraw Bay, Albany area 1975. Kingston area 1971–1973. Manhattan area collections initiated in 1979	Data collected prior to 1974 hand reduced, data from 1974–1980 in card image format. Haverstraw Bay data 1974–1980 in SAS/CHGE 1977; ORU 1977; NMPC 1976; Grabe 1978; QLM 1973h; LMS 1975d, 1978c, 1979d, 1979f, 1980d, 1980e, 1980f, 1981a, 1981c, 1981d
	1972–1975	1–Sex 2–Fecundity 3–Stomach contents		
Striped bass	1971–1980	1–Total weight 2–Weight 3–Age and growth 4–Sex	A representative subsample of all striped bass collected were analyzed for total length and weight. Scales to evaluate age, growth, time of annulus formation, and growth patterns of males and females collected from northern Newburgh Bay 1972–1979 and from Haverstraw Bay 1972–1980. Sex and fecundity analysis on scale analysis specimens.	See Atlantic tomcod

White perch	1971–1980	1–Total length 2–Weight 3–Age and growth 4–Sex	See Atlantic tomcod
	1972–1974	Fecundity	A representative subsample of all white perch collected are analyzed for total length and weight. Scales to evaluate age, growth, time of annulus formation, and growth patterns of males and females collected from northern Newburgh Bay 1972–1979 and from Haverstraw Bay 1972–1980. Sex and fecundity analysis on scale analysis specimens.
Blueback herring Alewife	1971–1973	1–Total length 2–Weight 3–Sex 4–Fecundity	See Atlantic tomcod Representative sub-sample of all specimens collected 1972–1980 analyzed for total length and weight. Sex composition and fecundity analysis during 1972–1973
Other species	1978–1980	1–Total length 2–Weight	See Atlantic tomcod Representative sub-sample of all specimens collected, analyzed for total length and weight.

Table 10 Summary of data sets related to physiology and behaviour of Hudson River fishes collected by Texas Instruments Incorporated. (Continued)

PARAMETER	YEARS	PRIMARY SPECIES	DATA	COMMENTS	DISPOSITION/ACCESS
Thermal preference	1972–1974	Striped bass White perch	1–Juvenile, adult 2–Vertical temperature gradient 3–Tested exposure range of 1.5–27.5 hr		Data in card imate format at CE; TI 1974c; 1975c; 1976g
Thermal avoidance	1972–1974	Striped bass White perch Atlantic tomcod	1–Juvenile adult 2–Horizontal temperature gradient		see Thermal Preference
Thermal tolerance (heat shock)	1972–1974	Striped bass White perch Atlantic tomcod	1–Juvenile, adult 2–Range of acclimation temperatures 3–Time to equilibrium loss and 96-hr TLm	Only single test on Atlantic tomcod	see Thermal Preference
Thermal tolerance (cold shock)	1972–1974	Striped bass White perch	1–Yearling, adult 2–10°, 15°, and 20°C acclimination temperatures 3–Varying rates of temperature decrease to 2°C 4–Time to equilibrium loss and 96-hr TLm		see Thermal Preference
Active respiration	1972–1974	Striped bass White perch	1–Juvenile, yearling, adult 2–Acclimated and tested at 8°–24°C 3–Current velocity range of 10–30 cm/sec	Unexplained low and negative metabolic rates	see Thermal Preference
Chemical discharge effects	1974	Striped bass White perch	1–Juvenile, yearling 2–96-hr acute tests 3–28-day chronic tests	Used prepared test chemical slurry	Data in card image format at CE; TI 1974d

in patchy or stratified media is reduced. In addition, the pump samplers generally provide more dependable and accurate sample volumes and eliminate the clogging encountered with plankton nets. The pumped samples were also used to obtain continuous 24-hour samples in order to evaluate diel patterns of fish abundance. Entrainment survival was estimated using intake samples to measure sampling-related mortality and using discharge samples to measure combined entrainment and sampling mortality. Several major modifications to the sampling procedure (Jinks et al. 1981) were made in order to reduce the potential for synergistic interactions between sampling and entrainment stresses. Therefore, caution must be exercised in comparing data from different years. Initial and latent mortality were observed, and length, life-history stage, and water chemistry data were recorded. At the Indian Point plant, plankton nets also were used to examine striped bass entrainment survival. Differences in "tow" speed between the intake and discharge result in differences in gear effects between stations; therefore, the intake probably underestimates sampling-related mortality and does not provide a good control for this source of mortality.

Most of the data sets for individual plants and years are in SAS format. Data are also available on phytoplankton and zooplankton survival at most plants and are tabulated in technical reports (Tables 12 and 13).

LMS Data

LMS conducted entrainment abundance studies (Table 14) using conical plankton nets at the intakes of Bowline Point (1973–1974), Lovett (1972–1974), Danskammer Point (1973–1979), and Roseton (1973–1979). Comparative samples were collected in a 1.0-m Hensen plankton net at different depths along several river transects during all plant studies to assist in the evaluation of plant operational impact. Entrainment studies conducted during 1973–1975 generally included samples from January (Atlantic tomcod) to mid-October and encompassed the spawning seasons of the majority of Hudson River fish populations. Ichthyoplankton specimens were identified to the lowest taxon possible, and a subsample was analyzed for total length and life-stage designation.

The Danskammer Point entrainment study conducted during 1973 employed two sets of nets vertically arrayed in the cooling water intake canal: 1.0-m Hensen nets (two nets) and 0.5-m conical nets (three nets), each with 571-μm mesh. A 1.0-m Hensen net was also

Table 11. *Summary of data sets related to physiology and behavior of Hudson River fishes and invertebrates collected by Ecological Analysts Incorporated*

PARAMETER	PRIMARY SPECIES	DATA	COMMENTS	DISPOSITION/ACCESSA
Thermal tolerance	Striped bass, white perch, Atlantic tomcod, alewife, white catfish, spottail shiner, *Gammarus*, *Neomysis*, *Chaoborus*	1–Thermal plume entrainment simulation 2–96-hour exposures 3–Yearling, adult, young-of-the-year 4–Range of acclimation temperatures 5–Upper and lower tolerance limits 6–Length effects	1–Data base not complete for all species & acclimation temperature combinations 2–Data not in general access computer files; all initial data tabulation and summarization by hand	EA 1978a
	Striped bass, white perch, Atlantic tomcod, alewife, white catfish, spottail shiner, invertebrates	1–Short term exposure—condenser entrainment simulation 2–Length effects 3–Larvae and early juveniles		EA 1978a, 1979a
Hatching success	Striped bass, white perch, alewife, Atlantic tomcod	1–Thermal exposure 2–Percent normal hatch		EA 1978a, Kellogg et al. 1978

Growth	Striped bass, white perch, alewife, white catfish, Atlantic tomcod, spottail shiner	1–Thermal exposure 2–Individual weight/total biomass/mortality 3–Larvae and early juveniles 4–Constant excess food supply		EA 1978a
Thermal preference	Striped bass, white perch, Atlantic tomcod, alewife, white catfish, spottail shiner	1–Vertical and horizontal test gradients 2–Length/age class effects		EA 1978a
Predator-prey interaction	White perch yearling Striped bass larvae	1–Vertical and horizontal test gradients 2–Preferential predation 3–Recovery time 4–Natural turbidity conditions and filtered water 5–Alternate food source	Larval density approx. 20 times peak river density	EA 1978a
Thermal effects literature	Striped bass, white perch, Atlantic tomcod, alewife, white catfish, spottail shiner, Atlantic and shortnose sturgeon, bay anchovy, weakfish, *Neomysis, Crangon, Gammarus, Chaoborus*	Tolerance, preference, avoidance data and life history relationships	Review of literature through 1977	EA 1978b

Table 12. Summary of fish entrainment abundance data sets for Hudson River generating plants collected by Ecological Analysts Incorporated

PLANT	YEARS	PRIMARY SPECIES	DATA	STRENGTHS	WEAKNESSES	DISPOSITION/ACCESS
Danskammer Point	1977	Striped bass White perch Clupeids Other spp.	1–Unit 3 discharge 2–Only Mar and Jun samples processed 3–Water chemistry 4–1 to 3 days/week		1–Only 1 year 2–Only during peak season 3–Life stages, no lengths	CHG&E; multiple record semi-hierarchical file; EA 1978h
Roseton	1976–1979	Striped bass White perch Clupeids Other spp.	1–Pumped discharge 2–Diel density distribution 3–Lengths—all years except 1976 4–Water chemistry 5–1 to 3 days/week 6–May to Jul	1–More even distribution in water column 2–More reliable flow determination 3–Eliminates net clogging and associated avoidance factors 4–Continuous sampling for 24-hr period w/diel record		CHG&E; SAS format; EA 1978h, 1980a
Indian Point	1978–1979	Striped bass Macrozooplankton	1–Nets, intake and discharge 2–12-hr period surrounding dusk		1–Only about 2 weeks of samples in 1979 (unit off line)	Con Edison; SAS format; EA 1980b, 1981c

	1979–1981	Striped bass, white perch, clupeids, bay anchovy, other spp.	1–Pumped discharge samples 2–Lengths 3–12-hr around dusk—1979 4–24-hr coverage—1980–1981 5–Water chemistry 6–1 to 2 days/week 7–Jun to Jul	1–More reliable flow records 2–Eliminates clogging 3–Comparison of 12-hr periods in 1979 showed pumps comparable or better than nets for YSL and PYS	2–Low filtration capacity at intake	Con Edison; SAS format; EA 1981a
Lovett	1975–1978	Striped bass, white perch, clupeids, Atlantic tomcod, bay anchovy, other spp.	1–1975–1976 Unit 5 intake nets 2–1977–1978 Unit 5 pumped samples at discharge—24 hr 3–Net samples taken at surface, mid-depth and bottom at dawn, noon, dusk, and midnight 4–Lengths 5–Generally 2 days/week	vertical and diel distribution for nets	1–Most 1978 samples voided by sampling problems 2–Net sample volumes estimated from velocity matrix constructed from electromagnetic flowmeter rotated among the three depths. Matrix set up by tide, depth, and plant flow	ORU; multiple record semihierarchical file; EA 1976c, 1977c, 1978i

Table 12. *Summary of fish entrainment abundance data sets for Hudson River generating plants collected by Ecological Analysts Incorporated (Continued)*

PLANT	YEARS	PRIMARY SPECIES	DATA	STRENGTHS	WEAKNESSES	DISPOSITION/ACCESS
Bowline Point	1975–1979	Striped bass, white perch, Atlantic tomcod, clupeids, other spp.	1–Intake net samples—3 depths 2–Electromagnetic flow measurement at all depths for 1978–1979 3–Velocity matrix by depth tide and plant operation used for 1975–1977—clogging monitored at one depth with electro-magnetic meter 4–Dawn, noon, dusk, and midnight 1975–1976 5–Noon and midnight 1977–1979 6–Length data 7–Water chemistry 8–1975 net clogging study to	1–5-yr data base 2–Depth and diel distribution	1–Sample durations variable in 1975 and 1976 2–Sample volumes by direct measure only in 1977–1979 3–Clogging may be problem for long duration tows in 1975 and heavy algal blooms in 1978	ORU; SAS format, EA 1976d, 1977d, 1978j; 1979e, 1978b

Bowline Point	1977–1981	Striped bass, white perch, Atlantic tomcod, clupeids, other spp.	1—Pumped discharge sampling 2—Diel distribution 3—Length data 4—Water chemistry 5—Primary station at Unit 1 6—Comparison of Units 1 and 2 1978 and 1980 7—Mar, May–Aug, or Sep determine optimal "tow" duration 9—2 days/week	1—Reliable flow record from in-line meters 2—24 hr sampling reduces effects of temporal patchiness 3—5-yr data base 4—3 to 5 days/week in 1978–1980 Some voided samples as a result of electrical supply and computer memory problems	ORU; SAS format; EA 1978j, 1979e, 1981b, d

YSL = yolk-sac larvae
PYS = post yolk-sac larvae
ORU = Orange and Rockland Utilities, Inc.

Table 13. *Summary of fish entrainment survival data sets for Hudson River generating plants collected by Ecological Analysts Incorporated.*

PLANT	YEARS	PRIMARY SPECIES	DATA	STRENGTHS	WEAKNESSES	DISPOSITION/ACCESS
Danskammer Point	1975, 1978	Clupeids Striped bass White perch	1–Water chemistry 2–Lengths 3–Unit 4 4–Collection gear 5–4 days/week 6–Mar, Jun		1978 low intake (control) survival	CHG&E; multiple record semi-hierarchical file; EA 1981e
	1981	Clupeids Striped bass White perch	1–Survival collection gear comparison 2–Intake only 3–4 days/week 4–Jun	1–Low sampling mortality 2–Large numbers collected 3–High egg abundance		CHG&E, ESEERCO; multiple record semi-hierarchical file; EA 1981e
Roseton	1975–1978, 1980	Clupeids Striped bass White perch	1–Water chemistry 2–Lengths 3–Collection gear calibration 4–Combined discharge sealwell 5–Complete system effect w/diffuser sampling in 1980 6–4 days/week 7–Mar, Jun	1–Multiple gear comparisons—1980 2–Diffuser discharge effects—1980	1–Lengths for 1975–1976 measured 2 years later 2–Gear changes between years, sampling mortality may not be comparable 3–Low intake survival operational problems w/collection gear accelerated flow	CHG&E, SAS format; EA 1976b, 1978f, g, 1980 report in preparation

	1977, 1978	Hatchery-reared striped bass	1–Direct release of fish at intake 2–Water chemistry 3–Plant operating mode 4–1980—all releases at <30°C to evaluate mechanical effects of entrainment	Provide larger sample sizes for evaluation of factors effecting entrainment survival	1–Precision of released population estimate 2–Low recovery efficiency for some releases in 1980 3–Not all stations operating for all releases in 1980	CHG&E; SAS format; EA 1978g
Roseton	1980	Hatchery-reared striped bass	Assess effects of conductivity on collection and holding stress			CHG&E; hand generated
	1980	Hatchery-reared striped bass	Assess synergy between thermal and mechanical stress	First study on Hudson River specifically designed to address synergy under natural water conditions	Small data base	Con Edison; hand generated; report in preparation
Indian Point	1977–1980	Striped bass, white perch, Atlantic tomcod, clupeids, bay anchovy	1–Length data 2–Water chemistry 3–Gear calibration 4–Larval table/pumped and reardraw 5–4 days/week 6–Mar, Jun	1980 gear modifications reduced sampling mortality to negligable levels	Lack of inter-year comparability; changes in gear design, differences in plant operation	Con Edison; SAS format; EA 1978d, 1979c, 1981a

Table 13. Summary of fish entrainment survival data sets for Hudson River generating plants collected by Ecological Analysts Incorporated. (Continued)

PLANT	YEARS	PRIMARY SPECIES	DATA	STRENGTHS	WEAKNESSES	DISPOSITION/ACCESS
	1977–1980	Hatchery-reared striped bass	1–Gear efficiency and calibration 2–Survival 3–Lengths 4–Jun	Larger single sample size for analyses	Precision of released population size estimate	Con Edison; SAS format; EA 1978d, 1979c, 1981a
	1977–1979	Striped bass, macrozooplankton	1–Net collections 2–Intake and discharge 3–Length data 4–1 day/week 5–Mar to Dec		Adequacy of control station; differences in velocity between intake and discharge	Con Edison; SAS format; EA 1980b, c, 1981c
Lovett	1975–1977	Atlantic tomcod, striped bass, white perch, clupeids, bay anchovy	1–Length data 2–Water chemistry 3–4 days/week 4–Mar, Jun	Comparable gear for all 3 yr		ORU; multiple record semi-hierarchical file; EA 1976c, 1977c, 1978k
Bowline Point	1975–1979	Atlantic tomcod, striped bass, white perch, clupeids, bay anchovy	1–Pumped samples in all years 2–Pumpless gear 1979, 1980 3–Total system effect with diffuser samples 1979–1980 4–Lengths 5–Water chemistry	Comparable data for 5-yr data base	Sample sizes for individual years—often too small for comparisons of temp., regimes, pump mode and length or life stage	ORU; SAS format; EA 1976d, 1977d, 1978l, 1979e, 1981b

	1977–1980	Hatchery-reared striped bass	1–Gear calibration and efficiency 2–Survival 3–Lengths 4–Transit time 5–2 to 4 days/week 6–Jun 6–Primarily Unit 1 7–3 to 5 days/week 8–Mar, Jun, Jul	Provides large single samples for analyses	1–Accuracy of population size estimate 2–Additional handling stress on control groups not adequately reflected in released fish collected at discharge	ORU; multiple record semi-hierarchical file; EA 1978l, 1979e, 1981b
Roseton, Danskammer Point Lovett Bowline Point	1975–1976 1975, 1976 1979	Microzooplankton *Gammarus* *Neomysis* *Crangon* *Chaoborus* *Monoculodes*	1–Density 2–Water chemistry 3–Microzoo—replicated samples for latent effects 4–Mar–Nov	1–Generally large sample size 2–Seasonal succession		CHG&E and ORU; multiple record semi-hierarchical file; EA 1976c, d, 1977c, d, 1979e, 1981b
Roseton, Danskammer Point, Lovett, Bowline Point	1975, 1976	Phytoplankton	1–ATP 2–Phytopigments 3–C^{14} primary productivity 4–Auto fluorescence 5–Density and biomass 6–Species indent. 7–Mar–Nov	1–Seasonal succession 2–Diversity of parameters to evaluate relative health of community	Parameters examined not consistent between years	CHG&E and ORU; multiple record semi-hierarchical file; EA 1976c, d, 1977c, d, 1981b

ESEERCO = Empire State Electric Energy Research Corporation

Table 14. *Summary of fish entrainment abundance data sets for Hudson River generating plants collected by Lawler, Matusky & Skelly Engineers.*

PLANT	YEARS	SPECIES	DATA	STRENGTHS	WEAKNESSES	DISPOSITION/ ACCESS
Bowline Point	1973–1974	Striped bass, white perch, Atlantic tomcod, *Alosa* spp., plus other common species	1–Intake net samples—3 depths 2–Through net velocity measured with a TSK meter 3–Day, night, dusk, and drawn sampling 4–Species identification, length and life-stage data 5–Water chemistry	1–Depth and diel data 2–Net mesh same as net used for simultaneous river samples	1–Intake velocity generally below TSK meter threshold 2–Long duration samples subject to net clogging 3–Inconsistent sample times	ORU; 1973 hand generated, 1974 card image format/ LMS 1974; ORU 1977
Lovett	1972–1974	Striped bass, white perch, *Alosa* spp., plus other common species	1–Intake net samples at Unit 5 2–Day, night, dusk, and dawn sampling periods 3–Species identification lengths 4–Water chemistry	1–Depth and diel data 2–Similar net used as at other Hudson River plants	1–Inconsistent sample times 2–Net clogging	ORU; all data hand generated/QLM 1974a; LMS 1975b
Danskammer Point	1973–1979	Atlantic tomcod, white perch, *Alosa* spp., striped bass,	1–Intake net samples located in intake canal.	1–Long-term study using same gear and similar meth-	1–Samples collected from discrete time periods	CHGE; 1973 and 1974 data maintained on hand

plus other common species	a. 1.0-m and 0.5-m nets in two vertical arrangements 1973–1974 b. Surface, mid and bottom, 0.5 m 1975–1978 c. Surface and bottom, 1.0 m, 1978–1979 2–Sample volume determined using net mounted TSK meters 3–Samples collected day, night, dusk, and dawn 1973–1978; replicate day and night samples 1979 4–All ichthyoplankton identified and representative sample measured for total length and life stage designation 5–Water chemistry parameters recorded with each sample	odology among years 2–All larvae identified; representative subsample measured for length 3–Net clogging evaluated by comparing velocity meter readings between net mouth and frame mounted TSK meters 4–Large volume of water filtered for each sample	2–Net clogging and avoidance especially by larger larvae 3–Sampling conducted only during peak abundance periods	reduced summary tables, 1975–1979 data in card image format/LMS 1975c, 1978e, 1978f, 1979e, 1979f, 1980d, 1980e

Table 14. *Summary of fish entrainment abundance data sets for Hudson River generating plants collected by Lawler, Matusky & Skelly Engineers. (Continued)*

PLANT	YEARS	SPECIES	DATA	STRENGTHS	WEAKNESSES	DISPOSITION/ ACCESS
Danskammer Point	1980	*Alosa* spp., white perch, striped bass, and other common species	1–One-hour pump samples collected over 24-hr period from three cross channel locations at intake, random site selection used during 24-hr sample period 2–All organisms identified to species and representative sample analyzed for total length and life stage 3–Water chemistry taken with each sample	1–Continuous sample covering 24-hr 2–Permits detailed explanation of diel activity patterns	1–Relatively small volume of water sampled from each one-hr period 2–More organism damage observed compared to net samples 3–Only conducted during peak organism abundance periods	Data in card image format/LMS 1981d
Roseton	1973–1979	White perch, *Alosa* spp., striped bass, Atlantic tomcod, plus other common species	1–Intake net samples at three depths 2–Volume of water filtered based on net and frame mounted TSK meters	1–Stratified sampling program concentrating efforts during peak larval periods 2–All larvae identified; representa-	1–Low intake velocity resulted in through net velocities below TSK meter threshold 2–Small net size increases poten-	CHGE; 1973 data hand generated, 1974–1979 card image format/LMS 1975c; 1978e, 1978f, 1979e, 1979f, 1980d, 1980e, CHGE 1977

Roseton	1980	White perch, striped bass, *Alosa* spp., and other common species	1—24-hr pump sample composited from 3-hr discharge collections 2—All organisms identified to species; representative sample analyzed for total length and life stage 3—Sampling conducted at time periods corresponding to mid-day, mid-night, dusk and dawn, 1973–1978 4—Day and night replicate samples from 3 depths during 1979 5—Discharge 24-hr pump samples collected at 3-hr intervals in 1980 6—Water chemistry parameters recorded with each sample	Continuous sample covering 24 hrs tive subsample measured for length 3—Stage of larval development determined using definable morphological characteristics 4—Long-term study utilizing same gear and methodology 5—Corresponding river collections	1—Relatively small volume of water sampled from each 3-hr period compared to discharge flow 2—Greater amount of organism damage observed compared to net samples 3—Only conducted tial net avoidance	Data in card image format/ LMS 1981d

Table 14. *Summary of fish entrainment abundance data sets for Hudson River generating plants collected by Lawler, Matusky & Skelly Engineers. (Continued)*

PLANT	YEARS	SPECIES	DATA	STRENGTHS	WEAKNESSES	DISPOSITION/ ACCESS
			3–Water chemistry taken with each sample		during periods of peak organism abundance	
Roseton	1980	*Alosa* spp., striped bass, white perch, and other common species	1–Floating induced flow larval tables at plant intake and discharge; intake table using pump and discharge table using siphon 2–All ichthyoplankton identified to species and life stage and measured for total length 3–Six samples collected per night at each location, four nights per week from 28 May–26 Jun 4–Water chemistry parameters recorded with each sample	1–Comparison study done in conjunction with EAI 2–Larval table calibration and survival tests conducted 3–Through plant survival of ichthyoplankton determined	Problems with set-up of both tables; limits the number of comparative samples	Data in card image format/LMS 1981e

located at the Unit 3 discharge. Discrete samples were collected continuously over twenty-four hours, with the sampling dates stratified to concentrate more sampling effort during the spring. The same sampling scheme was used in 1974. However, sampling was conducted four times, at six-hour intervals over a 24-hour period. In 1975, the sampling scheme was modified to include only three vertically arrayed 0.5-m plankton nets in the intake canal each year from 1973 through 1978. In 1979, the Danskammer Point sampling scheme was changed to two 0.5-m plankton nets (surface and bottom) sampled in replicate day and night.

At Roseton, three vertically arrayed 0.5-m plankton nets were deployed at the intake during all years from 1973 through 1978. During the period 1974–1978, four discrete samples were collected over a 24-hour period at approximately six-hour intervals. In 1979, two net holding frames were used to include replicate day and replicate night samples. In 1973, a single 0.5-m plankton net was placed in a frame at the thermal discharge and sampled during periods of plant operation (pre-startup year); discharge sampling was done in conjunction with intake sampling.

Four-inch trash pumps were employed for entrainment sampling at Roseton and Danskammer Point during the spring–summer season of 1980. Twenty-four-hour periods were sampled at the Danskammer Point intake in 1-hour subsamples and at the Roseton thermal discharge in 3-hour subsamples. A short-duration entrainment survival study was conducted at Roseton in conjunction with the entrainment abundance program using induced flow larval tables, one at the intake and a second at the discharge seal well. Entrainment programs conducted at Lovett and Bowline Point employed the same vertically arrayed 0.5-m plankton nets at the intakes and, during 1973, a single frame-mounted 0.5-m net at the thermal discharges. River ichthyoplankton programs using 0.5-m Hensen nets were conducted at Kingston during 1973–1974 and on a bimonthly schedule at Albany during 1975.

Entrainment and near-field ichthyoplankton data and survey results have been published in several technical reports (Table 14). These data have served as the basis for several predictive models related to estuarine distribution patterns and plant operational impact.

IMPINGEMENT

TI Data

TI monitored fish impingement daily on a year-round basis at the Indian Point plant from 1972–1980 (Table 15). This long-term con-

Table 15. *Summary of fish impingement data sets collected by Texas Instruments Incorporated at the Indian Point generating plant, Hudson River.*

PARAMETER	YEARS	PRIMARY SPECIES	DATA	STRENGTHS	WEAKNESSES	DISPOSITION/ACCESS
Total impingement rates/environmental influences	1972–1980	All, except where noted	1–Daily collections from intake screens 2–Lengths and weights (striped bass, white perch, Atlantic tomcod) 3–Water chemistry, tidal stage 4–Collection efficiency measurements 5–Plant operational variables 6–Distribution, residency period, movement of fish in discharge canal (1972) 7–Histopathology in impinged vs. river collections, white perch and Atlantic tomcod (1972,1974)	1–Long-term, continuous series of daily impingement counts 2–Provides data to examine role of environmental factors on impingement	1–Lack complete matrix of collection efficiency values between units and among species, seasons and years for adjusting daily impingement counts to estimate total impingement 2–White perch surrogate for collection efficiency tests 3–Many short-term studies with minimal replication	Most post 1973 data in SAS format at CE; format; TI 1972a, b; 1974b, c; 1975e; 1976b; 1977h; 1979c; 1980b, d; McFadden 1977; McFadden et al. 1978

Mitigation	1972–1978	All, except where noted	8—Gill parasites on impinged vs. river collections, white perch, striped bass, bluefish (1974) 9—Gas bubble disease in impinged fish (1974) 1—Impingement rates with and without air curtains (1972–1974) 2—Impinged fish survival from Unit 2 vs. Unit 3 intake screens (1973) 3—Impingement rates at approach velocities of 0.15 vs. 0.30 m/sec (1973) 4—Survival of white perch passed through fish pump (1973–1974) 5—Vertical distribution of impinged fish (1974)	Date useful for establishing priorities and rigorous experimental designs for future studies Short-term studies with limited ability to adequately test all important variables	Data in card image format at CE; TI 1974b; 1975e; 1978b, d; 1979f; 1980b

Table 15. *Summary of fish impingement data sets collected by Texas Instruments Incorporated at the Indian Point generating plant, Hudson River. (Continued)*

PARAMETER	YEARS	PRIMARY SPECIES	DATA	STRENGTHS	WEAKNESSES	DISPOSITION/ACCESS
			6—Collection efficiency and survival of fish impinged on fine mesh traveling screen (1977–1978) 7—Effects of submerged weir on impingement (1977)			

tinuous data set contains daily counts and total weights of all fish species impinged; individual length and weight measurements on subsamples of striped bass, white perch, and Atlantic tomcod; records of associated physicochemical parameters (water temperature, dissolved oxygen, and conductivity); and relevant plant operational variables. The data set also contains information on special impingement-related studies such as: (1) estimates of impingement collection efficiency; (2) effects of environmental variables on impingement rates; (3) extent of fish movement into the discharge canal; (4) effectiveness of air curtains in reducing impingement; (5) comparisons of impingement rates and survival of impinged fish at Units 2 and 3 intakes; (6) effects of approach velocity near the intake screen face on impingement rates; (7) effectiveness of a fish pump for use in a screen bypass and fish collection system; (8) comparisons of incidence of histopathological symptoms, condition factor, gas bubble disease, and gill parasites for impinged and river fish; (9) vertical distribution of fish impinged on intake screens; and (10) effects of a fine mesh, continuously operating traveling screen on initial and extended survival of impinged fish. The major strength of this data set is the daily count of fish impinged on the intake screens. The major weakness is the inherent variability in actual collection efficiency among units, species, seasons, and years, and the unknown effects of this variability on the accuracy of adjustments used to estimate total impingement at Indian Point. Limited replication in several of the special studies also is a weakness. Methods, results, and summarized portions of this data set were published in journals (Hogan and Williams 1976, Lieberman and Muessig 1978) and in data displays and technical reports (Table 15).

EA Data

Impingement survival studies were performed by EA at the Roseton, Danskammer Point, and Bowline Point plants from 1975 through 1980 (Table 16). The effects of screenwash frequency and spray pressure were evaluated. For some stations during some years, the frequency and duration of sampling controls are limiting factors in evaluating impingement survival. The rate of reimpingement, effects of fish-return systems, and modified screens and a barrier net for fish protection also were evaluated. Initial and latent mortality and the influence of water chemistry were evaluated. The specific studies and data contained in computer data files and technical reports are summarized in Table 16.

Table 16. *Summary of fish impingement survival data sets for Hudson River generating plants collected by Ecological Analysts Incorporated.*

PLANT	YEARS	PRIMARY SPECIES	DATA	STRENGTHS	WEAKNESSES	DISPOSITION/ ACCESS
Danskammer Point	1975–1977	White perch Clupeids Atlantic tomcod	1–Life stage 2–Water chemistry 3–Screen operation modes continuous 2-hr, 4-hr, and 6-hr holds 4–Fall period of peak impingement		1–Treatment of infection in 1975 with potassium permanganate 2–No sampling controls in 1975 3–Collection basket holding period was 30 min for continuous wash and 15 min for intermittent wash	CHG&E; multiple record semi-hierarchical file; EA 1976a, 1977a, b, 1978e; King et al. 1978
	1979, 1980	White perch Clupeids	1–Assessment of modified screens with baskets 2–Length data 3–Continuous and 2 hr intermittent	Comparison w/ standard screens and controls		CHG&E; multiple record semi-hierarchical file; EA 1976a, 1977a, b, 1978e; Clock and Huggins 1982
Roseton	1975–1977	White perch Clupeids Atlantic tomcod	1–Life stage 2–Screenwash pressure 3–Fish return line effects		1–Treatment of infection in 1975 w/potassium permanganate 2–No sampling con-	CHG&E; multiple record semi-hierarchial file; EA 1976b, 1977b, 1978e; King et al. 1978

Bowline Point	1975–1981	White perch Striped bass Clupeids	1–Life stage 2–Water chemistry 3–Screenwash pressure 4–Collection mortality vs. hold duration 5–Reimpingement 6–Barrier net efficiency 7–Fall, winter, spring peaks of impingement	4–Screenwash mode continuous, 2 hr, 4 hr, 6 hr intermittent 5–Fall and winter water chemistry 6–Water chemistry 7–Fall and winter impingement peak trols in 1975 3–Continuous mode wash was 2 hr, intermittent wash was 30 min. All controls held in collection basket for 2 hr 1–Limited number of controls during winter prior to 1979—controls frequently collected from intake screen during week prior to sampling 2–Small sample sizes in 1979 and 1980 when barrier net effectively diverted fish from intake structure	ORU; SAS format; EA 1979d; King et al. 1978

LMS Data

LMS has conducted impingement monitoring studies incorporating a minimum of one 24-hour collection per week from each operating unit at the Lovett and Danskammer Point plants since 1972 and at Bowline Point and Roseton since 1973. All organisms were identified as to species, total species biomass recorded, and a representative subsample measured for total length and weight. During the initial survey years (1972–1975), major species were examined for sex and gonad development. Physical and chemical parameters were measured for each impingement sample. A two-year impingement monitoring program was also conducted by LMS at the Albany Steam Station (May 1974–April 1976). Two 24-hour samples were collected each week from operating units during the first survey year, and a weekly alternating 12-hour/24-hour schedule was followed during the second year.

In addition to data on actual numbers of fish impinged at the various plants and related information on spatial and temporal distribution patterns, seasonal and annual impingement estimates were calculated based on cooling water flow at each plant. These impingement estimates were used in evaluating plant impact on the adult fish community. Special programs conducted in conjunction with impingement monitoring programs include: (1) evaluation of collection efficiency and traveling screen carryover; (2) recirculation of impinged fish; (3) predation on impinged organisms (Bowline Point); (4) deployment and evaluation of a barrier net during the fall–spring period (Bowline Point); (5) effectiveness of an air curtain in reducing impingement (Roseton 1979); and (6) effectiveness of a chain barrier in reducing impingement (Danskammer Point 1976–1977).

Information from all impingement monitoring studies has been used in the preparation of demonstrations for Section 316. Methods, results, and summary data were published in a series of technical reports (Table 17).

GEAR PERFORMANCE AND SPECIAL STUDIES

TI Data

TI conducted several studies of gear performance between 1972 and 1980 (Table 18). During the first three months that TI sampled the fish populations of the Hudson River (April, May, and June 1972),

Table 17. *Summary of fish impingement data sets for Hudson River generating plants by Lawler, Matusky & Skelly Engineers.*

PLANT	YEARS	PRIMARY SPECIES	DATA	STRENGTHS	WEAKNESSES	DISPOSITION/ ACCESS
Bowline Poant	1973–1980	White perch, bay anchovy, blueback herring, alewife, striped bass, and other common species (total species = 62)	1–24-hr for impingement collections three times per week, 1973–fall 1979, two 24-hr collection per week fall 1979, one 24-hr collection per week and addition 24-hr collection within 72 hr if ≥ 1000 fish collected, 1980 2–All organisms identified to species and enumerated; major species analyzed for total length and weight 3–Mark/recapture program conducted 1973–1975 4–Traveling screen collection efficiency	1–All organisms identified and counted 2–Long-term data base 3–River nekton sampling coincident with impingement collections	Increase in confidence limits on annual estimates of impingement owing to low frequency of sampling	1973 hand reduced; 1974–1980 card image format and SAS/QLM 1973c, 1974a; LMS 1975b, 1976a, 1977b, 1978c, 1979d, 1981a, 1981c

Table 17. *Summary of fish impingement data sets for Hudson River generating plants by Lawler, Matusky & Skelly Engineers. (Continued)*

PLANT	YEARS	PRIMARY SPECIES	DATA	STRENGTHS	WEAKNESSES	DISPOSITION/ ACCESS
			ciency and carryover observations during 1978–1980 5—Annual plant impingement total calculated from cooling water flows 6—Water quality parameters measured with each sample			
Lovett	1972–1980	Atlantic tomcod, alewife, blueback herring, white perch, American shad, striped bass, and other common species	1—One 24-hr collection per week from intake traveling screens at each operating unit 2—All organisms identified to species and representative sample of dominant species measured for total length and weight	1—Long-term data base with all organisms identified 2—Simultaneous monitoring of water quality parameters enables analysis of role of environmental variables on organism presence and abundance	Impingement estimates based on one 24-hr sample per week	1972–1973 data hand reduced, 1974–1980 data in card image format, 1974–1979 data in SAS format/ QLM 1973c; 1974a; LMS 1975b, 1976a, 1977c, 1978b, 1979c, 1980c 1981b

Danskammer Point	1972–1980	Atlantic tomcod, white perch, blueback herring, alewife, American shad, striped bass, and other common species	1—One 24-hr collection per week from intake traveling screens 2—Samples collected and analyzed in 6-hr intervals 3—Additional 24-hr impingement collection conducted within 72-hrs of regular impingement if	1—All organisms for 24-hr period identified and counted 2—Long-term data base 3—Monitoring of water quality parameters with each sample permits correlation analyses 4—Simultaneous col-	Annual impingement estimates generally based on one sample per week	1972–1973 data hand reduced; 1974–1980 in card image format/QLM 1973g, 1974b; 1978e, 1978f; 1979e, 1979f; 1980d, 1980e, 1981d

3—Generating unit and plant annual impingement calculated from cooling water flows
4—Traveling screen collection efficiency calculated for 1978–1980
5—Data on condition factor, spawning condition and sex collected during 1972–1975
6—Water quality parameters measured with each sample

3—Simultaneous river nekton sampling 1973–1977

Table 17. *Summary of fish impingement data sets for Hudson River generating plants by Lawler, Matusky & Skelly Engineers. (Continued)*

PLANT	YEARS	PRIMARY SPECIES	DATA	STRENGTHS	WEAKNESSES	DISPOSITION/ACCESS
			organism abundance ≥ 1000, 1977–1980 4–All organisms identified to species and representative sample of each species measured for total length and weight 5–Plant annual impingement calculated from cooling water flows 6–Water quality parameters measured with each sample	lection river samples for comparison, 1973–1979		
Roseton	1980	*Alosa* spp., white perch, striped bass	1–Induced flow larval table collections at intake and seal well discharge 2–Total length and life stage	1–Several through table velocities evaluated 2–Simultaneous entrainment abundance sampling	1–Low intake survival noted for striped bass 2–Overall larvae abundance low during study period	CHGE; data in card image format/LMS 1980h

			determinations 3–Four sample dates per week, six samples per date 4–Water quality parameters with each collection		3–Sampling problems delayed start-up of comparison program	
Albany	1974–1976	Blueback herring, alewife, American shad, striped bass, white perch, and other common species (total species = 53)	1–Two 24-hr impingement collections per week, May 1974–Apr 1975. One 12-hr per week alternating with one 24-hr per week, May 1975–Apr 1976 2–All organisms were identified, enumerated, and subsample measured for total length and weight 3–Water quality parameters measured with each sample 4–Major species examined for sex and gonad development	1–Large number of samples collected during first year 2–NY State water quality monitoring station on-site for chemical parameter correlations	1–Sampling program covered only2 years 2–Small number of samples collected during second year 3–Little information available on Albany section of river	Data in card image format/LMS 1975f; NMPC 1976

Table 18. Summary of fisheries gear performance data sets collected by Texas Instruments Incorporated in the Hudson River (Continued)

GEAR	TESTS	YEARS	PRIMARY SPECIES	DATA	COMMENTS	DISPOSITION/ ACCESS
0.5-m conical net 1.0-m^2 conical net 1.0-m^2 epi-benthic net 2.0-m^2 Tucker trawl	1–Compare conical nets over range of mesh sizes 2–Compare conical nets to sled and trawl	1973	Striped bass	1–Eggs and larvae per volume sampled 2–May-Oct	Mesh size increased as season progressed	Data in card image format at CE; TI 1973b
1.0-m conical net 2.0-m Hensen net 1.0-m^2 epi-benthic sled 1.0-m^2 Tucker trawl	1–Compare catch of four gear 2–Test effect of mouth size on sled and trawl catch 3–Test effect of tow speed on trawl catch 4–Measure sled & trawl filtration efficiency 5–Test effect of tow duration on sled & trawl filtration efficiency	1974–1975	Striped bass	1–Eggs and larvae per volume sampled 2–Jun–Aug 1974 3–May–Sep 1975	1–Also evaluated day/night effects 2–Tests conducted between km 54 and 102	Data in card image format at CE; TI 1977a

Gear	Objective	Year	Species	Details	Notes	Source
30.5-m beach seine	Measure catch efficiency day and night	1977–1978	Striped bass, White perch	1–Juvenile and yearling plus 2–Water chemistry 3–American shad, alewife, and blueback herring catches also recorded	1–Day tests conducted in 1977, night tests in 1978 2–Tests in lower river between km 54 and 62	Data in SAS format at CE; TI 1978c, 1979b
1.0-m² Epibenthic	Measure catch efficiency at night	1978–1979	Striped bass, White perch	1–Juveniles per volume sampled 2–Sep–Oct 1978 3–Nov–Dec 1979 4–Paired samples, two boats 5–Range of two speeds	Tests conducted between km 38 and 61	Data in card image format at CE; TI 1980c
7.8-m (headrope) otter trawl 3.7-m (headrope) otter trawl	1–Compare catch rates in two trawls 2–Compare sizes of fish caught in two trawls	1979	Striped bass, White perch, Atlantic tomcod	1–Juveniles, yearlings plus 2–Lengths for primary species 3–Nov–Dec 4–All species counted 5–Paired tows of about equal area swept	1–Tests conducted during day between km 43 and 120 2–Gear compared in deep (>10 m) and shallow (<10 m) areas 3–Very few striped bass collected	Data in card image format at CE; TI 1981a

Table 18. *Summary of fisheries gear performance data sets collected by Texas Instruments Incorporated in the Hudson River (Continued)*

GEAR	TESTS	YEARS	PRIMARY SPECIES	DATA	COMMENTS	DISPOSITION/ ACCESS
15.2-m (headrope) otter trawl	1–Compare catch (number and size distribution) with small vs. large mesh in cod end 2–Compare size distribution of trawl catch vs. catches in gill nets and haul seines	1979	Striped bass	1–Subadults, adults 2–May–Jun, Oct. Nov. Dec 3–Daytime	1–Two mesh sizes not fished simultaneously 2–Tests conducted between km 38 and 98 3–Effective for sturgeons without rollers on lead line	Data in SAS format at CE; TI 1981b
10.7-m wide × 5.5-m deep Atlantic western trawl	Develop mobile gear for collecting striped bass during spawning run at bottom and mid-depths	1980	Striped bass	1–Subadults, adults 2–Summer, fall 3–Daytime	Only preliminary data collected	Data in card image format at CE
3-m beam trawl	Develop quantitative bottom gear for juvenile fishes	1980	Striped bass White perch Clupeids	1–Juvenile 2–Fall 3–Night	1–Conducted prelim. study 2–Potential replacement for 1.0-m^2 epibenthic sled	Data on card image format at CE

they tested a beach seine, purse seine, bottom trawl, surface trawl, traps, gill nets, and electroshocker to find the most effective gear for mark-recapture studies on striped bass and white perch. In 1973, TI began a series of evaluations of ichthyoplankton sampling gear which included: (1) mesh size comparisons in 0.5-m and 1.0-m conical nets; (2) catch comparisons among several types and sizes of ichthyoplankton gear; (3) effect of net mouth size on the catch efficiency of a 1.0-m^2 epibenthic sled and a 1.0-m^2 Tucker trawl; (4) effect of tow speed on the catch efficiency of a 1.0-m^2 Tucker trawl; and (5) effect of tow duration on the filtration efficiency of a 1.0-m^2 epibenthic sled and a 1.0-m^2 Tucker trawl. TI also compared two bottom trawls and conducted studies to measure the catch efficiency of a 30.5-m beach seine and a 1.0-m^2 epibenthic sled for juvenile striped bass and white perch. Most recently, they tested two large trawls (a 15.2-m headrope otter trawl and a 10.7-m × 5.5-m Atlantic western trawl) for collecting adult striped bass, and a 3.0-m beam trawl for collecting quantitative samples of juvenile fishes. TI's gear performance data set includes tests on a variety of fisheries gear, but it has two major weaknesses. (1) The tests focused on only two or three selected species, usually striped bass and white perch. (2) The tests were conducted in the middle and lower portions of the study area; hence, test results in these habitats may not accurately represent gear performance in the upper estuary. Methods, results, and summarized portions of this data set were published in several technical reports (Table 18).

EA Data

Several special studies were conducted by EA that were related to sampling gear and fish entrainment (Table 19). These studies include evaluation of a prototype fine-mesh screen at the Indian Point plant for reduction of entrainment and a simulated fine-mesh angled screen for diversion of ichthyoplankton from the intake. Several studies were conducted to assess the relative efficiency of nets and pumps for ichthyoplankton sampling. An entrainment survival model was developed which integrates empirical estimates of mechanical-entrainment effects and laboratory estimates of thermally induced mortality with plant operating characteristics to provide historical or projected estimates of mortality rates for each plant. The disposition of these data sets are listed in Table 19.

Table 19. *Summary of gear performance and special studies related to fish entrainment at Hudson River generating plants conducted by Ecological Analysts Incorporated.*

PLANT	YEARS	STUDY	SPECIES/TEST CONDITIONS	DATA	DISPOSITION/ACCESS
Indian Point	1977–1978	Fine mesh screen ichthyoplankton entrainment/impingement survival	Hatchery-reared striped bass YSL, PYS, JUV	1–Length distribution 2–Recovery efficiency 3–Ichthyoplankton imping.	Con Edison; multiple record semi-hierarchical file; EA 1979b
		Pump vs. net comparison for ichthyoplankton sampling	1–79 paired discharge samples 2–0.5m, 571-μm mesh net 3–6-inch pump 4–Volume, net = 87–148m³ 5–Volume, pump = 29–67m³ 6–3-in and 4-in pumps vs. net (1977)	1–Density estimate for individual taxa/lifestage and total ichthyoplankton 2–1977 study had small sample volumes and abundance for pumps thus reducing the power of statistical analyses	Con Edison, multiple record semi-hierarchical file; EA 1979c; King et al. 1981

Site	Years	Study	Species	Variables	Source
Bowline Point	1978–1979	Simulated, angled fine-screen evaluation w/modular flume	Hatchery-reared striped bass	1–Diversion system 2–Length effects 3–Survival	ORU; hand generated data summary; Muessig and Hutchison manuscript in preparation
Multiplant	1975–1980	Entrainment survival modeling	Striped bass, white perch, Atlantic tomcod, clupeids	1–Hourly discharge temp. 2–Daily minimum intake temp. 3–Plant pumping rate 4–Median length, striped bass, white perch, Atlantic tomcod 5–Empirical mechanical mortality estimates 6–Thermal mortality equations 7–Hourly and weekly mortality factors (fc)	Con Edison, ORU, CHG&E; multiple record semi-hierarchical file

JUV = juvenile

LMS Data

Several gear-evaluation studies for both ichthyoplankton and older fishes have been conducted by LMS in relation to Hudson River programs.

One of the first major ichthyoplankton programs on the Hudson River was conducted during 1965–1968 and reported by Carlson and McCann (1968). Different net designs, net cloth, and mesh sizes were employed during this survey. In 1973, LMS conducted an ichthyoplankton gear-comparison study at Cornwall (km 92), which included exact replicas of the 1965–1968 nets: the 0.5-m nets used by New York University at Indian Point and the 0.5- and 1.0-m nets used by LMS in near-field studies. All organisms were identified, and a representative subsample of each species was measured for total length. The nets with large mouth openings generally collected more and a larger size range of larvae than nets with smaller mouth openings. Major weaknesses in this gear-comparison study were the lack of information on net velocity through the water and the statistical design of the study.

Two related ichthyoplankton gear-comparison studies designed to evaluate tow velocity on net efficiency and larval catches were conducted during 1977 (LMS 1978d) and 1978 (LMS 1980d). In 1977, 0.5-m and 1.0-m nets were towed together at different velocities. Fewer organisms were collected in the 0.5-m net than in the corresponding 1.0-m net when all 0.5-m net data were compared to all 1.0-m net data.

The 1978 study was designed to address the effect of tow velocity on ichthyoplankton abundance estimates, and it specifically considered plant entrainment sampling conditions. Because of space restrictions, 0.5-m nets were employed at plant locations under generally lower velocities than at river locations where 1.0-m Hensen nets were used. The study design included four 0.5-m nets towed at velocities comparable to intake velocities at the downriver power plants in direct comparison to four 1.0-m nets towed at 90 cm/sec^{-1} (LMS standard river tow speed). A significant difference in abundance estimates between the net pairings was observed, with the larger net catching more organisms. The results of the two 0.5-m/1.0-m gear comparison studies have a direct bearing on estimates of entrainment impact when using two different sampling gears at plant and river locations. The predominant ichthyoplankton life stage evaluated was post-yolk sac *Alosa* spp. and *Morone* spp. Different results could have occurred with other life-history stages and species. In 1979, a comparison study evaluating the difference in nekton catches for a 7.9-m headrope otter trawl and 11.9-m headrope Yankee trawl was conducted in northern Newburgh Bay. A total of eighteen day and eighteen night comparison tows were made at surface and mid-water depth locations during October. On a catch/volume basis, the two types of gear did not differ significantly.

A second trawl study was conducted in northern Newburgh Bay during October 1980 to compare fish catches in otter trawls towed for a constant time versus otter trawls towed over a constant distance. The primary factor related to efficiency of trawl collections is net speed over the bottom. The faster the net moves over the bottom, up to the point where hydrodynamics begin to lift the net, the greater the catch.

Gear-comparison study methods, results, and data summaries have been published in several technical reports (Table 20). LMS also conducted several special studies related to fish distribution and impingement mitigation that are described in Table 21.

CULTURE

TI Data

From 1973 through 1975, TI conducted a program to assess the feasibility of culturing and stocking Hudson River striped bass as a potential mitigation technique to offset predicted striped bass losses at a pumped-storage generating station proposed for construction at Cornwall. TI continued to operate the hatchery at Verplanck, New York, through 1980 and cultured striped bass eggs and larvae for experimental purposes. Studies were also conducted to compare first-year survival, stomach contents, growth, maturity, and movements of hatchery-reared and wild striped bass. Methods, results, and most of this data set were published in several technical reports (Table 22).

TI also conducted small-scale culture studies with white perch and Atlantic tomcod, but only portions of the results have been published (Watson 1978).

EA Data

EA has not compiled specific data sets on fish culture studies, but artificial spawning and rearing procedures were developed for many of the physiologyical and behavioral studies (Table 11). Some of these procedures were described in the Methods sections of technical reports (EA 1978a). The propagation and rearing techniques described for most species were appropriate for providing laboratory test organisms

Table 20. Summary of fisheries gear performance data sets collected by Lawler, Matusky & Skelly Engineers in the Hudson River. (Continued)

GEAR	TESTS	YEARS	PRIMARY SPECIES	DATA	COMMENTS	DISPOSITION/ACCESS
Ichthyoplankton Nets –18-in conical net –0.5-m conical net –3-ft × 3-ft square net –1.0-m Hensen net –18-in conical net with solid vinyl collar	1–Compare nets over range of mesh sizes and tow speeds 2–Evaluate tow speed on species catch and length frequency 3–Evaluate mesh size and net configuration on species composition and body length 4–Evaluate net filtration efficiency	1972	*Alosa* spp., striped bass, and white perch	1–Eggs and larvae per volume sampled 2–Jun-Aug 3–Total length on representative subsample for all species	1–Tests conducted at river kilometer 92.5 2–Conducted in conjunction with cross river distribution study	Data hand reduced/QLM 1973f; Matousek 1973
Ichthyoplankton Nets –0.5-m conical net –570-µm mesh –1.0-m Hensen net –570-µm mesh	1–Compare nets over range of velocities on diel schedule 2–Evaluate composition and length frequencies of collections in separate nets at different velocities	1977	*Alosa* spp., striped bass, and white perch	1–Eggs and larvae per volume sampled 2–Water quality measured with each sample 3–Total length and life stage designation for all species	Two sets of comparative nets towed from same boat in Haverstraw Bay (km 58)	Data in card image format/LMS 1978d

Gear	Year	Species	Objectives	Notes
Ichthyoplankton Nets –0.5-m conical net with 571-μm mesh –0.5-m conical net with 505-μm mesh –1.0-m Hensen net with 571-μm mesh	1980	*Alosa* spp., striped bass, and white perch	1–Compare 0.5-m nets towed at velocities similar to power plant intake with 1.0-m nets towed at 90 cm/sec 2–Test effect of different net velocities on catch 3–Monitor net filtration efficiency	1–Eggs and larvae per volume sampled 2–Total length and life stage designated for all species 3–Water quality monitored with each sample set	1–Diel difference in collection evaluated 2–Study conducted at river kilometer 94 3–Direct net comparison study with four nets of one configuration towed behind one boat and four nets of other configuration behind second boat	Data in card image format/LMS 1980d
Otter trawl and Yankee trawl	1979	Blueback herring, alewife, and white perch	1–Compare catch in otter and Yankee trawls at surface and mid-depth stations 2–Evaluate net filtration efficiency 3–Test catch parameters between two nets including abundance, species composition, and length distribution	1–Species abundance per sample time and volume filtered 2–Length frequency data for all species collected	1–7.9-m headrope otter trawl; 11.9-m headrope Yankee trawl 2–Simultaneous tows of otter trawl and Yankee trawl using two boats 3–Diel comparison of catches conducted 4–Study con-	Data in card image format/LMS 1980g

Table 20. *Summary of fisheries gear performance data sets collected by Lawler, Matusky & Skelly Engineers in the Hudson River. (Continued)*

GEAR	TESTS	YEARS	PRIMARY SPECIES	DATA	COMMENTS	DISPOSITION/ ACCESS
					ducted at river kilometer 107	
Oter trawl –7.9-m headrope	1–Compare catches in trawls towed for constant time vs. trawls towed for constant distance 2–Correlate number of fish collected with distance towed over bottom when tow duration was constant 3–Correlate number of fish collected with trawl duration when tow distance over the bottom was constant	1980	White perch, blueback herring, hogchoker, bay anchovy, Atlantic tomcod	1–Organisms per sample and volume filtered 2–Length frequency and biomass for all species 3–Water quality parameters measured with each sample	1–Study conducted at a river kilometer 107 2–Diel sampling schedule	Data in card image format/LMS 1981e

on a small scale. However, during 1980 the spawning and rearing methodology described for alewife was modified for hatchery-level production of approximately three million larvae and early juveniles for a large-scale testing program.

LMS Data

LMS has not conducted any fish culture studies as part of the Hudson River study program.

DISCUSSION

In combination, the TI, LMS, and EA fisheries data bases compiled between 1971 and 1980 include an array of a basic and problem-oriented information on Hudson River fishes. The largest component of the TI data includes basic information on fish distribution, abundance, and species composition, and contains at least sixty-five thousand samples collected between 1972 and 1980. The biological characteristics and impingement data sets are other major components of the TI data. The largest component of the EA fisheries data relates to entrainment abundance and survival and to impingement survival studies. Data sets on near-field distribution and abundance, physiology, and behavior, as well as impingement, are other major components of the EA data. Fisheries data collected by LMS are concentrated primarily in the near-field areas of electrical generating stations and include information on the spatial and temporal distribution of ichthyoplankton and finfish, entrainment abundance estimates based on net collections at the plant intakes, and traveling screen impingement samples. Information on fishery meristic and physiological parameters and water-quality data is also included in the LMS data sets.

The TI, LMS, and EA data sets cover a wide area of the estuary, include many fish species, and focus on most portions of the life cycles of selected species. The major emphasis of these studies, especially TI's, was on striped bass, with white perch and Atlantic tomcod as secondary species. Consequently, most far-field surveys were designed to key on these species, and much of the existing data for many nonfocal species were collected with suboptimal designs for those particular species. However, beginning in 1979, TI modified most of the far-field surveys and made them more generic. Effective use of a given TI or LMS or EA fisheries data set should include a clear understanding of the objective(s) and sample allocation scheme/ experimental design(s) for the studies which generated that data set.

Table 21. *Summary of special studies related to fish distribution and impingment mitigation for Hudson River generating plants conducted by Lawler, Matusky & Skelly Engineers*

PLANT	YEARS	PRIMARY SPECIES	DATA	STRENGTHS	WEAKNESSES	DISPOSITION/ACCESS
Danskammer Point —Chain Barrier Evaluation	1975–1976	Atlantic tomcod, White perch	1–Number of fish per volume sampled with and without chain barrier in place 2–Total length and weight on all species 3–Water quality parameters	Sampling corresponded to impingement monitoring at intake traveling screens	1–Fish were collected by trap nets located behind the plant intake 2–Mechanical problems with chain barrier 3–Short duration study with minimal replication	CHGE; data hand reduced/MS 1977g
Roseton —Air Bubble Curtain	1979	Blueback herring, White perch, Alewife	1–Number of fish per volume with and without bubbler operating 2–Total length and weight of each species	1–Sampling conducted in 2-hr blocks, over six 24-hr periods 2–Random selection of bubbler operation	1–Low species abundance during study period 2–Short duration	CHGE; data in card image format/LMS 1980e

| Hydroacoustic Survey in northern Newburgh Bay | 1978 | Blueback herring, White perch, Alewife | 1–Number of single targets and aggregates per volume insonified
2–Otter trawl data for species composition
3–27 transects monitored day and night encompassing all general habitats and depths
4–Water quality data monitored along each transect | 1–Large segment of river monitored during short period
2–Depth and spatial distribution patterns
3–Ground truth evaluation of hydroacoustic data collected along each transect
4–Comparative fishery data collected in same river area at same time of year during six previous years | 1–Minimal resolution of targets near the bottom
2–Abundance within aggregations unknown | CHGE; data in card format and magnetic tape/LMS 1980d; Matousek et al. 1981 |

3–Water quality information

Table 22. Summary of striped bass culture data sets collected by Texas Instruments Incorporated in the Hudson River.

PARAMETER	YEARS	DATA	COMMENTS	DISPOSITION/ACCESS
Culture techniques	1973–1980	1–Brood fish collection/transport and holding 2–Hormonal inducement of gonads 3–Egg viability 4–Fertilization rates 5–Hatching rates 6–Larval rearing and transport	1–Larvae and juveniles reared for stocking in 1973–1975 only 2–Eggs and larvae reared for several research groups in 1976–1980 (experiments)	Data in card image format at CE; TI 1974a, e, f; 1975f; 1977b, c, d, i; 1979d, e; McFadden 1977; McFadden et al. 1978
Stocking and evaluation	1973–1979	1–Fingerling harvesting, tagging, shipping, and stocking 2–Growth, survival, movements maturity, and stomach contents of recaptures	Striped bass stocked only in 1973–1975	Data in card image format and SAS formats at CE; TI 1974a, e; 1975f; 1977b, c, d, h; 1979c, 1980b, d; McFadden 1977; McFadden et al. 1978

As extensive as the fisheries data sets collected by TI, LMS, and EA since 1971 are, they comprise only a portion of the ecological data and literature base for the Hudson River estuary. Research sponsored by the five utilities mentioned in this paper (and others), and that conducted by New York University, NALCO, and other groups, have resulted in the compilation of additional fisheries data sets. Several studies sponsored or conducted by Boyce Thompson Institute (e.g., Dovel 1981b), U.S. Environmental Protection Agency, Federal Energy Regulatory Commission, New York's Departments of Environmental Conservation and Transportation, the Port Authority of New York and New Jersey, and others, have also generated ecological data on the Hudson estuary.

We hope that our initial efforts to "catalogue" the TI, LMS, and EA fisheries data sets will stimulate those individuals familiar with the numerous other Hudson River data sets to do likewise. Perhaps, eventually, Hudson River researchers will have access to a central depository of fisheries and other ecological data that will be maintained and expanded during the decades ahead and that will become the model for other such long-term multiorganizational efforts. We also encourage the use of these fisheries data sets as the "seeds" for an environmental atlas of the Hudson River estuary that would build upon the contribution of Dovel et al. (1977) and attain the comprehensive nature of Lippson et al. (1980).

ACKNOWLEDGMENTS

Our examination of the fisheries data sets would have been impossible without the dedicated efforts of the men and women who collected and processed the data. Morris H. Baslow, J. B. Hutchison, Jr., and Susan G. O'Connor reviewed the manuscript and offered several helpful comments. We thank Beverly Knee for her clerical assistance. Financial support for the Hudson River Ecological Study was provided by: Consolidated Edison Company of New York; Orange and Rockland Utilities, Inc.; Central Hudson Gas and Electric Corporation; Niagara Mohawk Power Corporation and the New York Power Authority.

Part II
Striped Bass and White Perch

2. Commercial Fishery for Striped Bass in the Hudson River, 1931–80

James B. McLaren, Ronald J. Klauda, Thomas B. Hoff, and Marcia Gardinier

INTRODUCTION

Commercial fisheries have traditionally served as a convenient source of data on the status of commercially important fish stocks. Data collected by state and federal regulatory agencies and containing annual landings, amount of licensed or registered fishing gear, dockside value of the catch, and so forth, are often the only long-term data available to fishery managers. Detailed information on growth, age structure, sex ratios, fecundity, and other indicators of stock status may also be obtained through special sampling programs of existing commercial fisheries. Using fisheries, a resource manager can minimize expenses and collect data on stock status and dynamics that could not be obtained in any other way.

Commercial fisheries data on striped bass have been used extensively in description of the periodicity of strong year classes (Koo 1970; Van Winkle et al. 1979) and coastal stocks (Mansueti 1961; Trent and Hassler 1968; Grant 1974; McGie and Mullen 1979). The commercial fishery data from the Hudson River may play a similarly important role in future research on striped bass. The New York State Department of Environmental Conservation (NYDEC) is presently conducting a research program which uses the commercial fishery to monitor population trends.

Interest in striped bass has recently intensified in the Hudson River and along the entire Atlantic Coast because the stocks appear to be declining, particularly in the Chesapeake Bay (Richkus et al. 1980; Boreman and Austin 1985). The decline in the Chesapeake system is evident in both the commercial fishery records and juvenile production indices. Juvenile production in the Hudson has not declined as much as in the Chesapeake Bay and other coastal populations (Klauda et al. 1980; Borman and Austin 1985). In this paper, we examine Hudson River commercial fishery records of 1931–1980 for

89

trends in landings (catch), fishing effort, and catch-per-effort for striped bass. The age and sex composition of striped bass catch by Hudson River commercial fishermen also are described from a four-year (1976–1979) sampling program. Finally, we estimate the exploitation rate for the commercial fishery from data compiled during a four-year tagging program.

HISTORICAL RECORDS, DATA COLLECTION

Commercial fishery records for the Hudson River were obtained through the courtesy of the Statistics and Market News Office of the National Marine Fisheries Service (NMFS) in Patchoque, Long Island. Details on the specific survey procedures used by NMFS personnel may be obtained from that NMFS office. General descriptions of the NMFS survey of the Hudson River commercial fishery were presented in Klauda et al. (1976) and McFadden et al. (1978). The data were compiled by NMFS personnel from interviews and correspondence with the major or representative fishermen operating on the river, license application reports submitted to New York's Department of Environmental Conservation, and "judgemental" increments to the compiled data by survey personnel based on insights gained from the interviews. Survey data were collected early in the calendar year for landings and effort in the previous year, and therefore were based on the memory or catch and effort records of the fishermen that were interviewed.

The primary purpose of the annual Hudson River Statistical Survey was to establish catch trends (catch, value, fishermen employed, and fishing gear used) for the American shad fishery. Data on other species such as striped bass were collected as a by-product of the survey. Striped bass have traditionally been the major by-catch in the American shad fishery, but in recent years (up to 1976) many commercial fishermen actively pursued striped bass.

DESCRIPTION OF THE FISHERY

Most commercial fishing in New York waters of the Hudson River occurred between river kilometer (km) 19 and km 188 (Figure 2). The commercial fishery for striped bass and American shad was conducted during spring spawning runs, mostly by part-time fishermen. Most striped bass were taken by stake-and-anchor gill nets on the

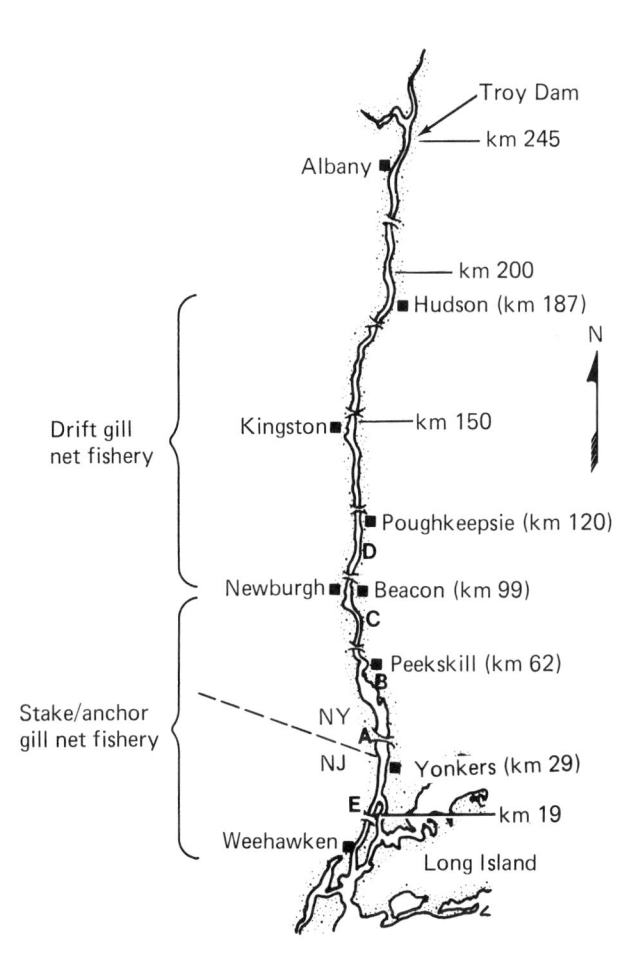

Figure 2. *Hudson River from Troy Dam to the Atlantic Ocean, showing location of commercial gill net fishing effort. Fishing locations for commercial fishermen A through E are shown.*

shallow flats between km 29 and 69 (Figure 2). Few striped bass were caught in drift gill nets, because this gear is used upriver from the major striped bass spawning areas (Texas Instruments 1981b).

Prior to 1950, commercial fishermen used natural fiber (cotton and linen) gill nets, but between 1950 and 1955 they gradually converted to mono- and multifilament nylon gill nets.

REGULATIONS

Striped bass has been a commercial species only in the New York waters of the Hudson River, and its fishery has been regulated by New York's Department of Environmental Conservation. After the 1975 season, the commercial fishery for striped bass in the Hudson River was closed and the sale of fish banned because of polychlorinated byphenyl (PCB) concentrations greater than 5 ppm in the edible flesh, the maximum concentration then allowed by the U.S. Food and Drug Administration (Hetling et al. 1978). This limit has since been lowered to 2 ppm. Prior to its closure, regulations included a fishing season, minimum size limit, maximum gill net length and minimum mesh size, restricted fishing areas, and escapement (net lift) period. Season and size-limit regulations applied specifically to striped bass. Other regulations were established primarily to control exploitation of American shad. The legal commercial fishing season for striped bass extended from 16 March through 30 November, even though most fish were taken prior to 1 June. An additional winter fishery (1 December through 15 March) existed through the 1948 season, but it was closed in early 1949. In 1938, the minimum legal size limit was increased from 12-in. (305-mm) to 16-in. (407-mm) fork length.

By law, individual stake-and-anchor gill nets had to be less than 1,200-ft (366-m) long. In practice, a stand could contain several individual nets staked end-to-end in a line perpendicular to the shoreline, particularly betweem km 40 and 62. Stake-and-anchor gill nets could not be set within 1,500 feet (457 m) of another licensee's net. Drift gill nets could not exceed 2,400 feet (732 m) in length. Minimum legal bar mesh size for all gill nets was 2.25 inches (5.72 cm). There were no restrictions on gill net material. The use of haul seines and fyke nets to take striped bass was outlawed in early 1949. Between 15 March and 15 June, commercial netting was forbidden on the Kingston Flats (km 146 to 154).

Commercial fishing for American shad remained legal in the Hudson River after 1976. During the 15 March–15 June American shad season, all commercial netting was restricted to only the weekday portion of the week to allow escapement of some spawners. Weekends were thus called "lift periods." Net lift periods have been in effect in the New York portion of the Hudson River for most years since at least 1868 (Hudson River Valley Commission 1966; Klauda et al. 1976). The lift period regulation was set annually and was designed to protect American shad, but the commercial fishery for striped bass was also affected. During the last year of legal striped bass commercial fishing in the Hudson River (1975), the weekend net lift period started at 6:00 A.M. EST Friday and ended at 6:00 A.M. EST Sunday.

LANDINGS

Reported landings of striped bass from the Hudson River varied by a factor of 35 between 1931 and 1980 (Table 23). Reported landings for 1976 through 1980, after the legal closure of the fishery and

Table 23. *Reported landings (kg) of striped bass in Hudson River commercial fishery, New York waters*

	GEAR TYPE[e]			
YEAR	STAKE AND ANCHOR GILL NETS	DRIFT GILL NETS	HAUL SEINES	TOTAL LANDINGS
1931	—	—	—	2,417
1932	—	—	—	2,044
1933	—	—	—	6,175
1934	—	—	—	4,946
1935	—	—	—	8,466
1936	—	—	—	9,125
1937	—	—	—	13,086
1938[a]	—	—	—	11,147
1939	—	—	—	13,577
1940	—	—	—	15,707
1941	—	—	—	9,676
1942	—	—	—	10,687
1943	—	—	—	14,009
1944	—	—	—	27,627
1945	—	—	—	35,986
1946	—	—	—	22,958
1947	—	—	—	21,974
1948	—	—	—	17,610
1949[b]	—	—	—	4,142
1950	—	—	—	4,326
1951	—	—	—	7,863
1952	—	—	—	13,536
1953	—	—	—	8,776
1954	—	—	—	25,397[f]
1955[c]	—	—	—	33,288
1956	—	—	—	42,097
1957	—	—	—	38,322
1958	—	—	—	34,966
1959	—	—	—	60,363
1960	—	—	—	60,272
1961	—	—	—	32,064

Table 23. *Reported landings (kg) of striped bass in Hudson River commercial fishery, New York waters (Continued)*

YEAR	STAKE AND ANCHOR GILL NETS	DRIFT GILL NETS	HAUL SEINES	TOTAL LANDINGS
	GEAR TYPE[e]			
1962	—	—	—	21,814
1963	—	—	—	21,179
1964	—	—	—	13,379
1965	15,873	771	0	16,644
1966	19,410	635	0	20,045
1967	24,392	389	0	24,781
1968	27,347	227	0	27,574
1969	34,991	0	0	34,991
1970	19,379	1,438	0	20,817
1971	11,223	0	0	11,223
1972	8,139	0	0	8,139
1973	22,592	5,406	2,404	30,402
1974	11,737	1,338	680	13,755
1975	18,685	1,782	476	20,943
1976[d]	71,361	27	0	71,388
1977[d]	8,118	2,041	0	10,159
1978[d]	6,240	184	27	6,451
1979[d]	15,283	409	0	15,692
1980[d]	11,417	322	0	11,739

[a] Minimum legal size limit increased from 12 in. to 16 in. (fork length).
[b] Winter fishing season closed; use of haul seines and fyke nets became illegal.
[c] Commercial fishermen completely switched over from natural fiber (cotton, linen) to synthetic fiber (nylon) gill nets.
[d] Striped bass commercial fishery closed indefinitely due to PCB contamination. Reported landings represent estimates of fish caught and released, collected for biological or chemical analyses, given away or consumed by the fishermen.
[e] 1931 through 1964 landings not reported separately by gear type.
[f] Landings estimated from total reported landings of striped bass for New York minus marine landings.

ban on sales in late 1975, represent estimates of striped bass caught and released, collected for biological or chemical analyses, and given away or consumed by the fisherman (personal communication, Fred Blossum, NMFS). Landings by gear type since 1965 show that most striped bass were taken in stake-and-anchor gill nets in Rockland and Westchester counties (Tables 23 and 24). Gill nets are generally reported to be effective for catching striped bass (Power 1962; Lyles 1965; Koo 1970). Reported haul seine landings from 1972 to 1973 were coincident with a change in statistical survey personnel and may reflect a spurious modification in survey policy. The landings may

Table 24. Reported landings (kg) of striped bass by county in Hudson River commercial fishery, New York waters

YEAR	ROCKLAND (WEST, KM 35–74)[a]	WESTCHESTER (EAST, KM 29–74)	PUTNAM (EAST, KM 74–91)	DUTCHESS (EAST, KM 91–163)	ULSTER (WEST, KM 109–170)	COLUMBIA (EAST, KM 163–170)	TOTAL LANDINGS
1965	6,032	9,887	0	0	725	0	16,664
1966	7,619	12,336	0	0	90	0	20,045
1967	5,601	18,976	0	0	204	0	24,781
1968	10,839	16,508	0	0	227	0	27,574
1969	15,556	19,435	0	0	0	0	34,991
1970	8,504	12,313	0	0	0	0	20,817
1971	5,215	5,896	0	112	0	0	11,223
1972	2,527	5,612	0	0	0	0	8,139
1973	14,224	11,801	2,540	522	907	408	30,402
1974	7,483	2,894	816	1,587	680	295	13,755
1975	11,291	7,668	816	340	510	318	20,943
1976	67,574	3,401	386	0	27	0	71,388
1977	6,576	1,542	0	0	2,041	0	10,159
1978	3,175	3,066	0	26	184	0	6,451
1979	12,154	2,676	454	0	408	0	15,692
1980	10,930	488	0	91	230	0	11,739

[a] km 0 is at the Battery, southern tip of Manhattan.

Figure 3. Reported landings (kg) of striped bass in Hudson River commercial fishery (New York waters) expressed as percent deviation from mean catch, 1931–1980.

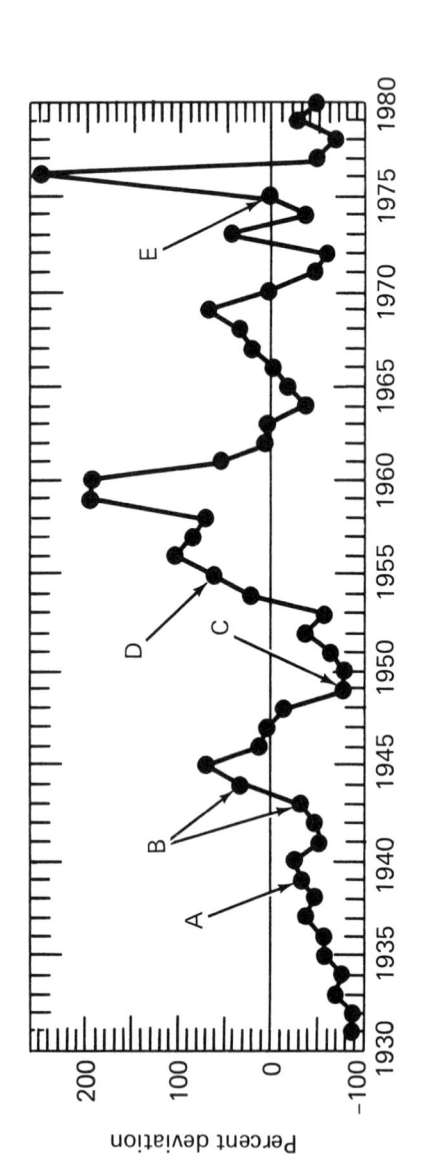

MAJOR REGULATION OR GEAR CHANGES

A - Minimum size limit increased from 12 in. to 14 in. FL
B - No weekend lift period
C - Haul seines and fyke nets outlawed, winter season closed
D - Change over from natural to synthetic fiber gill nets complete
E - Striped bass fishery closed after 1975 season due to PCBs

also have reflected an intensified interest in striped bass related to an increased market price (Strand et al. 1980). The reported dockside value of the striped bass commercial fishery was relatively low through the early 1970s (Table 25). The value of the fishery increased in response to the increase in market price beginning about 1973. The market value and general economic conditions during the late 1970s provided strong stimuli for illegal fishing after the 1976 closure of the fishery.

The pattern of reported striped bass landings since 1931 (annual landings expressed as percent deviation from the long-term mean) was irregular, with no discernible increasing or decreasing trends (Figure 3). Major peaks occurred in 1959–1960 and 1976 with minor peaks in 1945, 1956, 1969, and 1973. These peaks likely reflected, at least partially, annual variations in year-class success. The 1976 peak was likely exaggerated by some fishermen who tried to make the

Table 25. *Reported dockside value of striped bass catch to commercial fishermen, nominal and adjusted for inflation, New York waters of Hudson River*

YEAR	CATCH (kg)	VALUE (DOLLARS) NOMINAL (UNADJUSTED)	VALUE (DOLLARS) ADJUSTED FOR INFLATION[a]	ADJUSTED DOLLARS PER KG
1965	16,644	4,048	4,357	0.26
1966	20,045	5,260	5,412	0.27
1967	24,781	4,400	4,400	0.18
1968	27,574	11,674	11,203	0.41
1969	34,991	12,616	11,490	0.33
1970	20,817	7,843	6,744	0.32
1971	11,223	4,455	3,953	0.35
1972	8,139	4,965	3,963	0.49
1973	30,402	24,130	18,129	0.60
1974	13,755	12,132	8,242	0.60
1975	20,943	24,476	15,184	0.73
1976	71,388	94,446[b]	55,103	0.77
1977	10,159	18,592[b]	10,244	1.01
1978	6,451	16,502[b]	8,541	1.32
1979	15,692	30,448[b]	15,760[c]	1.00
1980	11,739	45,817[b]	23,715[c]	2.02

[a]Based on consumer price index with 1967 as base year = 1.000 (Statistical Abstracts 1978).
[b]Value if fishing season was open and catch had been sold on the market.
[c]Based on consumer price index for 1978 (1,932).

striped bass fishery appear more important in order to emphasize the economic losses as a result of the closure of the fishery. The minor peak in 1973 may have been influenced to an unknown degree by the change in statistical survey personnel. Landings were below average prior to 1944, during the late 1940s to early 1950s, and during the early 1970s. Since 1976, reported landings have also been below the historic mean catch of 20,278 kg.

FISHING EFFORT

Trends in commercial landings may not necessarily reflect changes in the size of the Hudson River striped bass stock. Fishing effort trends must also be evaluated, because if fishing effort declined while stock size remained roughly stable, a decline in reported landings would erroneously suggest a decline in stock size. The converse is also true. Therefore, trends in landings may reflect trends in actual stock size or simply trends in fishing effort.

Detailed effort data were not available from the Hudson River commercial fishing records. An estimate of effective fishing effort for the striped bass gill net fishery would include the number of hours fished by a specified area of gill net, net material, and mesh size. However, adequate information was available from license records and NMFS statistical survey data for a description of nominal effort trends.

Fishing effort statistics pertinent to the striped bass (and American shad) fishery obtained from the NMFS survey are listed in Table 26. The commercial fishery declined since 1951, especially the drift gill net component. Since the early 1960s, the total number of operating gill nets closely approximated or exceeded the total number of operating fishermen (all gear), indicating that commercial fishing is a dying trade on the Hudson River and that the remaining effort primarily has been a gill net fishery targeted on the larger anadromous species such as American shad and striped bass. The commercial fishery was seasonal and concentrated into about nine to eleven weeks during the spring spawning run; therefore, we assumed that actual fishing hours per week closely approximated the maximum permitted by law (Table 26).

The pattern of fishing effort for stake-and-anchor gill nets (Figure 4) was similar to the overall pattern of striped bass landings, except for the early 1940s. Major peaks in effort occurred during the mid-1940s,

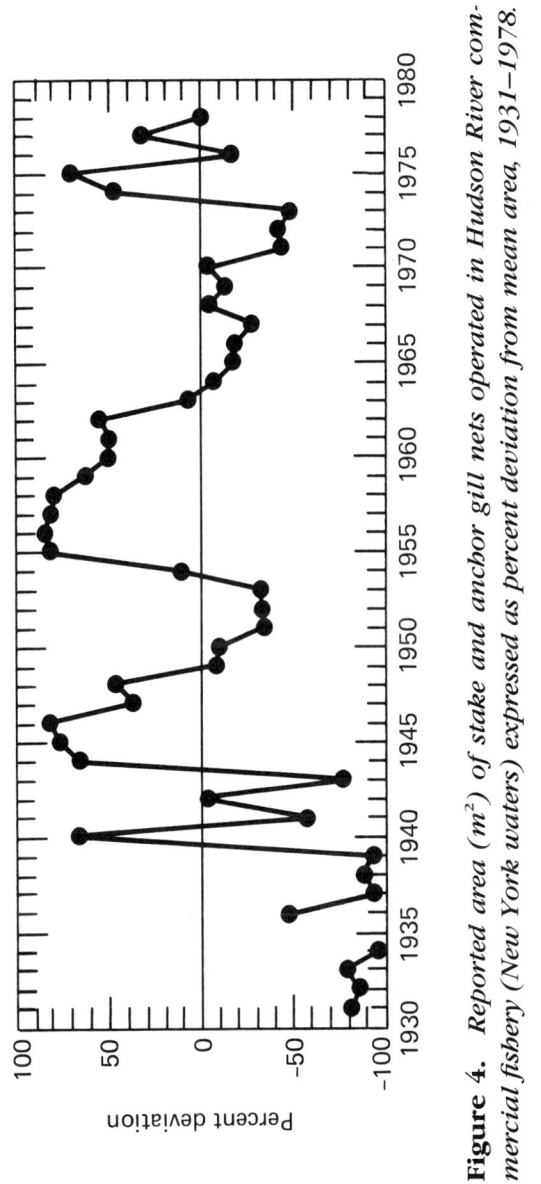

Figure 4. *Reported area (m^2) of stake and anchor gill nets operated in Hudson River commercial fishery (New York waters) expressed as percent deviation from mean area, 1931–1978.*

Table 26. *Effort statistics for Hudson River commercial fishery taking striped bass in New York waters (except where noted)*

YEAR	NUMBER OF OPERATING FISHERMEN[a] (ALL GEAR)	OPERATING GILL NETS TOTAL NUMBER[b]	DRIFT (m²)	STAKE AND ANCHOR (m²)[c]	RATIO OF DRIFT TO STAKE AND ANCHOR AREA	LEGAL FISHING HOURS FOR STRIPED BASS[d] PER WEEK
1931	252	123	262,748	12,167	21.6	108
1932	274	126	314,070	9,072	34.6	108
1933	317	146	227,294	13,370	25.2	108
1934	322	154	420,875	2,410	174.6	108
1935	498	301	—	—	—	108
1936	476	711	307,705	32,240	9.5	108
1937	613	247	496,249	3,599	137.9	108
1938	875	599	606,459	6,993	86.7	108
1939	647	417	459,670	4,632	99.2	108
1940	648	584	335,708	103,176	3.4	108
1941	650	332	358,412	25,142	14.3	132
1942	549	527	156,917	59,628	2.6	132
1943	608	275	384,067	13,280	28.9	168
1944	533	489	395,417	166,211	2.4	168
1945	545	383	562,917	69,400	8.1	132
1946	936	526	567,083	112,800	5.0	132
1947	1,172	533	671,667	85,500	7.9	132
1948	959	476	661,667	91,000	7.3	132
1949	845	468	517,500	57,125	9.1	132
1950	522	313	448,750	55,628	8.1	132
1951	419	197	325,333	40,275	8.1	96
1952	374	180	283,417	41,025	6.9	108
1953	363	173	263,521	41,025	6.4	108
1954	391	198	315,667	68,670	4.6	108
1955	322	176	209,314	113,340	4.6	108
1956	308	175	174,756	114,526	1.5	108

Year				
1957	276	160	165,969	113,425	1.5	108
1958	229	147	151,937	112,385	1.4	108
1959	234	143	121,241	101,359	1.2	120
1960	211	121	120,346	92,960	1.3	120
1961	191	112	104,894	92,639	1.1	120
1962	168	105	94,216	96,329	1.0	120
1963	142	83	84,820	66,434	1.3	120
1964	125	65	86,237	57,423	1.5	120
1965	83	55	79,018	52,006	1.5	120
1966	62	48	62,067	50,553	1.2	120
1967	62	41	61,457	44,618	1.4	120
1968	85	74	73,100	58,512	1.2	120
1969	68	53	59,492	53,630	1.1	120
1970	57	55	78,304	59,547	1.3	120
1971	39	42	63,596	34,133	1.9	120
1972	46	45	63,597	36,662	1.7	120
1973	41	66	76,687	32,253	2.4	120
1974	56	96	55,290	92,150	0.6	120
1975	100	145	110,580	106,033	1.0	120
1976	68	113	70,898	51,554	1.4	0
1977	65	148	58,643	82,985	0.7	0
1978	67	205	48,603	62,324	0.8	0
1979	88	178	NR	NR	—	0
1980	94	184	NR	NR	—	0

[a]Numbers for 1931 through 1964 include commercial fishermen operating in New Jersey waters of the Hudson River. Numbers represent regular (full-time) and casual (part-time) fishermen combined, but exclude common dip net fishermen because they were reported in 1974–1977.

[b]Numbers for 1931 through 1964 include gill nets licensed by commercial fishermen operating in New Jersey waters of the Hudson River. Numbers represent all types of gill nets.

[c]Area values for 1931 through 1964 have been multiplied by 1.2 to account for gill nets omitted by NMFS that primarily caught striped bass (McFadden et al. 1978).

[d]Obtained from New York Department of Environmental Conservation, Albany.

NR = not reported by NMFS.

1950s, and 1970s. Below average effort was expended prior to 1940, during the early 1940s, the early 1950s, and the early 1970s.

RELATIVE ABUNDANCE

It would be misleading to simply divide reported landings by reported effort and express relative abundance as landings (catch) per unit effort because of potential biases in the data. Biases may have been introduced by regulatory and gear changes through time, the practice of combining catches from several gear in the records, and changes in statistical survey procedures (Klauda et al. 1976). Subsets of the commercial fishery records, however, yielded reasonable indices of relative abundance for striped bass in the Hudson River. Restricting landings and effort data to 1955–1975 minimized the comparability problems introduced by regulation changes between 1931 and 1980. Restricting landings and effort (gill net area) data to stake-and-anchor gill nets eliminated multiple gear problems. Use of landings and effort records obtained only from interviewed commercial fishermen minimized inaccuracies from the statistical survey.

The proportion of total striped bass landings and total stake and anchor gill net area attributable to interviewed fishermen was available for 1977 and 1978 via direct testimony submitted to the Federal Energy Regulatory Commission in October 1979 by Fred C. Blossum of NMFS. Based on these surveys, interviewed commercial fishermen caught 93% and 98% of all reported Hudson River landings of striped bass in the respective years. These fishermen also accounted for 41% and 53% of the total reported stake-and-anchor gill net effort. By assuming that these percentages encompassed the range of survey results in earlier years, we adjusted the total reported landings and stake-anchor gill net effort in 1955–1975 downward to approximate data sets collected only from fishermen interviewed during the NMFS survey (Table 27).

This subset of the commercial fishery records was used to calculate two sets of relative abundance indices for striped bass from 1955 through 1975 (Table 28). Each index was calculated in the following manner.

Relative abundance indices of striped bass in the Hudson River varied by a factor of about 8 from 1955–1975. Major peaks in abundance closely tracked the trend in reported landings (Figure 3). Relative abundance was lowest in 1974 just prior to the closure of the

Table 27. *Total reported landings of striped bass in stake and anchor gill nets (Table 23) and operating stake and anchor gill net effort (Table 25) adjusted to reflect only reported landings and effort data from gill net fishermen interviewed by NMFS personnel during Hudson River statistical survey*

YEAR	ADJUSTMENTS BASED ON 1977 INTERVIEWS		ADJUSTMENTS BASED ON 1978 INTERVIEWS	
	ADJUSTED REPORTED LANDINGS (kg) COL. 1	ADJUSTED GILL NET EFFORT (m^2) COL. 2	ADJUSTED REPORTED LANDINGS (kg) COL. 3	ADJUSTED GILL NET EFFORT (m^2) COL. 4
1955	30,958	46,469	32,622	60,070
1956	39,150	46,956	41,255	60,699
1957	35,640	46,500	37,556	60,110
1958	32,518	46,078	34,267	59,564
1959	56,138	41,557	59,156	53,720
1960	56,053	38,114	59,067	49,269
1961	29,820	37,982	31,423	49,099
1962	20,287	39,495	21,378	51,054
1963	19,697	27,238	20,755	35,210
1964	12,443	23,543	13,111	30,434
1965	15,479	21,322	16,311	27,563
1966	18,642	20,727	19,644	26,793
1967	23,046	18,293	24,285	23,648
1968	25,644	23,990	27,023	31,011
1969	32,542	21,988	34,291	28,424
1970	19,410	24,414	20,454	31,560
1971	10,437	13,995	10,999	18,090
1972	7,569	15,031	7,976	19,431
1973	28,274	13,224	29,794	17,094
1974	12,792	37,782	13,480	48,840
1975	19,477	43,474	20,524	56,198

fishery. These abundance indices suggest that the size of Hudson River stock of striped bass fluctuated from 1955 to 1975, but with no discernible long-term trend.

DISCUSSION

Observed trends in commercial landings, effort, and relative abundance in the Hudson River support the inferences drawn from juvenile surveys that the Hudson River stock of striped bass is relatively healthy (Klauda et al. 1980). Apparently, the Hudson stock has not experienced the dramatic decline in reproductive success and relative stock

Table 28. *Indices of relative abundance[a] for striped bass spawning population in Hudson River based upon commercial fishery records obtained during NMFS interviews with New York commercial fishermen*

YEAR	INDEX A	INDEX B
1955	6.17	5.03
1956	7.72	6.29
1957	7.10	5.79
1958	6.53	5.33
1959	11.26	9.18
1960	12.26	9.99
1961	6.54	5.33
1962	4.28	3.49
1963	6.03	4.91
1964	4.40	3.59
1965	6.05	4.93
1966	7.50	6.11
1967	10.50	8.56
1968	8.91	7.26
1969	12.33	10.05
1970	6.63	5.40
1971	6.21	5.07
1972	4.20	3.42
1973	17.82	14.52
1974	2.82	2.30
1975	3.73	3.04

[a] See text for details on calculations of each catch-per-effort index of abundance.

abundance in the 1970s that has been exhibited by the Chesapeake Bay and Roanoke River-Albemarle Sound stocks (Johnson et al. 1979; Richkus et al. 1980; Boreman and Austin 1985). The Hudson River Statistical Survey conducted annually by NMFS contains data which provide a rough measure of long-term variations in striped bass stock abundance. The survey could be improved, particularly because fishing effort records are too crude for detailed analyses. Maryland's approach to obtaining commercial fishery records (Hamer et al. 1948) is an example of the means for improving the records on Hudson River landings and effort. All licensed commercial fishermen in Maryland have been required to maintain and submit confidential records of daily fishing activities. These records include the weight of each species landed, the location fished, the amount of fishing gear used, and the catch by gear type. To improve the Hudson River data on commercial fishing effort, total records on gill net surface area, mesh size, and total hours that each fisherman's nets are fished each week should also be compiled.

CATCH COMPOSITION, DATA COLLECTION

A sampling program to determine the size, age, and sex of striped bass exploited by the gill net fishery within the Hudson River began in 1976. Unfortunately, its beginning coincided with the closing of the fishery due to PCB concentrations in striped bass. Therefore, arrangements through the New York State Department of Environmental Conservation allowed representative commercial fishermen to continue fishing for striped bass, specifically to furnish collections of fish for scientific studies.

For 1976 and 1977, we chose four fishermen as representative of selected fishing regions in the river (Table 29). Two (labeled A and B) were fishermen who would contribute a large share of the total landings during a normal year. Fisherman A fished staked gill nets at km 43 in the vicinity of the Tappan Zee Bridge (Figure 2). Fisherman B fished staked-and-anchored gill nets for Atlantic sturgeon, as well as striped bass and American shad, at km 59–62. Fisherman C fished staked gill nets primarily for striped bass at km 83, near their major spawning grounds (Dey 1981; McLaren et al. 1981). Fisherman D was a drift gill net fisherman at km 104–107 whose chief target species was American shad, with only incidental catches of striped bass. In 1978 and 1979, fisherman E, a staked gill net fisherman from New

Table 29. *Gill nets used by sampled Hudson River commercial fishermen, 1976–1979*

FISHERMAN	GEAR TYPE	MESH SIZE (cm)*	LENGTH (m)
A	Staked	7.6–14.0	61–915
	Staked	30.5	61
B	Staked	11.7–15.2	12–183
	Anchored	13.6–35.6	23–183
C	Staked	15.2	91–183
D	Drift	14.0	30–439
	Staked	14.0	49–256
E	Staked	13.3	274

*Stretch

Table 30. Spatial-temporal distribution of striped bass catch by five† commercial fishermen in the Hudson River, 1976–1979

1976 SEASON (TOTAL CATCH = 1,223)

LOCATION (km)	MARCH 1	MARCH 2	MARCH 3	APRIL 4	APRIL 5	APRIL 6	APRIL 7	APRIL 8	MAY 9	MAY 10	MAY 11	JUNE 12	JUNE 13	JUNE 14	JUNE 15
104–107				A	A	*	A	A	*	A	*	*	*	*	*
83			A		A	A	C	A		A	A	B	A	A	
59–62		C	E	E	D	C	C	A	A	C	B	B	B	A	
43		D	B	C	C	D	C	D	C	B	C	D	C	C	A
19															

1977 SEASON (TOTAL CATCH = 1,499)

LOCATION (km)	MARCH 1	MARCH 2	MARCH 3	APRIL 4	APRIL 5	APRIL 6	APRIL 7	APRIL 8	MAY 9	MAY 10	MAY 11	JUNE 12	JUNE 13	JUNE 14	JUNE 15
104–107				A	B	B	B	A	B	A	*	*	A	C	E
83				A	A	A	A	A	A	B	B	A	A	A	*
59–62			E	C	D	B	C	B	C	B	B	A	A	*	*
43			D	E	C	C	D	C	C	B	A	*	*	B	A
19															

1978 SEASON (TOTAL CATCH = 1,351)

LOCATION (km)	MARCH 1	MARCH 2	MARCH 3	APRIL 4	APRIL 5	APRIL 6	APRIL 7	APRIL 8	MAY 9	MAY 10	MAY 11	JUNE 12	JUNE 13	JUNE 14	JUNE 15
104–107															
83															
59–62	C	B	B	C	C	C	C	C	A	B	B	B	A	*	*
43	C	D	C	C	C	B	C	C	C	D	C	C	C	D	E
19				C	C	B	C	C							

1979 SEASON (TOTAL CATCH = 865)

LOCATION (km)	MARCH 1	MARCH 2	MARCH 3	APRIL 4	APRIL 5	APRIL 6	APRIL 7	APRIL 8	MAY 9	MAY 10	MAY 11	JUNE 12	JUNE 13
104–107			A	A	A	A	B	A	A	A	A	A	
83	E	D	B	C	B	B	C	B	B	B	A	B	A
59–62	C	D	B	B	B	C	B	D	C	D	C	A	D
43				C	C	D	D						
19													

A = 0.1–0.5% of Total Catch per year
B = 0.6–1.5% of Total Catch per year
C = 1.6–5.0% of Total Catch per year
D = 5.1–10.0% of Total Catch per year
E = >10.0% of Total Catch per year
Blank space = No Effort
*No catch

†Only four fishermen fished in any one year; each location indicates a fisherman.

Jersey waters near the George Washington Bridge (km 19), was substituted for Fisherman D.

Each fisherman was encouraged to fish in his usual manner, and was accompanied twice per week during net tending to examine and record the striped bass catch. All striped bass were measured and scale samples were removed from slightly below and posterior to the insertion of the second dorsal fin. Live fish that were in good condition were tagged with an internal anchor tag (Floy FD-67c) and released. Dead and dying fish were taken to the laboratory for weighing and sex determination. Age was determined from scales (Mansueti 1961).

Sampling began as soon after the 15 March opening date as possible (weeks beginning: 29 March 1976; 3 April 1977; 19 March 1978; 18 March 1979). Weeks are numbered consecutively for data summaries (beginning at approximately 15 March).

TIMING AND LOCATION OF CATCH

The striped bass catch (Table 30) was consistently heavy at the start of the season (mid-March to early April) for fishermen A and B in the downriver locations (Tappan Zee and Croton and Haverstraw Bays). Fish in the river at this time consist of mature striped bass preparing to spawn further upriver and overwintering immature fish (McLaren et al. 1981). Fishermen A and B had the largest catches of striped bass throughout the remainder of the season. Fisherman A caught 41%–66% of each season's total sample. The combined catch of the four fishermen was 1,223 fish in 1976, 1,499 in 1977, 1,351 in 1978, and 865 in 1979. Fisherman B caught 23%–44% of each season's total sample. The proportionally greater catch by these two fishermen was consistent with historical records, indicating that the region between km 29 and km 69 produces the most striped bass landings.

Beyond the expectation of large catches early in the season, there is little predictability concerning the timing of large and small catches. The catch usually remained substantial into early to mid-June for fishermen A and B, with the exception of 1977 when catches dropped sharply by early June (week 12). Fisherman E (km 19) had large catches, but he restricted his fishing to April and early May, a time when late arriving spawners enter the river and immature nonspawners leave (McLaren et al. 1981). Fisherman D (km 104–107) caught few striped bass, except for several two-year-old, sublegal size striped bass in June 1977.

SIZE COMPOSITION

The size composition of the commercial striped bass catch depends on mesh sizes used in the nets, since gill nets are strongly size selective (Hamley 1975). Years of experience by each fisherman have directed their choice of mesh sizes. For example, fisherman C, who fishes near the major spawning grounds and expects to encounter large spawners, uses a larger mesh size (15.2 cm) than that used by fisherman A downriver. As a result, the modal lengths of striped bass caught by fisherman C were slightly larger than those caught by others (1979 data are presented in Figure 5 as an example).

The modal lengths of striped bass in the combined catches from all four fishermen varied somewhat from year to year, but they fell within the interval 451 to 650 mm total length (TL) (Figure 6). Some sublegal fish (less than 406-mm fork length) were vulnerable to the gill nets used, but their number were comparatively small. Striped bass larger than 950 mm TL (approximately one yard) were not uncommon in the catch.

AGE COMPOSITION

Age composition for the entire year's sample was estimated by sorting the catch into length groups of 20-mm increments beginning with 200 mm TL. Fish in each length group that were both sexed and aged were then used to estimate the proportion of the total catch in the length group that would have fallen into each category of sex and age.

The sampled striped bass ranged from two to fifteen years of age. The most common ages were IV–VI (Figure 7), which coincided with the modal lengths of 451 to 650 mm TL. Ages V and VI predominated in 1976, as did ages IV and VI (males only) in 1977, age V in 1978, and ages IV–VI in 1979.

The variation in modal ages among years can best be explained by year-class strength (Figure 7). Abundance indices for juvenile striped bass in the Hudson River (Klauda et al. 1980) have indicated strong year classes for 1969 and 1973. In 1976, the 1969 year class was age VII; females were quite abundant at age VII in 1976 but males were not. Gear selectivity against age VII fish or higher than usual mortality for this strong year class prior to age VII may have caused the diminution of this group in the catch. The progression of the strong 1973 year class through time is more apparent in the catch (Figure 8). The

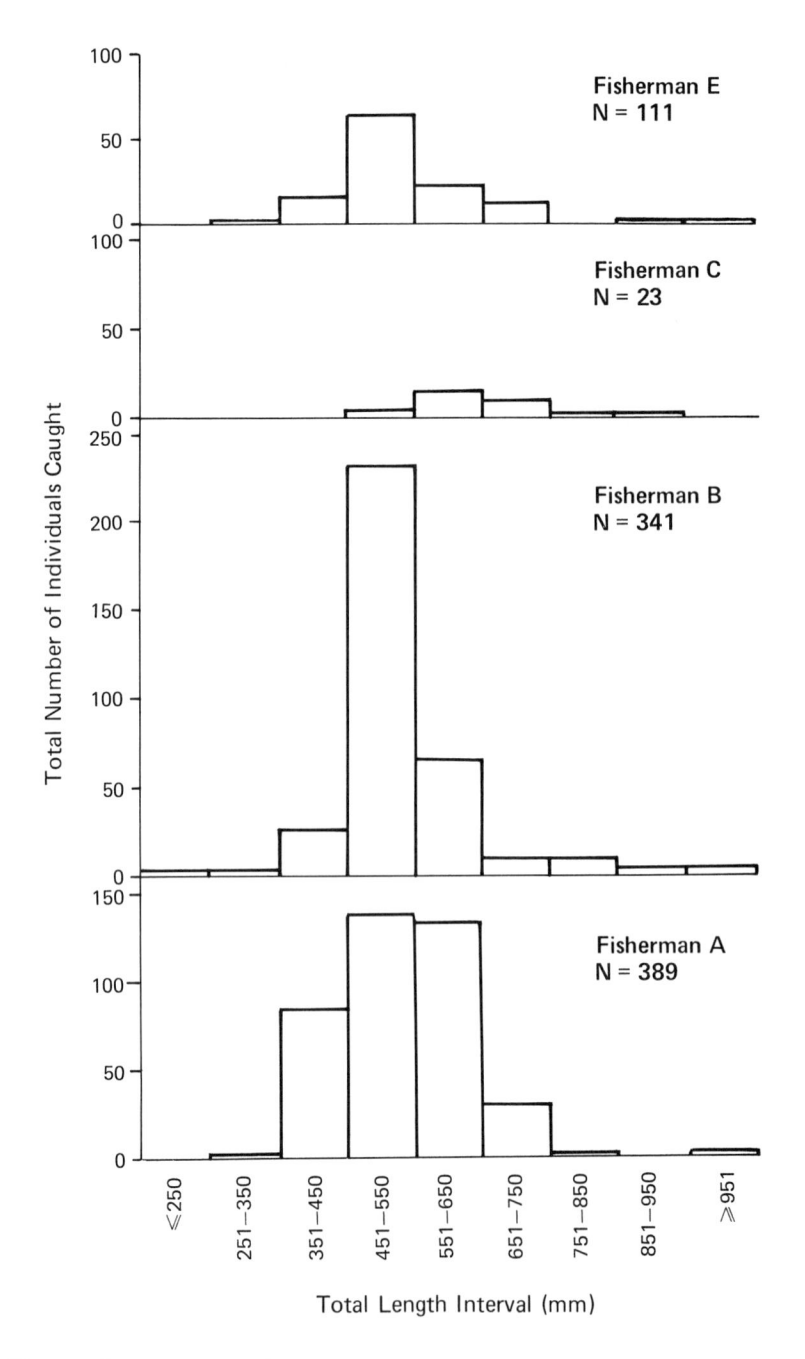

Figure 5. *Size distribution of commercial catch of striped bass by four fishermen in Hudson River estuary during 1979.*

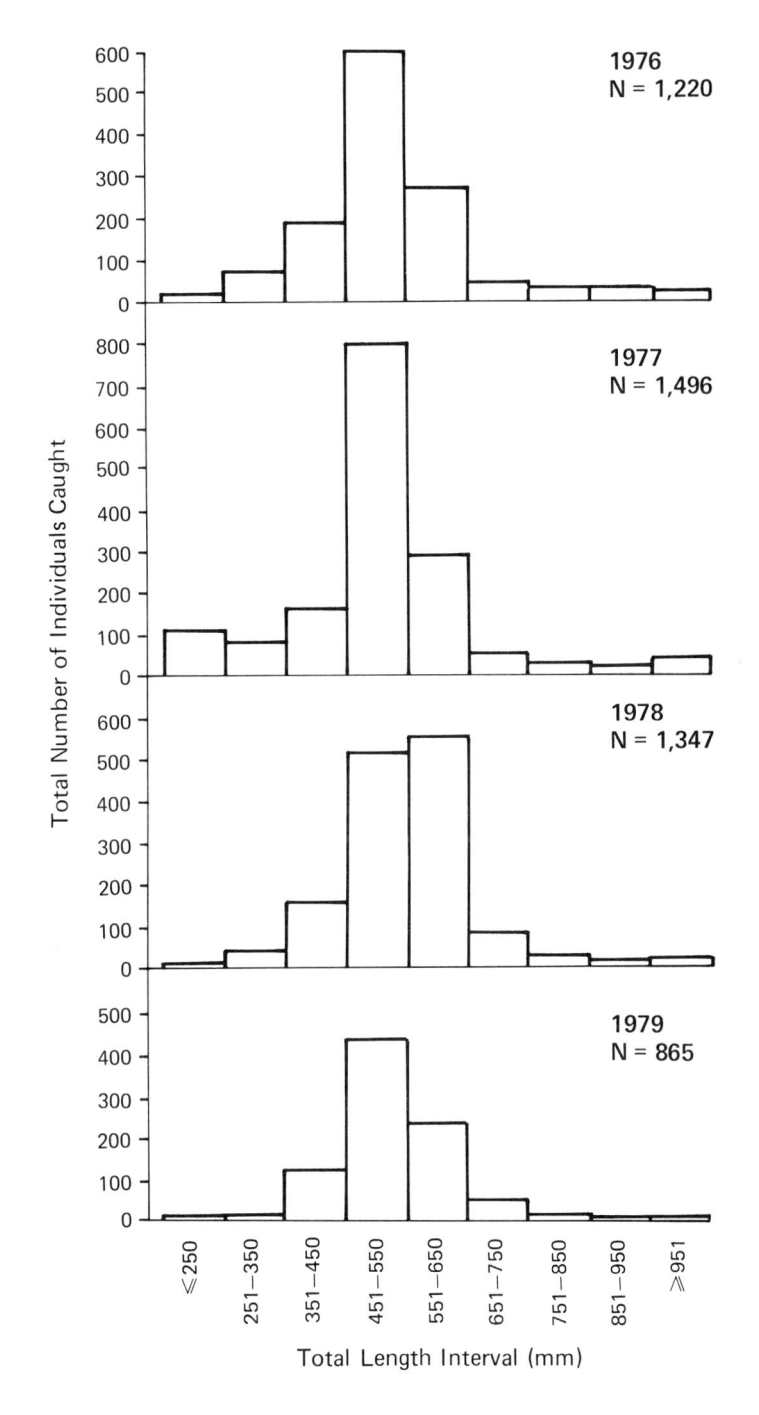

Figure 6. *Size distribution of commercial catch of striped bass by fishermen in Hudson River estuary during 1976 through 1979.*

111

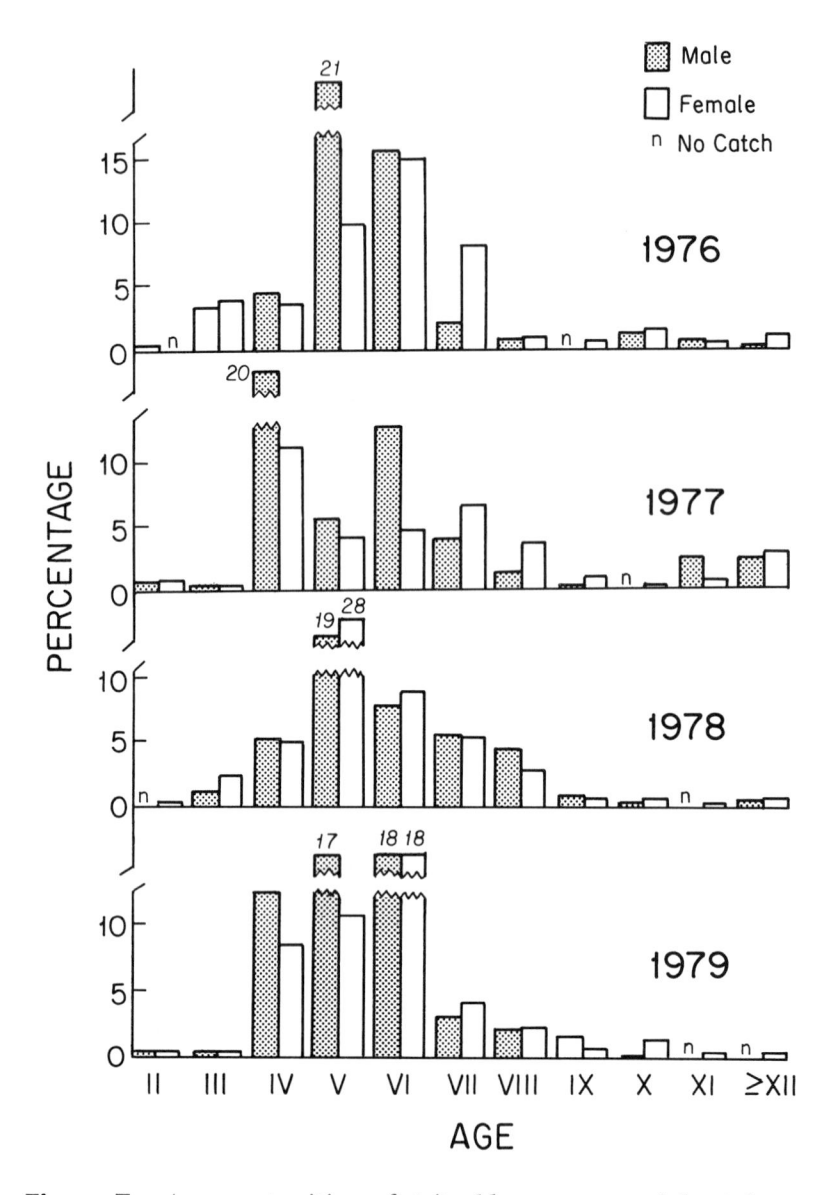

Figure 7. Age composition of striped bass commercial catch sampled in Hudson River estuary, 1976–1979.

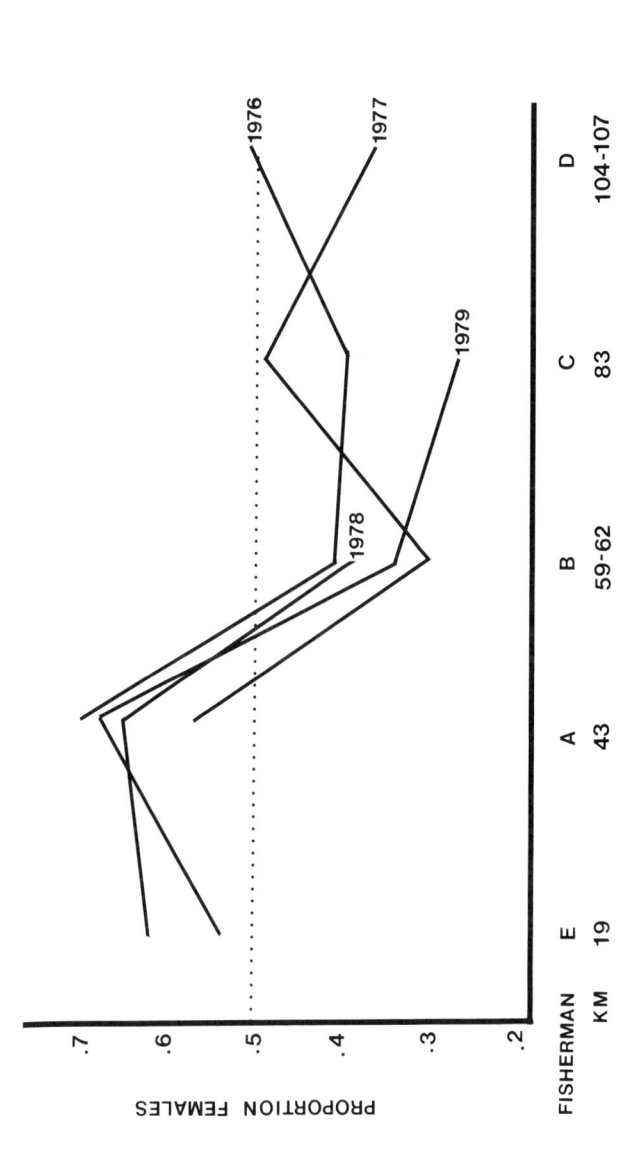

Figure 8. *Proportion of females in the striped bass catch of Hudson River commercial fishery, 1976–1979.*

strength of this year class first became obvious as age III in 1976. By age IV in 1977, it was the most abundant year class and continued to be so through 1979.

SEX RATIO

Movements associated with spawning behavior appear to account for sex ratio differences at distant points on the river. Females consistently outnumbered males in the catch of the two downriver fishermen, A and E (Figure 8). Dew (1981) also observed this tendency in the commercial catch of fisherman A during 1973, 1974, and 1975. Males almost as consistently outnumbered females in the catch of upriver fishermen, B, C, and D; the two exceptions (C in 1977 and D in 1976) of nearly 1:1 ratios occurred for small sample sizes (5 and 33 fish, respectively).

Spawning generally occurs above km 62. The proportion of females in the population below km 62 increased during April and May, and was attributed to the earlier departure of males to the spawning grounds (McLaren et al. 1981). Males mature at a younger age than females in the Hudson River (Klauda et al. 1980); therefore, males should outnumber females in the spawning population. The predominance of males in the catch of fisherman B was probably related to the early season residency of males in this area and movement of late arriving males through this area to reach the spawning grounds. In the combined catch of all sampled fishermen, males tended to outnumber females at ages V or younger, with the reverse true for ages VII and older (Figure 7).

SELECTIVITY OF THE FISHERY

To evaluate comparability of the gill net catch age composition to that of the population in the river, we compared the commercial catch (sexes combined) for 1976–1978 with age composition estimates for the spring population. Estimates for the spring population were derived from a related study on adult striped bass population parameters (TI 1980a, 1980b).

The commercial fishery appears to select for ages V and VI (Figure 9) and against ages III and IV. Ages VII and older were usually caught in proportion to their abundance in the population. Selectivity in favor of ages V and VI is not surprising, since 12.7-cm and 15.2-cm stretch mesh gill nets (approximating mesh sizes most commonly

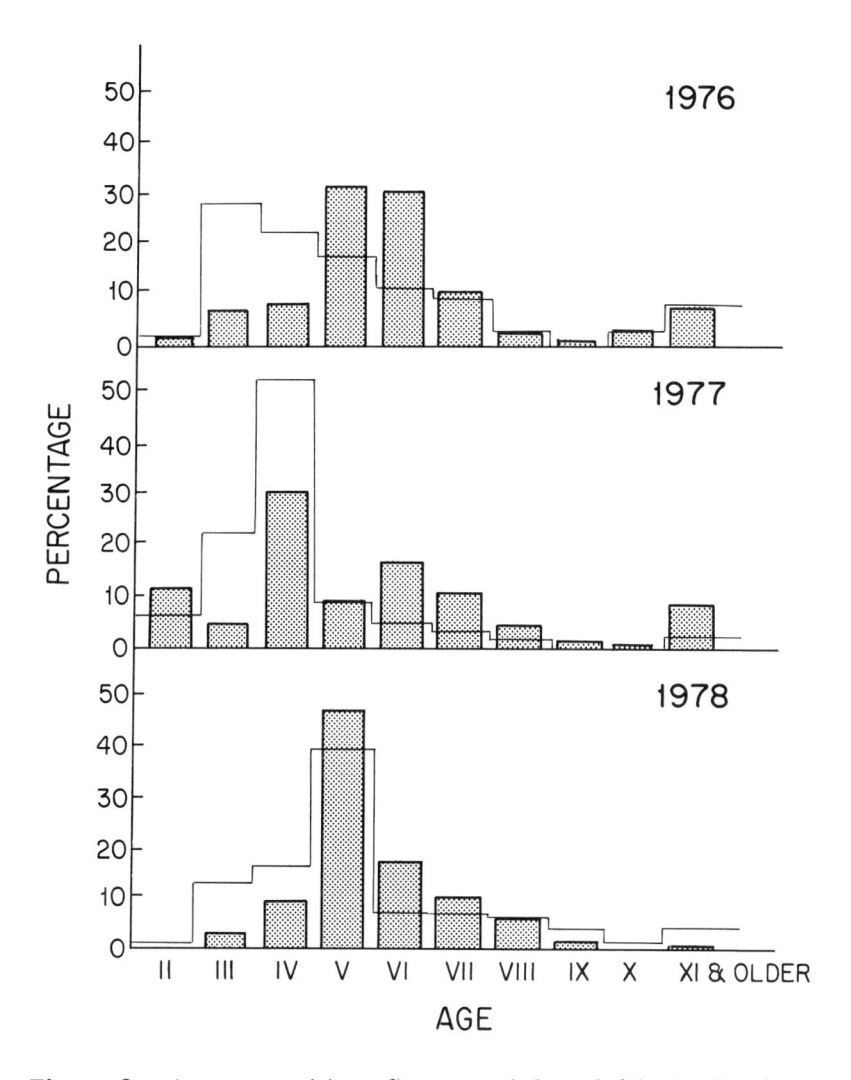

Figure 9. *Age composition of commercial catch (shaded bars) and spring population (solid line) of striped bass in Hudson River estuary, 1976–1978.*

used by commercial fishermen) select for striped bass with modal lengths of 501 to 550 and 601 to 650 mm TL, respectively (TI 1981b). The mean lengths of age V fish during 1976–1979 have ranged from 479 to 558 mm TL, and of age VI fish from 568 to 642 mm TL (TI 1981b).

The mesh sizes selected by commercial fishermen effectively avoided the capture of large numbers of sublegal-size striped bass in New York (16 in. = FL 406 mm), which was slightly less than the mean length of age IV fish: 412 to 425 mm FL for males and 428 to 441 mm FL for females (McLaren et al. 1981; FL = 0.9692 (TL) − 13.313). Age III fish were below legal size. Although age IV striped bass could be exploited more by use of slightly smaller mesh sizes (e.g., 10.2 cm), the capture of sublegal-size fish would increase substantially.

The commercial fisheries of the Chesapeake Bay in Maryland and the Roanoke River in North Carolina differed in terms of legal size and ages exploited from the Hudson River fishery during the years studied. Legal size for striped bass in Chesapeake Bay increased from 11 in. FL (279 mm) to 12 in. FL (305 mm) in June 1957. Ages II and III were the primary ages of exploited fish in the Chesapeake (Mansueti 1961), compared to ages IV to VI in the Hudson River. The modal length of age II striped bass caught in 7.6-cm to 10.2-cm stretch mesh gill nets, commonly used in the Chesapeake, fell just above the minimum length of 305 mm FL, unlike the Hudson fishery where up to 100 mm separates the minimum legal length and the modal length of fish caught in 12.7-cm mesh nets. In the Roanoke River, striped bass exploited by the fishery (Trent and Hassler 1968) fell well above the minimum legal size of 12 in. FL (305 mm). The dominant age groups were III and IV for males and IV and V for females. The mean fork length of age III males was 424 mm. Younger age groups participated very little in the spawning run up the Roanoke River from the Albemarle Sound, and therefore were not subject to the fishery within the river.

EXPLOITATION RATES, DATA COLLECTION

The tagging program conducted concurrently with the commercial catch composition study served to describe the annual migrations of striped bass spawning within the Hudson River and to estimate the rate of exploitation on the stock from the commercial and sport fisheries within the river and along the Atlantic coast. Results of the migration study for 1976–1978 were described by McLaren et al. (1981).

Striped bass were collected from gill nets of the sampled commercial fishermen and from gill nets operated by Texas Instruments, from approximately the Tappan Zee Bridge (km 42) to 11 km above the Newburgh-Beacon Bridge (km 99). Anchored 91-m gill nets were

Table 31. Number of Hudson River striped bass >250 mm total length collected, released, and recaptured annually, 1976 through 1979

	1976 [5]	1977 [5]	1977 [5]	1978 [10]	1978 TOTAL	1979 [5]	1979 [10]	1979 TOTAL	YEARS COMBINED
Number collected	5,607	5,921			6,708	2,228		5,130	23,336
Number released	2,406	2,813	3,307	259	3,566		93	2,321	11,106
Number Recaptured*									
Sport Fishermen	119	145	164	11	175	85	6	91	530
	(39.0)†	(43.8)	(31.5)	(25.0)	(31.0)	(31.1)	(50.0)	(31.9)	
Commercial	23	108	246	24	270	120	6	126	527
Fishermen									
	(7.5)	(32.6)	(47.3)	(54.5)	(47.9)	(44.0)	(50.0)	(44.2)	
Total	142	253	410	35	445	205	12	217	1,057
Percent Recapture	5.9	9.0	12.4	13.5	12.5	9.2	12.9	9.3	9.5

*Only those collected, released, and recaptured during same years.
†Numbers in parentheses indicate percentage of total recapture.
[ ]Dollar value of tag reward.

set at depths of 3 to 10 m and primarily had mesh sizes of 10.2-cm to 15.2-cm stretch in 1976 and 10.2-cm to 17.8-cm stretch in 1977–1979. Placement of the nets is described in McLaren et al. (1981). A 274-m haul seine (10.2-cm to 4.2-cm stretch mesh) and a 61-m haul seine (19.2-cm to 0.6-cm stretch) provided additional catch. All fish caught were measured to the nearest millimeter (TL), and scale samples were removed for age determination. From 1976 through 1979, the number of tagged fish released per year ranged from 2,321 to 3,566 (Table 31). A publicity program, including posters and advertisements in fishing periodicals, encouraged the public to return tags and the capture information (date, location, and method). All tags applied during 1976 and 1977 offered a $5 reward; during 1978 and 1979, a $10 reward was applied to 7% and 4% of the released fish, respectively.

ESTIMATES OF EXPLOITATION RATE

The percentage of released tags returned by commercial fishermen within the Hudson River (Table 32) increased from 0.8% in 1976 to 3.8% in 1977, 6.4% in 1978, and 5.0% in 1979. Overall, the percentage return from all sport and commercial fisheries ranged from 5.8% to 12.5%. These return rates represent minimum exploitation rates, since we made no adjustments for mortality associated with handling and tagging, tag shedding, or failure of fishermen to return tags ("nonresponse").

Laboratory and field observation of tagging mortality and tag shedding is currently underway (personal communication, Dennis Dunning, New York Power Authority), and preliminary results have

Table 32. *Percentage return of tags from Hudson River striped bass fisheries during 1976–1979**

	1976	1977	1978	1979
Commercial				
Inside River	0.8	3.8	6.4	5.0
Outside River	0.1	0.1	1.2	0.4
Sport				
Inside River	0.7	0.8	0.3	0.7
Outside River	4.2	4.4	4.6	3.2
Total Percentage	5.8	9.1	12.5	9.3
Total Number of Returns	142	253	445	217

*Totals include only tagged fish caught during the calendar year of release.

Table 33. *Return rates for $5 and $10 tags for Hudson River striped bass during 1978 and 1979*

	1978		1979	
	$5	$10	$5	$10
Number Released	3307	259	2228	93
Number Recaptured				
Sport	164	11	85	6
	(5.0%)	(4.2%)	(3.8%)	(6.5%)
Commercial	264	24	120	6
	(7.4%)	(9.3%)	(5.4%)	(6.5%)
TOTAL	410^a	35^a	205^b	12^b
	(12.4%)	(13.5%)	(9.2%)	(13.0%)

^{a}Calculated X^2 = 1.0, critical value = 3.84
^{b}Calculated X^2 = 0.3, critical value = 3.84

indicated an appreciable loss of the Dennison-type tag used during 1976–1979. Chadwick (1968) reported a nonresponse rate of 38% in the striped bass sport fishery in California. He obtained his estimate by assuming 100% return for tags that offered a $5 reward, and by comparing the return of reward tags with that of tags offering no reward. In the Hudson River, we only released reward tags, but we found no significant (α = 0.05) difference (1978, χ^2 = 1.0, 1d.f.; 1979, χ^2 = 0.3, 1d.f.) in the return of $5 and $10 tags (Table 33). We believe that the return rate for our tags was considerably less than 100% and that returns could not be appreciably enhanced without much larger rewards. Nonresponse was likely the result of fishermen losing tags or forgetting to return the tags, failure to observe the tags (approximately six tags per year were discovered on fish sold in the Fulton Fish Market in New York City), fear of the consequences of returning a tag (e.g., association of Hudson River fish with PCB contamination in the coastal fishery, which had remained open), and loss during mail delivery.

In the absence of quantitative estimates of tag loss, tagging mortality, and nonresponse rates under our specific conditions, we doubled the minimum exploitation rate to account for these sources for underestimation of exploitation. With this adjustment, total annual fishing exploitation during the first year at large for tagged fish would range (using our most reliable years of data, 1977–1979) from 0.18 to 0.25. Commercial exploitation within the river would range from 0.08 to 0.12. Commercial exploitation within the river may be slightly exaggerated owing to the proximity of the fishermen's nets to the area of tag releases, primarily km 42–99. We acknowledge that nonresponse rates could differ widely among the four fisheries (inside versus outside river, sport versus commercial) for purely sociological reasons. Although the striped bass fishery within the river was officially closed in 1976, total gill net fishing effort (Table 26) had not appreciably declined since the closure. Therefore, our exploitation estimates may reasonably represent an open fishery if the recent fishing effort did not effectively avoid striped bass as a by-catch.

CRITICAL AGE

Ricker (1975) defines the critical size or age for fish in an exploited population as the size or age at which instantaneous natural mortality (M) and instantaneous growth rate (G) are equal. At the critical age the biomass of a year class is at its maximum, and fishing for maximum yield could best be accomplished. Essentially, a fishery operates most efficiently when the mean age of exploited fish approximates the critical age.

The critical age for Hudson River striped bass was computed from age-specific instantaneous growth rates and from best estimates of instantaneous natural mortality rates for each sex. The data for natural mortality were derived from estimates of total annual mortality (A) of 0.55 for males and 0.40 for females, based on age composition estimates for the population during 1976–1978 (Klauda et al. 1980; TI 1980a). Assuming a rate of exploitation (u) of 0.25, the annual rate of natural mortality (v) for males and females would be 0.30 and 0.15, respectively ($A = u + v$). The instantaneous natural mortality rate (M) for males and females (assuming a Type 2 fishery, Ricker 1975) would be 0.44 and 0.19, respectively, using the formula: $M = vZ/A$ where $Z =$ instantaneous total mortality rate and $A = 1 - e^{-Z}$. Sex and age-specific instantaneous growth rates were calculated from mean weight data from 1976 and 1977 provided by McFadden et al. (1978) and Texas Instruments (1980b). Instantaneous growth rate was calculated as: $G = \log_e W_1 - \log_e W_1$ where W_1 and W_2 are the mean weights (g) for males (or females) of a particular year class in 1976 and 1977, respectively. Instantaneous growth rates were then regressed against age for each sex, and the simple linear regressions were found to be significant ($\alpha = 0.05$): males $G = 0.5638 - 0.0559x$, females $G = 0.9225 - 0.0875x$, where $x =$ age in years.

COMMERCIAL FISHERY FOR STRIPED BASS

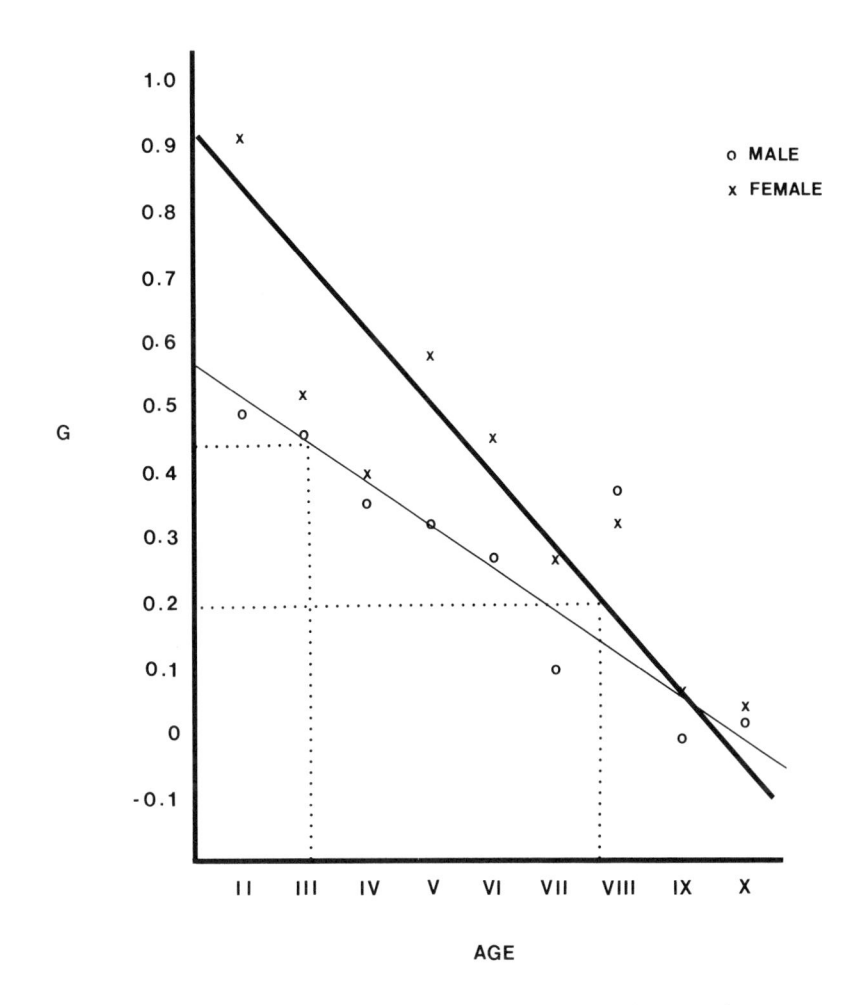

Figure 10. *Critical ages of male (Age III) and female (approximate Age VIII) striped bass in Hudson River commercial fishery derived from instantaneous growth rate (G) and instantaneous natural mortality rate (horizontal dotted lines) with associated regression lines of growth rate on age (female, heavy line; male, light line).*

Instantaneous growth equalled instantaneous mortality for age III males and for age VII–VIII females (Figure 10); thus, these were the critical ages for the Hudson River fishery. As shown earlier, the modal ages of exploitation by the river commercial fishery were IV, V, or VI, depending on year-class strength variations. Ages IV, V, and VI fall

121

midway between the critical ages for males and females. Since the commercial fishery did not exploit males and females independently, it would be logical to expect a compromise between the two sexes. It appears that this compromise (approximately age V, mean age of exploitation) has allowed the most efficient exploitation for the population under existing conditions of growth and mortality.

SUMMARY AND CONCLUSIONS

1. An increase in market price of striped bass beginning in 1973 stimulated interest in commercial exploitation despite the 1976 closure of the striped bass fishery in the river owing to PCB contamination.

2. Commercial fishing for striped bass was primarily by stake and anchor gill nets in Rockland and Westchester counties.

3. Reported commercial landings of striped bass in the Hudson River varied irregularly by a factor of 35 between 1931 and 1980, with no discernible increasing or decreasing trends.

4. Indices of relative abundance derived from reported landings and fishing effort data varied by a factor of 8 from 1955 through 1975, but they displayed no discernible long-term trends.

5. Commercial landings and relative abundance indices did not indicate a dramatic decline in stock abundance for the Hudson River since the early 1970s, as happened in the Chesapeake Bay and Roanoke River-Albemarle Sound stocks.

6. Sampling of the catch of four commercial fishermen annually from 1976 through 1979 revealed that striped bass between 450 and 650 mm TL, ages IV–VI, were most frequently caught.

7. Sex ratios within the catch of individual fishermen were determined by movements associated with spawning behavior such that males outnumber females as fishing location progressed upriver.

8. The commercial catch was not an accurate representation of the age composition of the river population but rather selected for ages V and VI as against ages III and IV.

9. The annual exploitation rate by Hudson River commercial fishermen likely fell between 0.04 and 0.12.

10. The commercial fishery appeared to be operating near maximum efficiency as determined by analysis of critical ages (ages of equal instantaneous rates of natural mortality and growth).

ACKNOWLEDGEMENTS

We are grateful to the staff at the Statistics and Market News Office of the National Marine Fisheries Service in Patchoque, Long Island, especially Fred Blossum, for providing summaries of the Hudson River Statistical Survey data. New York's Department of Environmental Conservation personnel in Albany and New Paltz provided the information on commercial fishing regulations. Paul Baumann assisted us in compiling and analyzing the NMFS commercial fishery records. Andrew Kahnle, John Merriner, and Laurence King provided helpful comments and critical reviews of the manuscript. Funding for these studies was provided by: ConEd, NYPA, ORU and CHGE.

3. Age-Specific Variation in Reproductive Effort in Female Hudson River Striped Bass

John R. Young and Thomas B. Hoff

INTRODUCTION

Life-history theory is currently one of the most studied topics in ecology. The basic premise of life-history theory is that biological populations allocate resources to growth, maintenance, and reproduction in response to environmental selective pressures so as to maximize the probability of successful reproduction. Thus, the wide range of life-history strategies—from redwoods that do not mature for decades but live for centuries, to invertebrates that hatch, mature, reproduce, and die within a few weeks—are all adaptive for the environments in which the particular organisms live.

As is common in rapidly developing branches of biological science, theories often proliferate at a much faster rate than do empirical studies which can support or refute the theories. Theoretical aspects of life-history evolution have been notably advanced by the works of Williams (1966), Gadgil and Bossert (1970), Giesel (1974), Schaffer (1974a, b), and Pianka and Parker (1975), to name only a few of the earlier authors. Many of the papers have advanced theories of optimal life-history strategies and some have used computer simulations to examine the behavior of populations under varying conditions. Stearns (1976) has been one of the few people to take a more critical look at life-history theory. He has pointed out that empirical evidence, both through documentation of life histories of wild populations and carefully controlled experimental studies, is needed to support the theoretical work.

Although empirical studies of reproductive strategy are becoming more common, few have been conducted on fish populations. Murphy (1968) examined a fish population whose age structure was artificially truncated by fishing mortality. Schaffer and Elson (1975) and Schaffer (1979) used optimal life-history theory to explain differences in reproductive characteristics among spawning stocks of Atlantic salmon from different rivers. Leggett and Carscadden (1979) examined variation in reproductive characteristics among American shad populations along a latitudinal gradient. Garrod and Knights (1979), Gundarson (1980), Kawasaki (1980), and Adams (1980) have used basic life-history characteristics to examine relationships between production and exploitation in diverse species of fish.

A common problem in attempting to apply life-history theory to wild fish populations is that the difficulty of capturing and identifying individuals, and measuring reproductive characteristics of live individuals, precludes repeat observations of the same individuals. As a result, researchers are forced to base their conclusions on age-specific averages derived from a single observation per fish. Such averages can be misleading if the variation that exists among individuals of the same age is not considered.

The object of this study was to answer two questions: first, do morphometric and fecundity data from a species with a relatively complex life history provide usable indices of reproductive effort; second, do striped bass increase their reproductive effort as they age, as predicted by optimal life-history theory (Gadgil and Bossert 1970)?

MATERIALS AND METHODS

Female striped bass captured during the 1978 spawning season were used for these analyses. By using data from a single year we hoped to eliminate variation caused by year to year changes in environmental conditions.

From March through June, striped bass caught in gill nets and haul seines were counted and subsampled. Sampling was concentrated in the vicinity of the Tappan Zee Bridge and Croton-Haverstraw Bay early in the spawning season (March and April), as far upriver as Newburgh Bay as the season progressed (May), and downriver again at the end of the spawning season (June) (McLaren et al., this volume).

Length and weight measurements were taken in the laboratory. Total length (TL) was measured to the nearest millimeter; weight was recorded to the nearest gram for fish 400 mm or less TL, and to the nearest 50 grams for larger fish. Scale samples for age determination were taken from each fish, and age was determined according to the criteria developed by Mansueti (1961). Sex and state of maturity were determined for up to thirty fish per size group per sample. Assessment of maturity was based on visible characteristics, according to the system developed by Texas Instruments (TI 1980e), and on an index based on the ratio of body weight to gonad weight. Ripe fish, defined as those with well-developed gonads but not yet ready to spawn, were found to have ratios less than 235 for males and less than 70 for females.

Ovaries of females classified as ripe were removed and preserved in 10% formalin for at least one month before processing for fecundity estimation. After obtaining the total gonad weight to the nearest 0.1 g, a section of the right ovary was removed and weighed. Eggs were then separated from the ovarian tissue of the sample and counted. Fecundity was estimated by multiplying the count of eggs in the sample by the total gonad weight and dividing by the weight of the sample.

Mean egg diameter was determined by counting a random sample of eggs arranged in a single row along a measured distance (approximately 30 mm) and dividing that number into the distance.

The variable of most interest in this study was reproductive effort, that is, the proportion of the total resources available to an individual that are committed to reproduction rather than growth or maintenance (Giesel 1976). Unfortunately, this allocation of resources cannot be measured directly without extremely rigorous experimental procedures; therefore, it must be estimated from more easily measured variables.

We used two indices:

$$GI = \frac{\text{Gonad wt} \times 100}{\text{Body wt.} + \text{Gonad wt.}}$$

and

$$EI = \frac{\text{Egg Count} \times \text{Gonad wt.}}{\text{Gonad sample wt.} \times \text{Body wt.}}$$

The egg index is the fecundity divided by body weight and contains more information than the gonad index, which is gonad weight as a percentage of the total body weight. However, the egg index also adds potential variability due to differences in egg size in different fish and inherent measurement errors associated with the additional data.

RESULTS

Eighty-five ripe females were used for fecundity studies, and age was determined for eighty-one of these. Their ages ranged from four to eighteen years; however, the sample of older fish was small, as 89% of the specimens were between five and nine years old.

The gonad index generally ranged between 3% and 14% of total weight (Figure 11), but near the end of May some individuals had

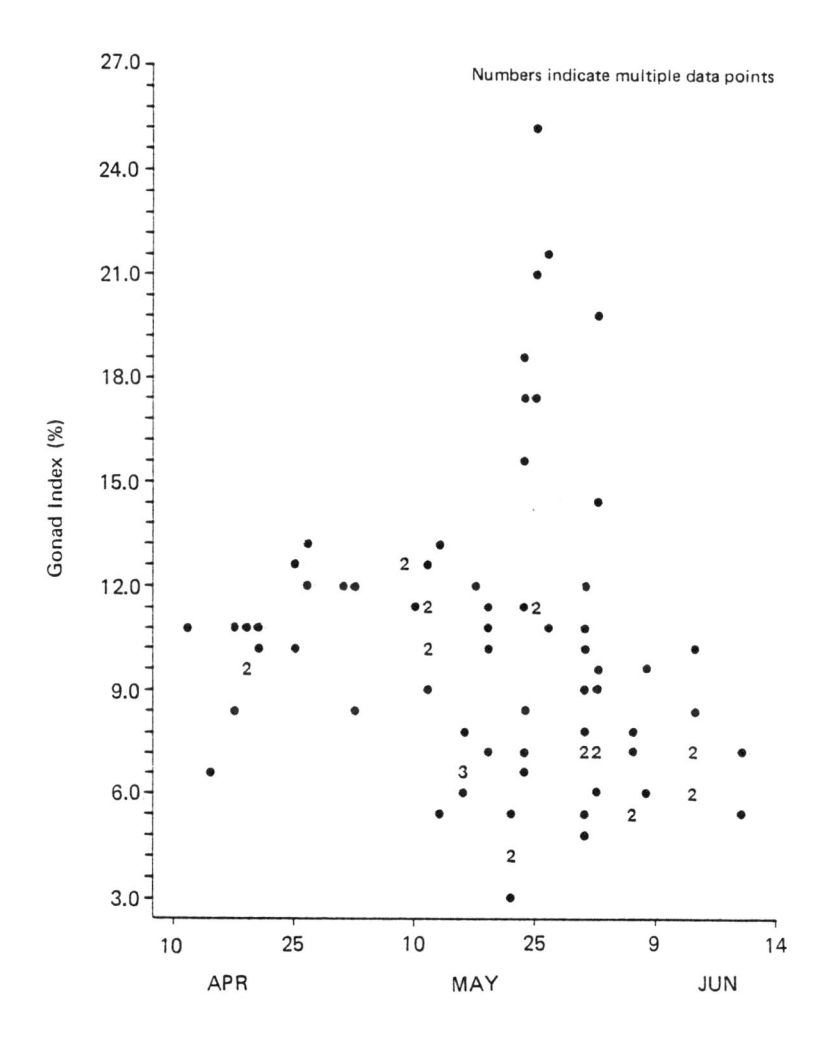

Figure 11. *Seasonal variation in gonad index of female Hudson River striped bass during the 1978 spawning season.*

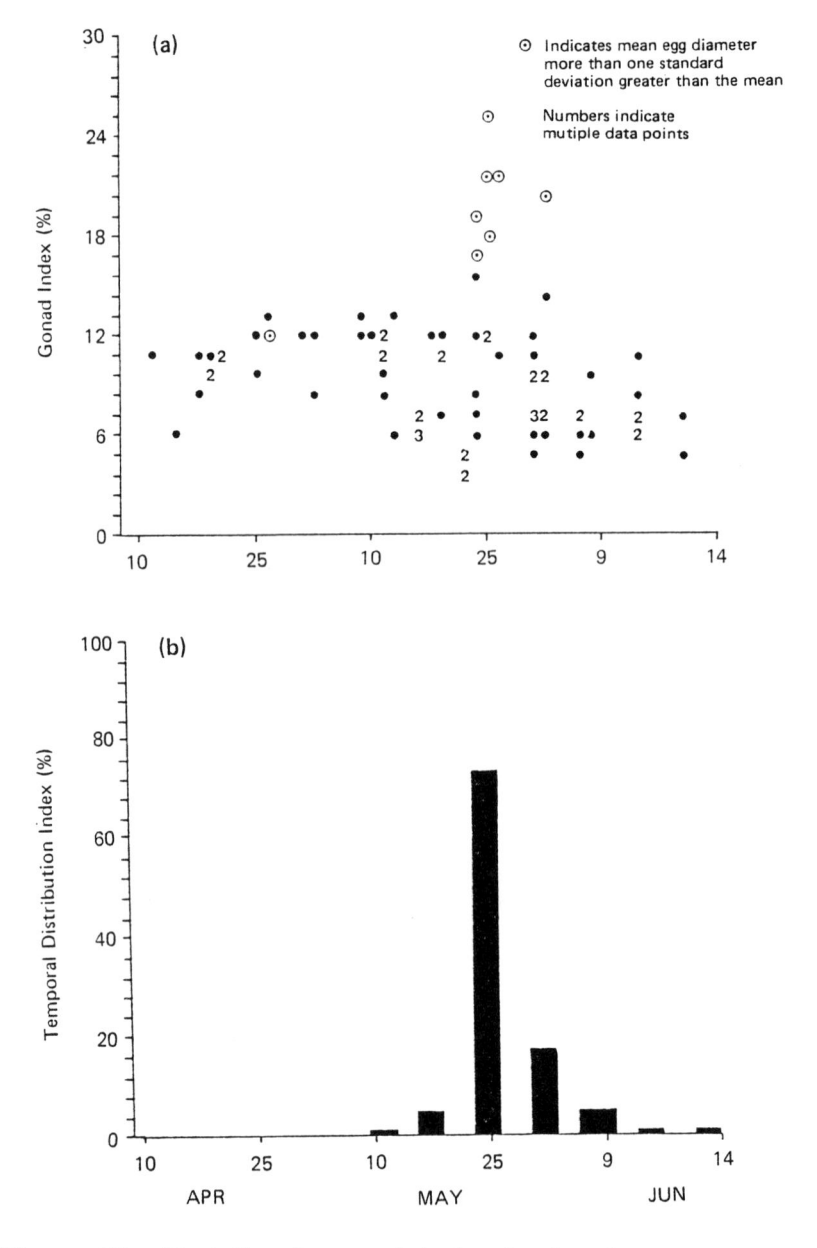

Figure 12. *Variation in gonad index for female Hudson River striped bass (a) and temporal distribution of spawning (b) during the 1978 spawning season.*

values up to 25%. To determine the significance of these unusually high values, we used a multidimensional plot to examine simultaneously gonad index, mean egg size, and date of capture (Figure 12a). The highest gonad index values were from fish that also had exceptionally large eggs. Seven of eight fish with mean egg diameters more than one standard deviation greater than the overall mean (i.e., mean egg diameter greater than 1.11 mm) were included in the group with the high gonad index. Since eggs and gonads enlarge rapidly just before ovulation (Bayless 1972), these fish may have been captured just before spawning and therefore were not representative of the majority of ripe fish. Studies by TI (1980e) indicated that more than 90% of the spawning in 1978 occurred during the same two weeks in which the high gonad index values were found (Figure 12b), and the first spent females were also captured in that same period. In

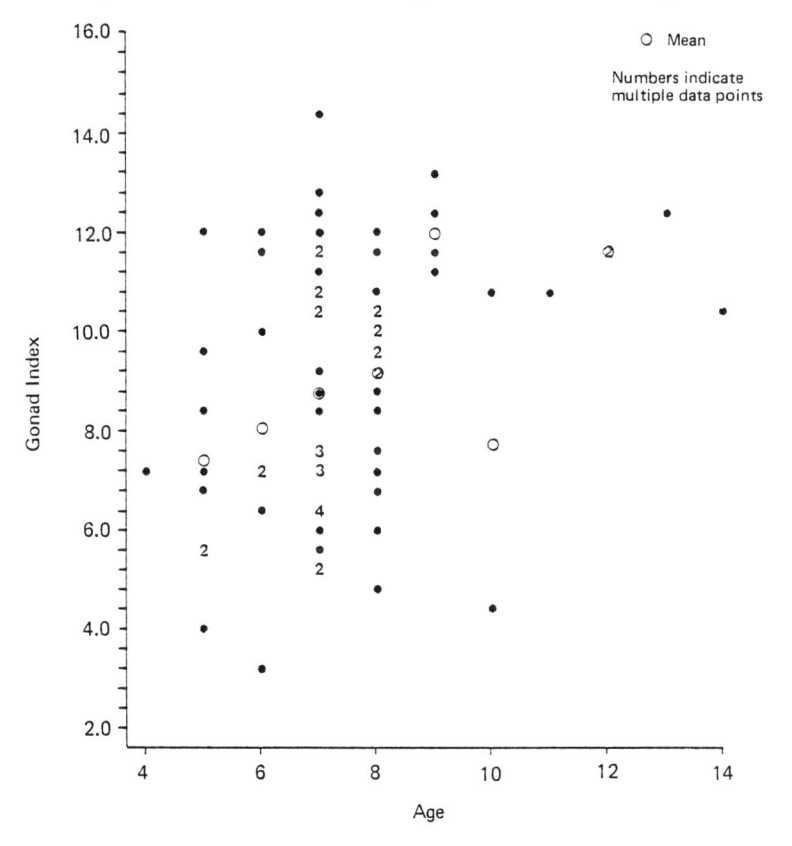

Figure 13. *Variation in gonad index with age for female Hudson River striped bass during the 1978 spawning season.*

order to reduce variation in gonad index resulting from the inclusion of fish that were probably within hours of ovulation, we removed from further consideration all fish with gonad indices of more than 15%. Although the egg index should not have been affected by their inclusion, we wanted to base our comparison of the two indices on the same data set.

Gonad index values for the remaining seventy-two fish varied widely but showed a general increase with increasing age (Figure 13). Mean values increased almost linearly from 7.2% at age five to 8.9% at age eight. Sample sizes were small for ages nine through fourteen, but only one fish in this range had a gonad index value below 10%.

The mean egg index, however, remained relatively constant for ages five through eight but increased in the older fish (Figure 14). The trend toward higher egg index values was not as pronounced as

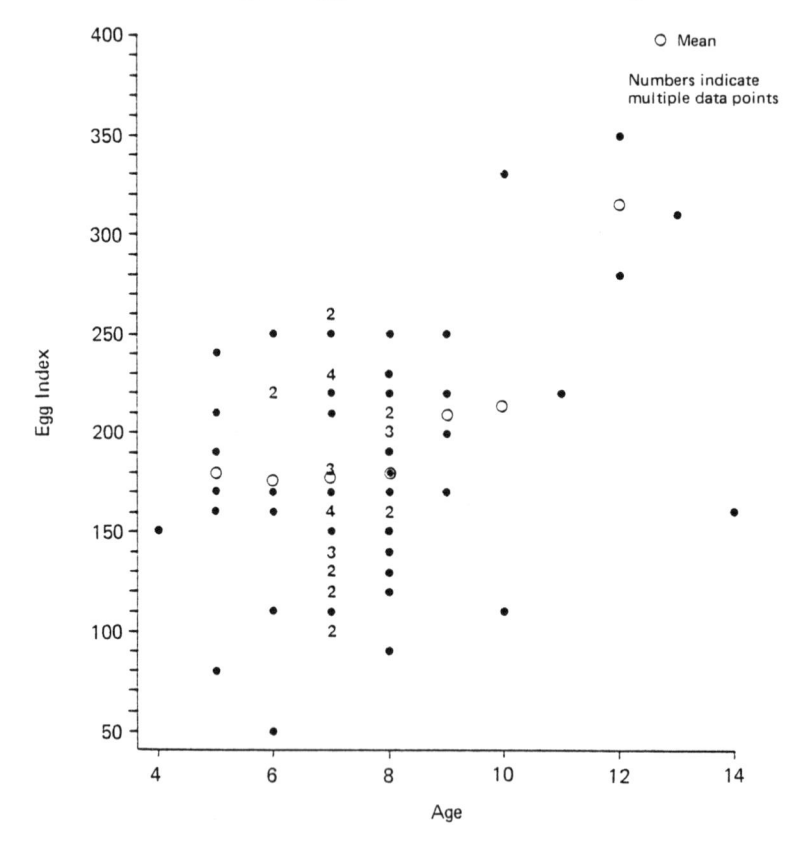

Figure 14. *Variation in egg index with age for female Hudson River striped bass during the 1978 spawning season.*

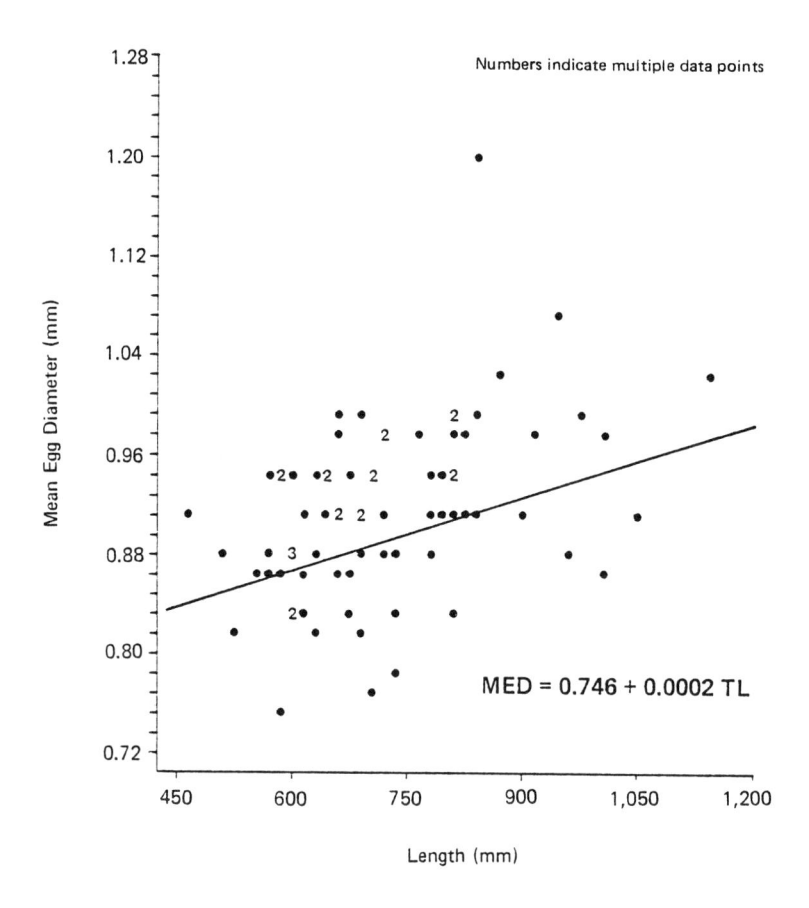

Figure 15. *Relationship between mean egg diameter and length for female Hudson River striped bass during the 1978 spawning season.*

the trend toward higher gonad index values, and several older fish had values near 200 eggs per gram or less.

The differences in trends of the two indices can partly be ascribed to changes in egg size with increasing age and size of the females. Although older fish may allocate relatively more resources to the gonads (higher gonad index), increasing egg size may result in a relatively smaller increase in the egg index. Egg size increased significantly ($p < 0.05$) with fish size (Figure 15), although the slope of the relationship (0.0002) was not nearly as great for the data set used here (with fish near ovulation removed) as it was when all ripe females were included. A study by TI (1980e) which did include all females found a slope of 0.0033.

Table 34. *Results of analysis of variance of egg index and gonad index for female striped bass, ages 5 to 9, during the 1978 spawning season*

		EGG INDEX			
SOURCE	DF	SS	MS	F	$\frac{\sqrt{\text{ERROR MS}}}{\text{MEAN}}$
Age	4	5,273	1,318	0.55	
Error	60	143,994	2,400		0.267
Total	64	149,267			

		GONAD INDEX			
SOURCE	DF	SS	MS	F	$\frac{\sqrt{\text{ERROR MS}}}{\text{MEAN}}$
Age	4	61.56	15.39	2.63*	
Error	60	350.60	5.84		0.270
Total	64	412.16			

*$P < 0.05$.

Further statistical analysis was done on ages five through nine, which provided a minimum of four observations for age group nine and a maximum of twenty-eight observations for age group seven. This eliminated the other ages which had only one or two observations each. A one-way analysis of variance performed separately for the two indices indicated a significant difference among ages for the gonad index but not for the egg index (Table 34). The F value for the gonad index was just significant at the 5% level. The analysis of variance also provided a measurement of the relative precision of the two indices. Dividing the square root of the error mean square (pooled standard deviation) by the overall mean provides a measure of the relative amount of unexplained variation associated with each index (Snedecor and Cochran 1967). The two indices had essentially identical relative precisions—0.267 for the egg index, 0.270 for the gonad index—indicating that there is no loss of statistical precision in spite of the increased number of variables in the egg index.

CONCLUSIONS

Indices to measure reproductive effort can be developed from standard fisheries sampling techniques. These indices are relatively precise statistically, but their ability to estimate reproductive effort accurately has not been established and may vary widely from species to species.

Indices based on gonad weight and fecundity do not account for the costs of reproduction other than those that are reflected in the sexual products. Fish that make arduous spawning migrations, build nests, or guard the young divert additional energy to reproduction that cannot be measured with these indices. Another variable that influences reproductive success is the quality of the eggs as reflected in the amount of energy reserves provided for the early larvae.

In striped bass, the egg size increases as the fish grow, and the two indices used here may show different trends in reproductive effort as the fish age. The gonad index is probably a more accurate indicator of effort in this case, since the changes taking place are not obscured by the changes in egg size. Gundarson (1980) also noted that the gonad index may be preferable to an index based on fecundity because of differences in egg size, although his emphasis was on different species and higher taxonomic groups. The gonad index is also much simpler to calculate since only gonad and total weight need be measured.

The rapid increase in egg size and gonad weight that occurs just prior to ovulation can bias estimates of effort, and this must be considered when examining either of these indices. The gonad index, however, is most likely to be affected; but for striped bass, a combination of mean egg diameter, gonad index, and timing of spawning are criteria for identifying fish which would have biased the estimates of reproductive effort.

Both indices show increasing reproductive effort with increasing age, which fits the predictions of Gadgil and Bossert (1970). These trends are apparent even though the indices are less than perfect estimators of reproductive effort. Unfortunately, our data set was inadequate to test for differences beyond age nine. Either larger samples of older fish in one year or pooling of fish across years (after checking for inter-year differences) could provide data to determine whether both indices depict similar patterns of change in reproductive effort.

ACKNOWLEDGEMENTS

We thank the Consolidated Edison Company of New York for permission to use their data, and EA, Inc., for use of computing facilities and support in preparing the manuscript. We are also indebted to Dennis Dunning for a thorough review of the manuscript, and to Susan Voor and Michele Fredmore for secretarial assistance.

4. Feeding Selection of Larval Striped Bass and White Perch in the Peekskill Region of the Hudson River

Douglas A. Hjorth

INTRODUCTION

Fisheries biologists have for some time suspected that year-class strength often hinges on the feeding success of the larval stages. In 1926, Hjort hypothesized that one factor that could contribute to a rich year-class of fish was "the contemporary hatching of the eggs and the development of the special sort of plants or nauplii which the newly-hatched larva needed for its development." The importance of temporal matching of prey with the transition of larvae to exogenous feeding has been confirmed for certain marine species by Cushing (1972) and by Dowd and Houde (1980). Kernehan et al. (1981), studying striped bass larvae in the vicinity of the Chesapeake and Delaware Canal, found little correlation between year-class strength and the abundance of striped bass eggs or yolk-sac larvae, but they did find that in years when post yolk-sac larve were abundant, juveniles were also abundant later. They believed that the availability of suitable size zooplankton during the critical period of transition at the end of the yolk-sac stage was important for year-class success. Other investigators (Hunter 1976, TI 1980e) have reached similar conclusions.

Clearly defining the early dietary preferences of striped bass and white perch is an important step in determining whether optimal food conditions exist in any given year. Numerous studies have attempted to quantify the food of striped bass larvae in culture ponds (Humphries and Cummings 1973), the laboratory (Doroshev 1970, Eldridge et al. 1977, 1981), and natural environments (Kretzer 1973, Beaven and Mihursky 1980), yet few have related the quantity of food in the stomachs to the availability of prey in the environment. Meshaw (1969) studied the feeding selectivity of striped bass larvae in hatchery ponds in North Carolina, but these ponds were fertilized to promote artificially large quantities of zooplankton, and how this may relate to natural feeding selectivity is uncertain.

The feeding habits of white perch larvae have been ignored almost totally. This is surprising, since white perch and striped bass frequently have overlapping nursery areas and interspecific competition for food could limit the population of one or both species. Certainly juvenile striped bass and white perch in the Hudson River feed on similar prey, including cladocerans, copepods, and insect larvae (TI 1981b). Thoits (1958) added little when he stated that there "appears to be good agreement between various investigators to the effect that white perch fry feed on plankton."

The objectives of this study were to examine prey selectivity by striped bass and white perch larvae within the study area, and to relate predator size to prey size for both species in the hope of discovering mechanisms that enable these similar predators to coexist in the same habitat.

STUDY AREA

Striped bass and white perch larvae and their prey were collected from Peekskill (RM 43.7) to below Stony Point (RM 40.0) (Figure 16), an area that is usually part of a major nursery for both species (TI 1980e). Seven stations were sampled biweekly for microzooplankton and weekly for ichthyoplankton during late spring and summer, though not all samples were used for this study. The water depth at each station was approximately 15 m.

METHODS

Field and Laboratory Procedures

Microzooplankton samples were collected by towing a 75-μm mesh conical plankton net (0.5-m diameter opening, 1.8 m long) through the upper 10 m of the water column. Thus, the volume sampled by each tow was approximately 7.8 m^3. The organisms and detritus were washed into a sample container and preserved in 10% formalin. In the laboratory, two 1 ml aliquots from each sample were placed in a Ward's plankton disk and the microzooplankton identified and counted.

Ichthyoplankton samples were collected immediately after the microzooplankton. Two 575-μm nets were deployed simultaneously and towed for ten minutes at constant engine speed. One net sampled

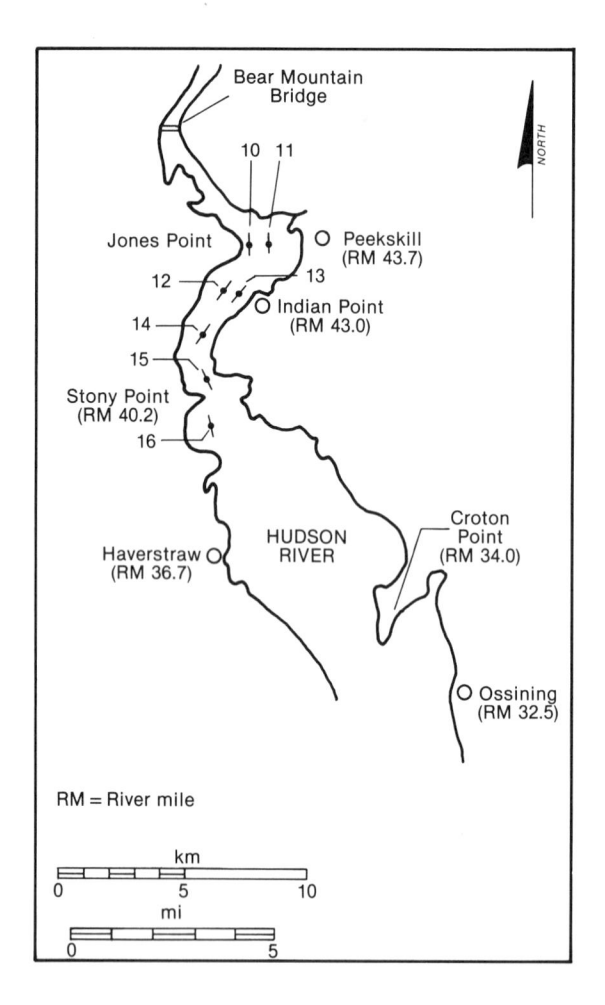

Figure 16. *Location of sampling stations (adapted from Ecological Analysts, Inc., 1980).*

just below the surface and the second net sampled at a depth of approximately 10 m. At the end of the tow, the nets were retrieved and the contents preserved. Each net was equipped with a standard vane General Oceanics digital flowmeter to determine the volume of water sampled. Striped bass and white perch larvae were identified on the basis of characters given by Mansueti (1958, 1964) and Bath (1974). Samples in which either species was present were used for the food habit analysis. These samples were collected between noon and 3 A.M. on 30–31 May, 14 June, and 26–27 June 1978.

Larvae were measured individually to the nearest 0.1 mm with an ocular micrometer, and unmeasurable larvae were not used. The contents of each digestive tract were gently removed and food items identified to the lowest possible taxon, using available literature (Ward and Whipple 1959, Eldridge 1976, Pennak 1953). When possible, the maximum length of each food item was measured to the nearest 0.01 mm with an ocular micrometer. Microzooplankton with spines were measured to the tip of the spine. Copepods were measured to the distal portion of the caudal rami, exclusive of setae.

Analytical Procedures

Prey selection by larval fish is undoubtedly based on several factors. Among these are the shape of the prey, its color and movement patterns, the hunger level of the predator, and the relative sizes of the prey and predator (Humphries and Cummings 1973, Pyke 1979). To account for as many of these variables as possible, two different analytical approaches were used. The first addressed the selection of particular taxa by striped bass and white perch larvae. The second analyzed the size of prey relative to the size of the predators. Both approaches have strengths and weaknesses, but the combined results give a reasonably complete picture of the feeding habits. Because individual stations had only small samples of the prey, the data were pooled to provide a more meaningful regional interpretation of the results.

Prey selectivity was determined by Strauss's Linear Feeding Selectivity Index (L) which is:

$$L = r_i - p_i$$

where r_i is the proportion of a particular prey in the diet and p_i is the proportion of a particular prey in the environment.

Potential values of this index range from -1.0 (indicating maximum avoidance of the prey item) and $+1.0$ (indicating maximum selection for the item). Zero values indicated opportunistic feeding. Significant variations from zero were tested by the t-test ($p\,0.05$), when the samples were large enough to ensure normality (Strauss 1979). Determining the adequacy of available or utilized prey sample size followed the method suggested by Strauss. If one or both sample sizes were not normal, the t-statistic was not calculated.

Prey availability was determined for the zooplankton samples that corresponded to the particular ichthyoplankton samples containing the predators. The portion of each prey category (taxa or size) (p_i) in the local environment was obtained by dividing the number of the prey of concern in both aliquots by the sum of the total counts in each aliquot. The portion that each prey category contributed to the diet of the predator (r_i) was determined by dividing the total number of the prey category in the stomachs by the total number of all prey consumed by striped bass or white perch at that location.

The predator-prey size analysis required certain liberties to be taken in assigning prey items that were not measured to a particular size category. A serious bias would be created if damaged food items were ignored, since larger prey were more frequently unmeasureable. Size characteristics for each taxa were tabulated from the literature (Ward and Whipple 1959, Hutchinson 1967, Pennak 1953). Based on this review and the lengths of the food items that were measured, three size intervals were established: < 0.29 mm, 0.30–0.49 mm, and > 0.50 mm. The smallest class included all rotifers and copepod nauplii. The intermediate interval included the cladocerans *Bosmina* spp. and *Chydorus* spp. The largest interval included gastropod velifers, annelid larvae, cirripedia larvae, all remaining cladocerans, nearly all copepodites and adult copepods. Two species of copepod, *Halicyclops fosteri* and *Ectinosoma cuticorne*, did not fit conveniently into one specific size category, and those occurring in the microzooplankton samples or that were unmeasureable in the stomachs, were divided equally between the intermediate and largest prey size categories. Measured food items were placed in categories warranted by their lengths, regardless of taxa. If all of the stomachs from larvae from a particular station were empty, the microzooplankton from that station were not used in the prey-size analysis.

RESULTS

The digestive tracts of 316 striped bass and 588 white perch were analyzed for this study. Of these, 36% of the striped bass stomachs and 22% of the white perch stomachs were empty. The maximum number of food items present in any one larva, a gluttonous white perch, was thirty-three. Normally, fewer than ten food items were present in each gut.

Virtually the only food consumed by either species was microzooplankton. The exceptions were two fish eggs and an unidentified fish larvae in striped bass (total number of prey consumed = 491), and four fish eggs in white perch larvae (total prey = 1253). Traces

FEEDING SELECTIVITY

of phytoplankton were occasionally found, but these most likely had been inadvertently consumed or were introduced from the ruptured stomachs of consumed prey. Neither fish eggs, larvae, nor phytoplankton were considered in the calculations.

White perch consumed far more species of rotifers than did striped bass. Certain seasonal taxa of microzooplankton (gastropod veligers, annelid larvae, and cirripedia larvae) indicative of brackishwater intrusion, were not utilized at all. Most available cladoceran and copepod taxa were preyed upon to some degree (Table 35).

Table 35. *Prey utilization of striped bass and white perch larvae in the Hudson River near Peekskill, New York from 30 May to 27 June, 1978*

PREY	AVAILABILITY	UTILIZATION STRIPED BASS	WHITE PERCH
ROTIFERA			
Unidentified rotifers	*	*	*
Asplanchna priodonta	*		
Euchlanis spp.	*	*	*
Synchaeta spp.	*	*	*
Polyarthra spp.	*		*
Trichocera spp.	*		
Brachionus spp.	*	*	*
Ploesoma truncatum	*		*
Notholca acuminata	*		*
Keratella cochlearis	*		*
K. quadrata	*		*
Kellicottia spp.	*		
Monostyla spp.	*		*
Filinia longiseta	*		
Elosa worrali	*		*
GASTROPODA VELIGER	*		
ANNELIDA LARVAE	*		
CERRIPEDIA LARVAE	*		
OSTRACODA			*
DIPTERAN LARVAE			*
CLADOCERA			
Unidentified cladocera		*	*
Ceriodaphnia spp.	*	*	*
Leptodora kindti	*	*	*
Diaphanosoma brachyurum	*		
Bosmina spp.	*	*	*
Daphnia spp.	*	*	*
Chydorus spp.	*		

Table 35. *Prey utilization of striped bass and white perch larvae in the Hudson River near Peekskill, New York from 30 May to 27 June, 1978 (Continued)*

		UTILIZATION	
PREY	AVAILABILITY	STRIPED BASS	WHITE PERCH
COPEPODA			
Copepod nauplii	*	*	*
Unidentified copepod		*	*
Unidentified calanoid copepod		*	*
Acartia tonsa	*		*
Eurytemora affinis	*	*	*
Unidentified cyclopoid copepod	*	*	*
Paracyclops fimbriatus poppei	*	*	*
Cyclops spp.		*	*
C. vernalis	*	*	*
C. varicans		*	*
C. bicuspidatus	*	*	*
Tropocyclops prasinus			*
Canuella canadensis	*		
Mesocyclops edax	*	*	
Halicyclops fosteri	*	*	*
Canthosamptus staphylinus	*		
Unidentified harpacticoid adult	*		
Onychocamptus mohammedi	*		
Ectinosoma cuticorne	*	*	*

Many taxa were available or consumed in trace amounts. Therefore, the analysis of prey selectivity by station was confined to major components of the diet of the two species that were studied. However, less common prey taxa are included in the size analysis.

Copepodids and adults of the calanoid copepod *Eurytemora affinis* were the prey most consistently selected, especially by striped bass larvae (Table 35). At eleven out of fourteen stations with striped bass containing food, *E. affinis* constituted 60% or more of their diet. This selection is even more striking in view of the fact that this taxon constituted less than 10% of the total microzooplankton. Selective values were always positive, ranging from 0.176 to 0.963, and, when the sample size was adequate, always significantly greater than zero. This same trend is evident for white perch collected on the same sampling days. However, on 30–31 May, the degree of selectivity for *E. affinis* was considerably less than on later dates. This may be related to the mean density of the prey, 0.38 per liter on 30–31 May and 2.16 per liter on 14 and 26–27 June.

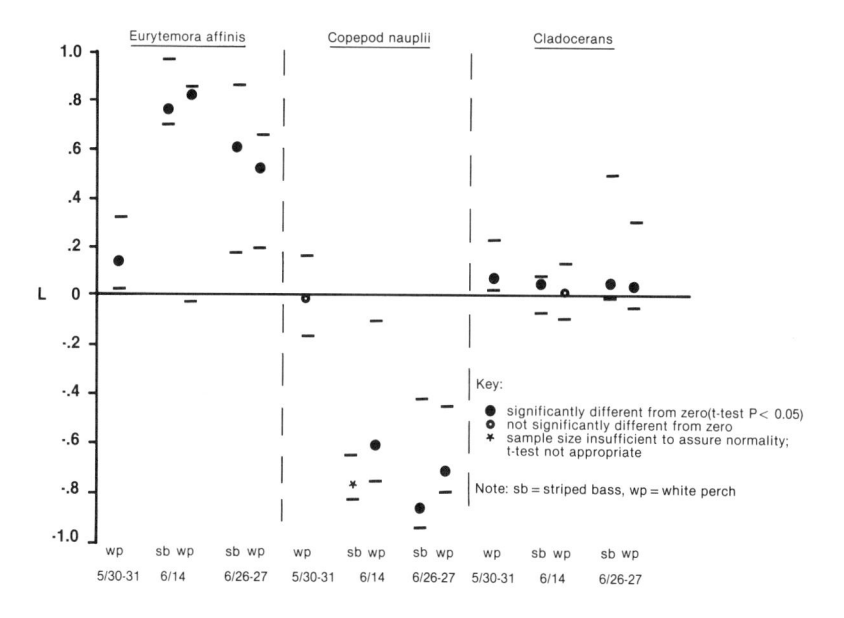

Figure 17. *Feeding selectivity of striped bass and white perch larvae for selected prey taxa (pooled daily data with range).*

Although the density of copepod nauplii in the water column was consistently high (the mean density was 17.44–53.10/l), they were infrequently consumed by striped bass and all selectivity values were strongly negative (Figure 17). White perch showed similar negative selection for copepod nauplii for corresponding days. An important deviation from this pattern was evident on 30–31 May, when copepod nauplii were opportunistically consumed and constituted 11% to 31% of the diet of white perch larvae.

Another persistent element of the diet of striped bass and white perch larvae was cladocerans, predominantly *Bosmina* spp. Striped bass generally selected for cladocerans (Figure 17) but much less strongly than for *E. affinis*. Similarly, white perch often selected for cladocerans, though opportunistic consumption was indicated for the pooled 14 June data.

On 30–31 May, not only were *Eurytemora affinis* uncommon and copepod nauplii abundant (mean density 17.44/l) but rotifers were especially abundant (mean density 55.06/l, compared with mean daily densities ranging from 0.20 to 19.59 on the other two sampling days). Therefore, selection for rotifers by white perch was analyzed for this date in order to determine whether these smaller prey were selected in the absence of the preferred prey *E. affinis* (Figure 18). When all rotifers were analyzed as a group, there appeared to be strong negative selection. However when two of the most dominant taxa were analyzed separately, *Keratella* spp. continued to be fairly strongly rejected while *Notholca acuminata* was, for the most part, selected.

The feeding habits of striped bass and white perch larvae seem to be strongly influenced by the size of the prey. They altered their foraging habits when *E. affinis*, a relatively large prey, usually 0.8–1.4 mm long, was not abundant. Striped bass larvae curtailed feeding while white perch switched to smaller prey such as copepod nauplii (usually 0.20 mm) of certain species of rotifers (usually 0.30 mm).

Recently transformed striped bass larvae (6 mm TL) are able to utilize small prey (0.30 mm) to some degree (Figure 19). On 26–27 June, this size prey constituted 32% of the diet. Yet, utilization of

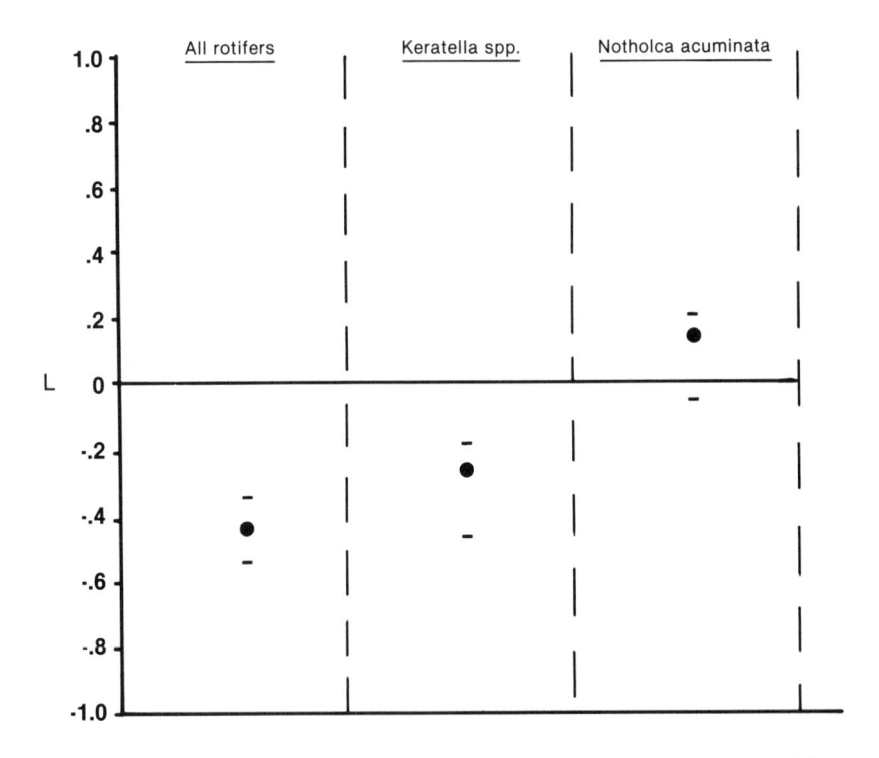

Figure 18. *Feeding selectivity of white perch larvae for all rotifers,* Keratella *spp. and* Notholca acuminata *in the Hudson River near Peekskill, New York, 30–31 May 1978 (pooled data with range).*

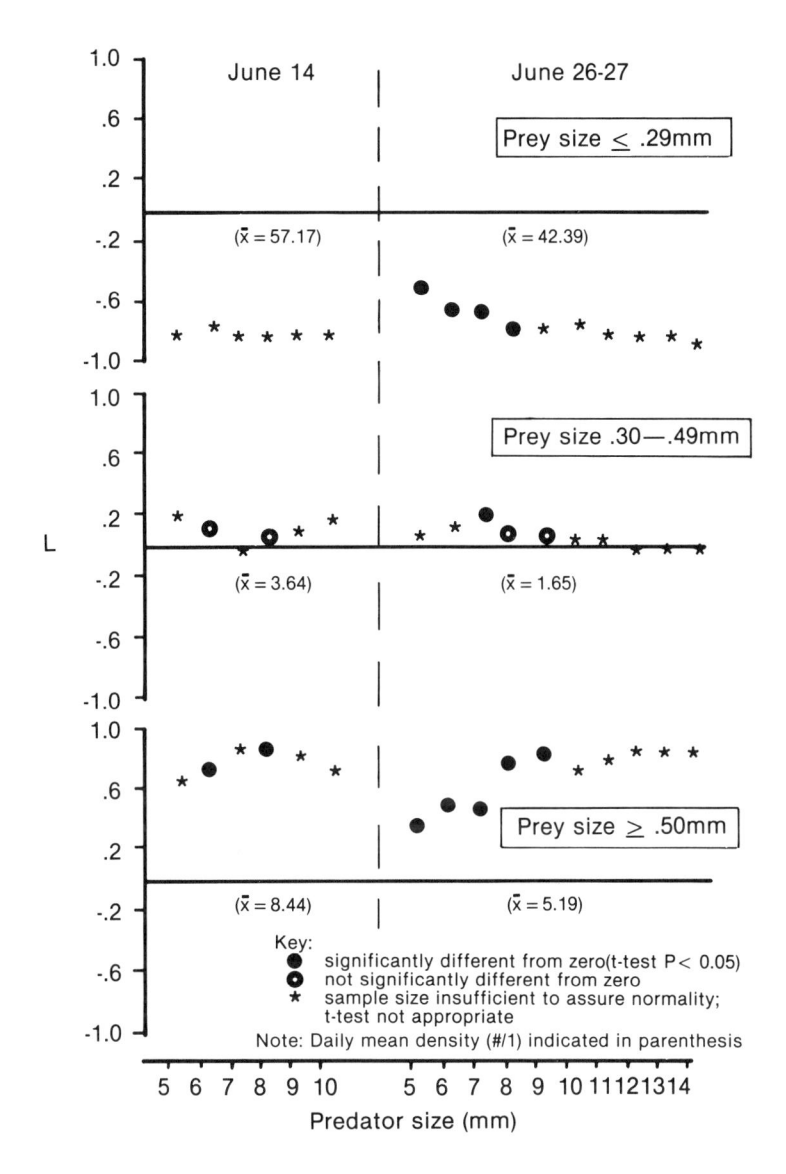

Figure 19. *Feeding selectivity of striped bass larvae based on prey size in the Hudson River near Peekskill, New York, 1978.*

small prey rapidly diminishes with increasing length, and by the time the larvae are 8 mm long, the pattern of selection for prey 0.50 mm long is established. Intermediate-sized prey is usually consumed opportunistically by all sizes of striped bass larvae.

Small white perch larvae make much greater use of small prey (Figure 20). The diet of larvae 4 mm was 38% to 67% small prey, the

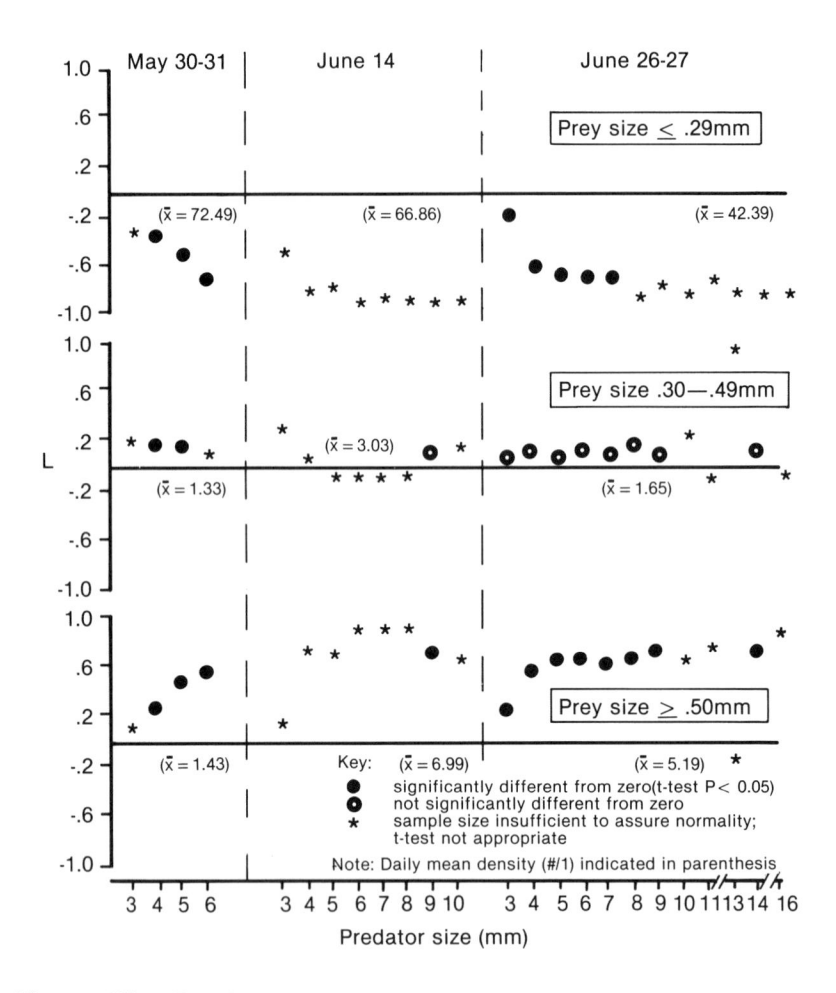

Figure 20. *Feeding selectivity of white perch larvae based on prey size in the Hudson River near Peekskill, New York, 1978.*

degree probably dependent to a large extent on the density of large prey. It is important to note that on the two later sampling days, when large prey were available in substantial quantities, white perch more than 4 mm long selected the larger prey. Like striped bass, white perch larvae consume intermediate-size prey opportunistically.

DISCUSSION

Although adults and copepodids of *Eurytemora affinis* are the preferred food of both striped bass and white perch larvae, when the density of this preferred species falls below a certain level, as it did on 30–31 May, white perch relied more heavily on alternate food sources while striped bass larvae stopped feeding. Previous studies of the food habits of striped bass larvae suggest that diet preferences are influenced by local conditions. *Bosmina* spp. were selected in some situations and avoided in others.

Prey size is an important determinant of the diet. Selection of larger prey presumably enables more energy to be devoted to growth. White perch and striped bass are both capable of feeding on smaller prey, yet for the most part they chose not to. Recent studies by Gardner (1981) showed that bluegills ignored smaller *Daphnia pulex* even when they were the only prey visible to the fish. Gardner also suggested that the size of prey selected by planktivores was based on the fish's previous experience. When previous encounter rates were low, the fish tended to eat zooplankton as they encountered it. When previous encounter rates had been high, the fish tended to select the larger prey, apparently because they had been able to "learn" the size range of the potential prey. The greater utilization of smaller prey by the smallest larvae of both species in this study may also have resulted from a lack of exposure to larger prey. Furthermore, the ability of larvae feeding for the first time may be limited by the size of the gape of the mouth. Similar mechanisms were hypothesized by Hansen and Wahl (1981) for yellow perch larvae. However, the small white bass and white perch larvae did take some relatively large prey; for example, one 5.4-mm striped bass had a 1.14-mm *E. affinis* in its gut, and a 1.05-mm *E. affinis* was found in a 3.8-mm white perch. It appears that experience plays a major role in determining the size of prey preferred by these larvae.

Doroshev (1970) pointed out that when the gas bladder of striped bass larvae is fully developed (at approximately 8 mm), energy expended on feeding activity is sharply reduced and the fish become more efficient at capturing large prey.

Planktivores generally prefer prey of the largest possible size. Optimal foraging theory states that modes of foraging evolve to maximize net energy intake or minimize searching time (Hansen and Wahl 1981). Fishes seeking to maximize energy intake would be expected to select larger prey, since caloric value is more or less proportional to size (Griffiths 1975). The results of this study generally confirm the selection of large prey by striped bass and white perch larvae. However, if the encounter rate for the preferred larger prey is such that searching time cannot be minimized, the predator must either use smaller but more abundant prey or spend more time searching for larger prey. The results of this study, especially the 30–31 May data, suggest that white perch are more inclined to consume smaller prey than striped bass. This may be a function of the smaller size of the white perch at commencement of exogenous feeding (3.5–4.0 mm versus 5.0–6.0 mm for striped bass). Presumably, small prey would appear larger to a smaller predator. Yet in laboratory situations, striped bass feed quite readily on prey less than 0.3 mm in length (Doroshev 1970), which indicates that they can perceive and capture small prey in certain situations. The feeding strategy of striped bass to ignore smaller prey during low densities of large prey serves as a type of niche partitioning, which reduces competition with white perch larvae for a suboptimal food source.

How can striped bass larvae in natural situations afford to wait for the proper-sized food to become available while white perch larvae consume both large and small prey? Recent work by Eldridge et al. (1981) suggests that the oil globule, which is prominent in striped bass but not in white perch, is a high-energy reserve which is utilized when the proper food is not available. In fact, striped bass larvae that were given no food at all survived twenty-eight days after hatching. This removes the urgency of finding food and could help to explain why all forty-eight striped bass larvae collected on 30–31 May, when there were few large prey available, had empty stomachs. As one might expect, Eldridge et al. found that starved striped bass larvae did not grow.

White perch, on the other hand, do not have the luxury of waiting for the proper-sized food, since without a large built-in energy reserve, rapid mortality would result if they ceased feeding. Small prey are consumed, but large prey are always preferred. By consuming small, as well as large prey, white perch gain additional experience in capturing prey. This may enable them to become efficient predators at a relatively smaller size than the striped bass. This is suggested by the 26–27 June data in Figures 19 and 20. The level of selection for large prey is stabilized by the time white perch are 4 mm long, but not until striped bass are 8 mm long.

The results of this study shed some light on how two congeneric species with preferences for the same prey can cope with unpredictable environmental conditions. When that prey is scarce, white perch switch to smaller prey but striped bass curtail their feeding. Murdock et al. (1979) reported that preference may break down at low prey density because of increased hunger. Presumably, striped bass are not as hungry at low prey density, so they stop feeding, thus leaving more of the alternate prey for the white perch. When the density of the preferred larger prey increases, as it did in mid- and late June when the copepods matured, enough preferred food is available to support both species.

Which of these two feeding strategies is most efficient from a competitive standpoint? Striped bass larvae would seem to be fairly well protected from starvation by their high-energy oil globule. However, growth is not possible unless food is consumed, and larval mortality due to predation is high and inversely related to the growth rate of the larvae (Houde 1978). In other words, the longer a striped bass remains a larva, the greater the probability that it will be eaten. On the other hand, white perch expend considerable energy to attack and consume small prey, and this energy must be equalled or exceeded by the usable caloric content of the prey if maintenance or growth is to occur.

White perch also gain additional experience in capturing prey, even if they only succeed in capturing enough prey for maintenance, which may give them a competitive advantage later. Thus, the high-energy oil globule of the striped bass may actually place them at a competitive disadvantage later if they rely on internal reserves at the expense of growth and increased predatory skills.

Striped bass and white perch have adopted different life-history strategies to ensure the survival of at least some offspring. These strategies were called into play in 1978 when transitional larvae of both species were common in the Peekskill region of the Hudson River, before the preferred adults and copepodids of *Eurytemora affinis* became abundant.

ACKNOWLEDGEMENTS

Ichthyoplankton samples were provided by ConEd. I am indebted to the staff of EA, especially John Young, Bill Dey, and Becky Ligotino for general support and constructive advice. The staff of Chas. T. Main, Inc., provided graphical and clerical support, and Li Suloway provided additional help in graphics.

5. Patterns of Movement of Striped Bass and White Perch Larvae in the Hudson River Estuary

Thomas L. Englert and David Sugarman

INTRODUCTION

During the past decade, consultants to the utilities and regulatory agencies have developed a number of mathematical models to predict the impact of power plant operations on fishes, in particular striped bass, in the Hudson River (Lawler 1972, Englert and Aydin 1975, Eraslan et al. 1976, Boreman et al. 1978). With the exception of Boreman's Empirical Transport Model, all of these efforts have involved the simulation of the effects of hydrodynamics on the spatiotemporal distribution of the early life-history stages of striped bass. The most complex of these simulation models is the Real Time Life Cycle Model (RTLC) of Englert and Aydin. Because it explicitly simulates the principal factors expected to influence the distributions of early life stages, it is a useful tool for investigating the effects of various mechanisms on these distributions.

MODEL SIMULATION OF STRIPED BASS LARVAE MOVEMENTS

The RTLC simulates the movement of striped bass yolk-sac and post yolk-sac larvae in two dimensions, longitudinal and vertical. The model's defining equation illustrates how the principal mechanisms are simulated.

$$\frac{\partial c^k(x,a,t)}{\partial t} + \frac{\partial c^k(x,a,t)}{\partial a} + \frac{1}{A}\frac{\partial}{\partial x}[c^k(x,a,t) \cdot e(a) \cdot Q^k(x,t)]$$

$$= \frac{1}{A\partial x}\left[EA\frac{\partial c^k(x,a,t)}{\partial x}\right] + M^k(x,a,t) \qquad (1)$$

$$- E_R^k(x,a,t) \cdot c^k(x,a,t) - KD(c,a) \cdot c^k(x,a,t)$$

where

$c^k(x,a,t)$ = the concentration (number of organisms per volume) at a point in the population space defined by x, longitudinal dimension; a, age; t, time; and in layer k.

$\dfrac{\partial c^k}{\partial t}$ = accumulation

$\dfrac{\partial c^k}{\partial a}$ = maturation

$\dfrac{1}{A}\dfrac{\partial}{\partial x}(c^k \cdot e \cdot Q^k)$ = convection

$\dfrac{1}{A}\dfrac{\partial}{\partial x}\left(EA\dfrac{\partial c^k}{\partial x}\right)$ = the dispersion term

$Q^k(x,t)$ = real time flow

$e(a)$ = advective transport factor (ATF); this is the effective fraction of flow experienced by a given life stage

E = the dispersion coefficient obtained from Elder's equation (Elder 1959)

A = the river cross-sectional area

$M^k(x,a,t)$ = the larval migration rate in the vertical direction

$KD(c,a)$ = the natural mortality rate

$E_R^k(x,a,t)$ = the entrainment or impingement rate

By including explicit simulations of the diurnal migration of the larval stages and the tidal flows in the upper and lower layers of the Hudson, the model accounts for the possible synergistic effect of these two mechanisms on larval distribution. Observations from field data collected from 1974 through 1978 provide evidence of diurnal migration of both yolk-sac and post yolk-sac stages (LMS 1981f). These field observations, which were used to evaluate the M_k term in equation (1), show a consistent pattern of larval movement to the lower layer during daylight hours and a more uniform distribution at night. Interaction of this diurnal migration pattern with the partially stratified flows in the Hudson can result in less net downstream movement of the larvae than would occur if the larvae were completely passive.

Comparisons of the output from the RTLC with the 1975–1978 field data suggest that other phenomena may be acting to reduce downstream movement of the larvae. Generally, such comparisons show that when the RTLC is run with the migration parameters and tidal flows computed from field data (ATF = 1.0), the model output shows a faster downstream movement of eggs and larvae than that seen in the field data. To get closer agreement between field data and the model output, it was necessary to reduce the influence of hydrodynamics on larval movement. This was accomplished through calibration of the advective transport factor, that is, the ATF. Calibration of the model based on field data collected in 1974–1978 gave ATF factors of approximately 0.1 or less for both the yolk-sac and post yolk-sac stages. Thus, field data suggest that the influence of hydrodynamic transport on the larval stages is only one-tenth the theoretical value, even when the interaction of diurnal migration patterns and the stratified flows is included in the simulation.

As shown in Figures 21–24, with ATF at approximately 0.1, the spatial distributions of striped bass larvae computed from model output and field data are in reasonably good agreement. In these figures and others discussed below, the shaded area represents the 95% confidence interval about the mean of the field data (McFadden and Lawler 1977, LMS 1975g, 1976). However, for 1978, the model output does not replicate the upstream peak in the field data near mile point 70.

Figures 25–28 show similar comparisons of model output and field data for striped bass post yolk-sac larvae. The 1975 and 1978 graphs (Figures 25 and 28) indicate that the model does not reproduce the upstream peaks in the distributions. There are several possible explanations for these differences. These include the following:

1. Since the starting point for the simulation is the distribution of eggs, any inaccuracy in field measurements could affect the ability of the model to reproduce the distribution patterns of later life stages. An undersampling of the upstream peak in egg distribution in 1978 could explain the model's failure to reproduce the upstream peak in yolk-sac and post yolk-sac stages (Figures 27 and 28).

2. Higher natural mortality of eggs and larvae transported into more saline water could result in the apparent lack of downstream movement. The RTLC has the capacity to consider differential mortality, but this has not yet been used because of the difficulty of determining the mortality rates.

3. Organisms may be transported downstream of the sampling region. Data collected in 1979 as part of the Westway study indicate that this is unlikely, since few striped bass larvae were found below milepoint 10 (LMS 1980f).

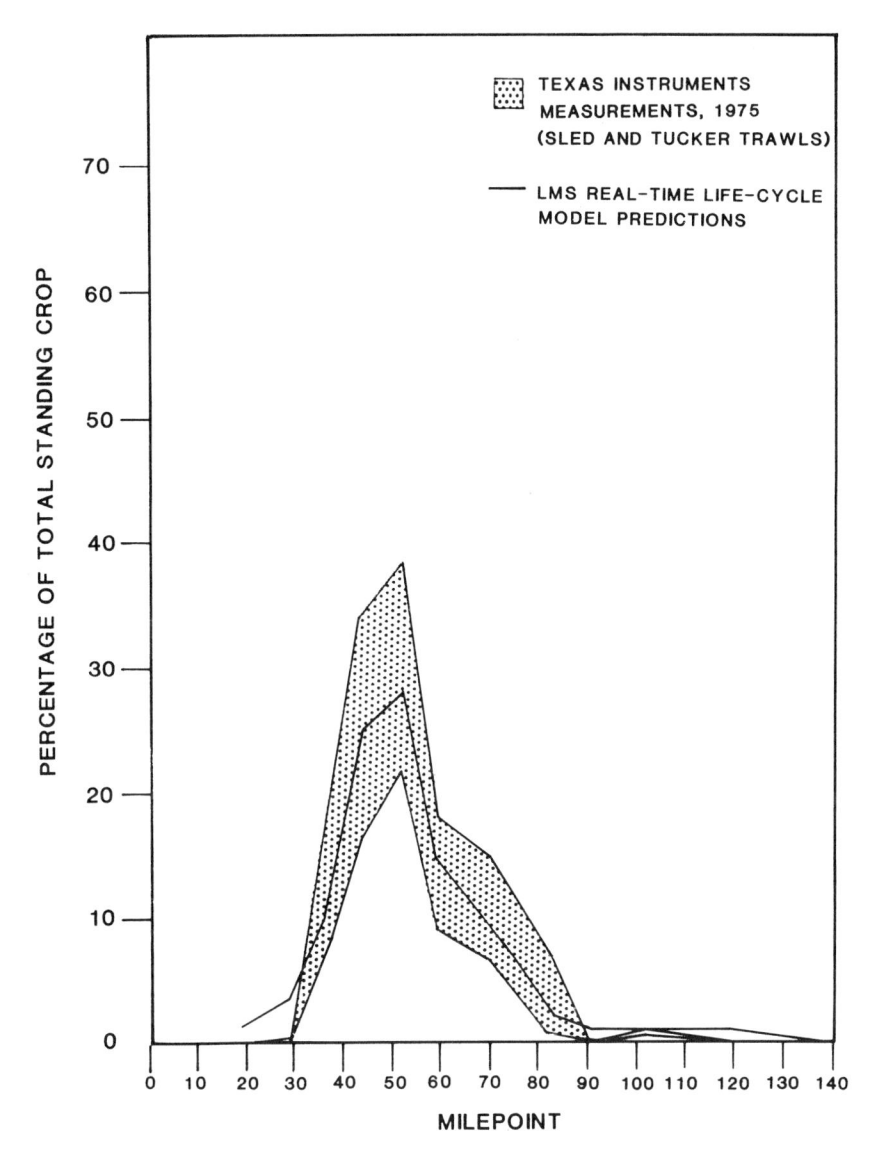

Figure 21. *Comparison of real-time model life cycle, predictions with field measurements of the spatial distribution of striped bass yolk-sac larvae 1–7 June 1975.*

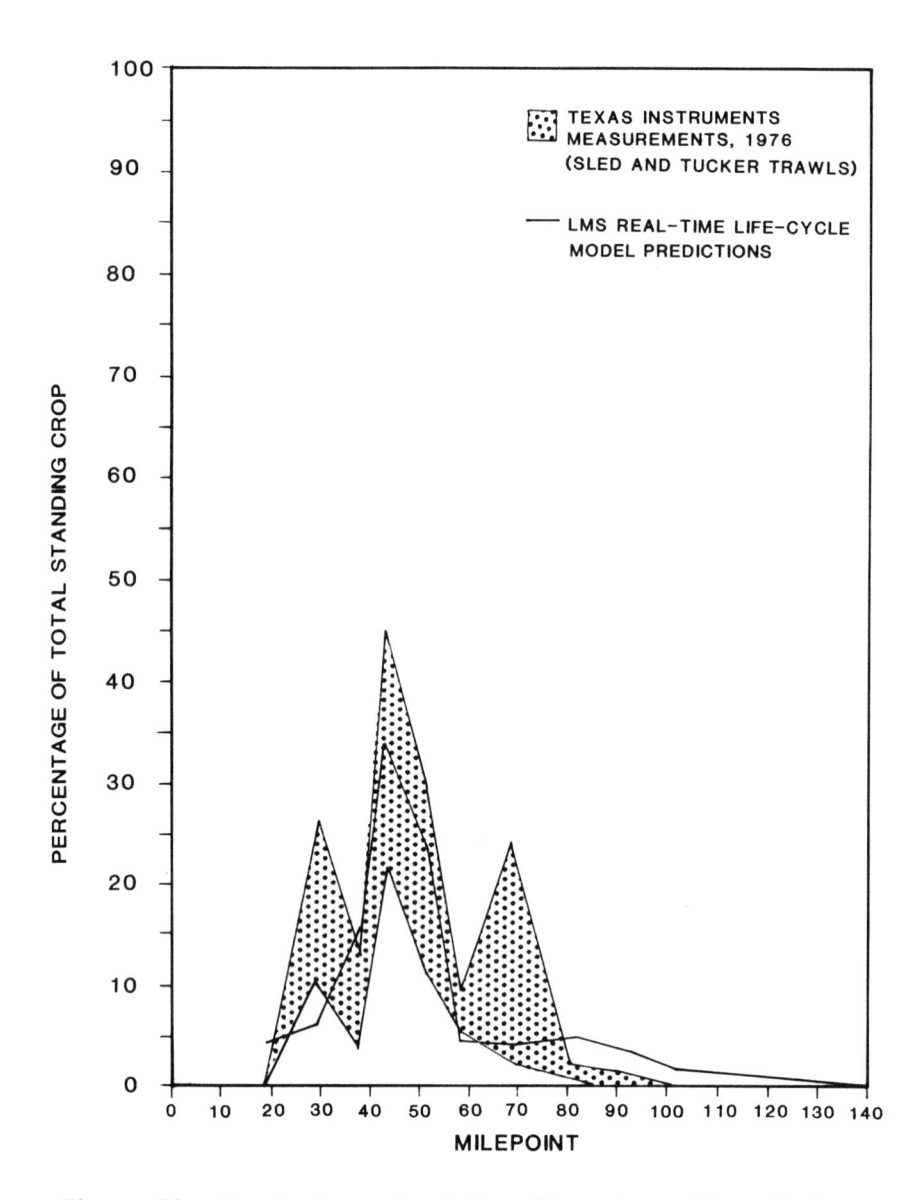

Figure 22. *Comparison of real-time life cycle model predictions with field measurements (bimodal egg distribution) of the spatial distribution of striped bass yolk-sac larvae 16–22 May 1976.*

152

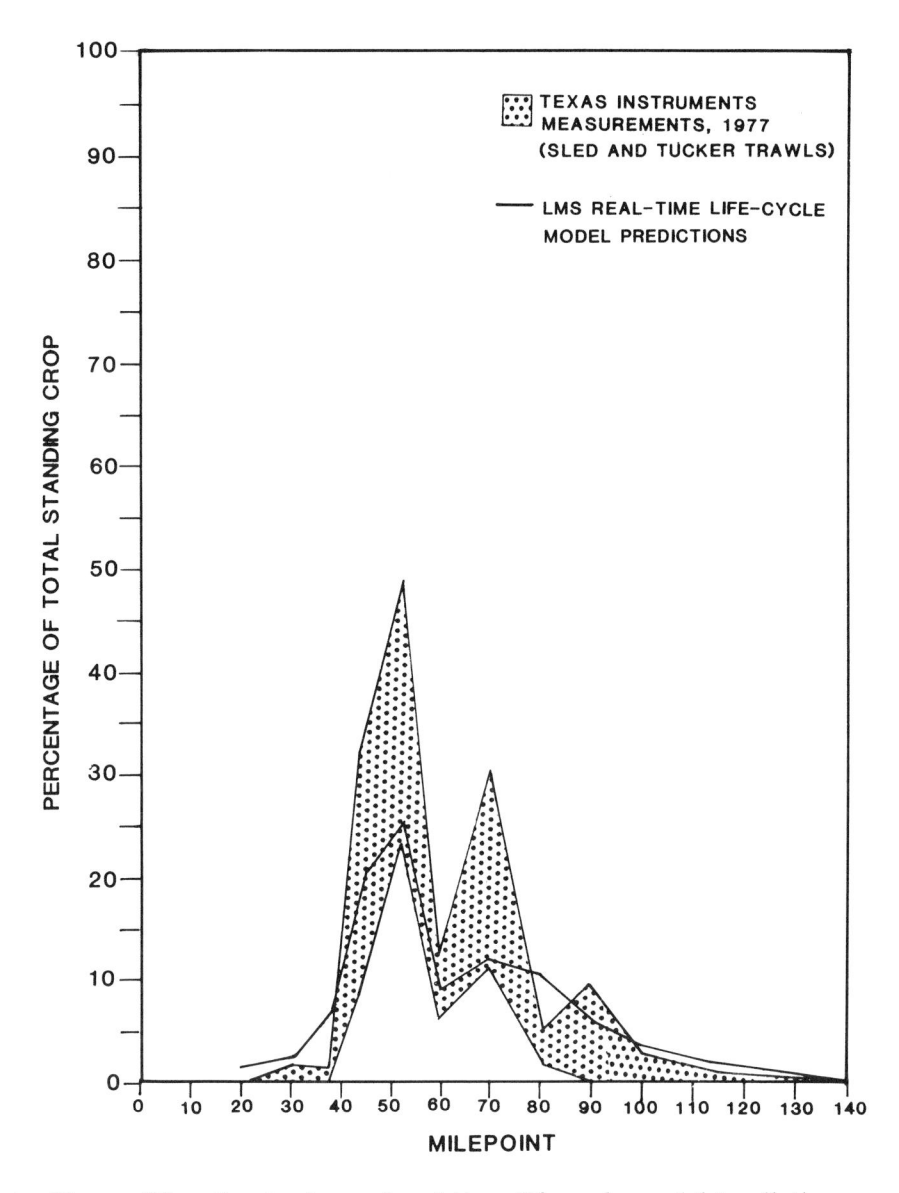

Figure 23. *Comparison of real-time life-cycle model predictions with field measurements of the spatial distribution of striped bass yolk-sac larvae 29 May–4 June 1977.*

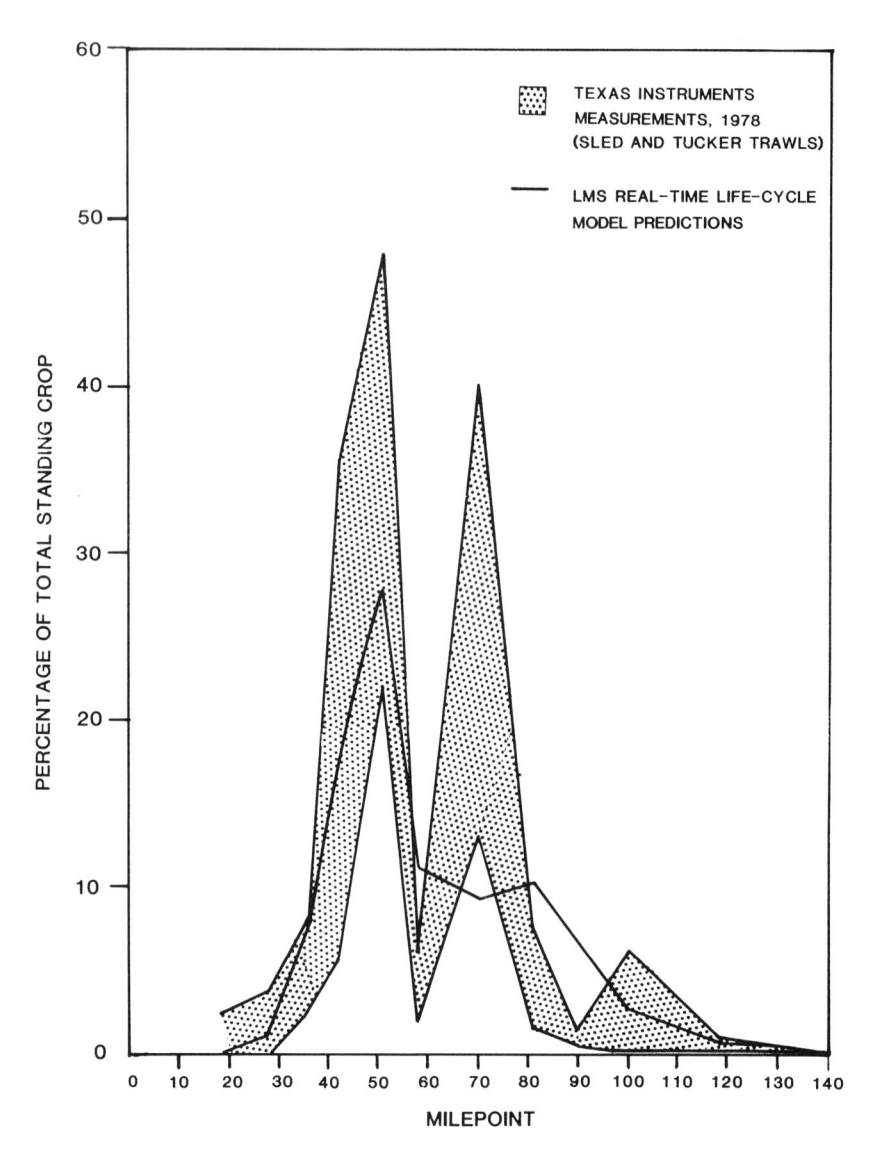

Figure 24. *Comparison of real-time life-cycle model predictions with field measurements of the spatial distribution of striped bass yolk-sac larvae 4–10 June 1978.*

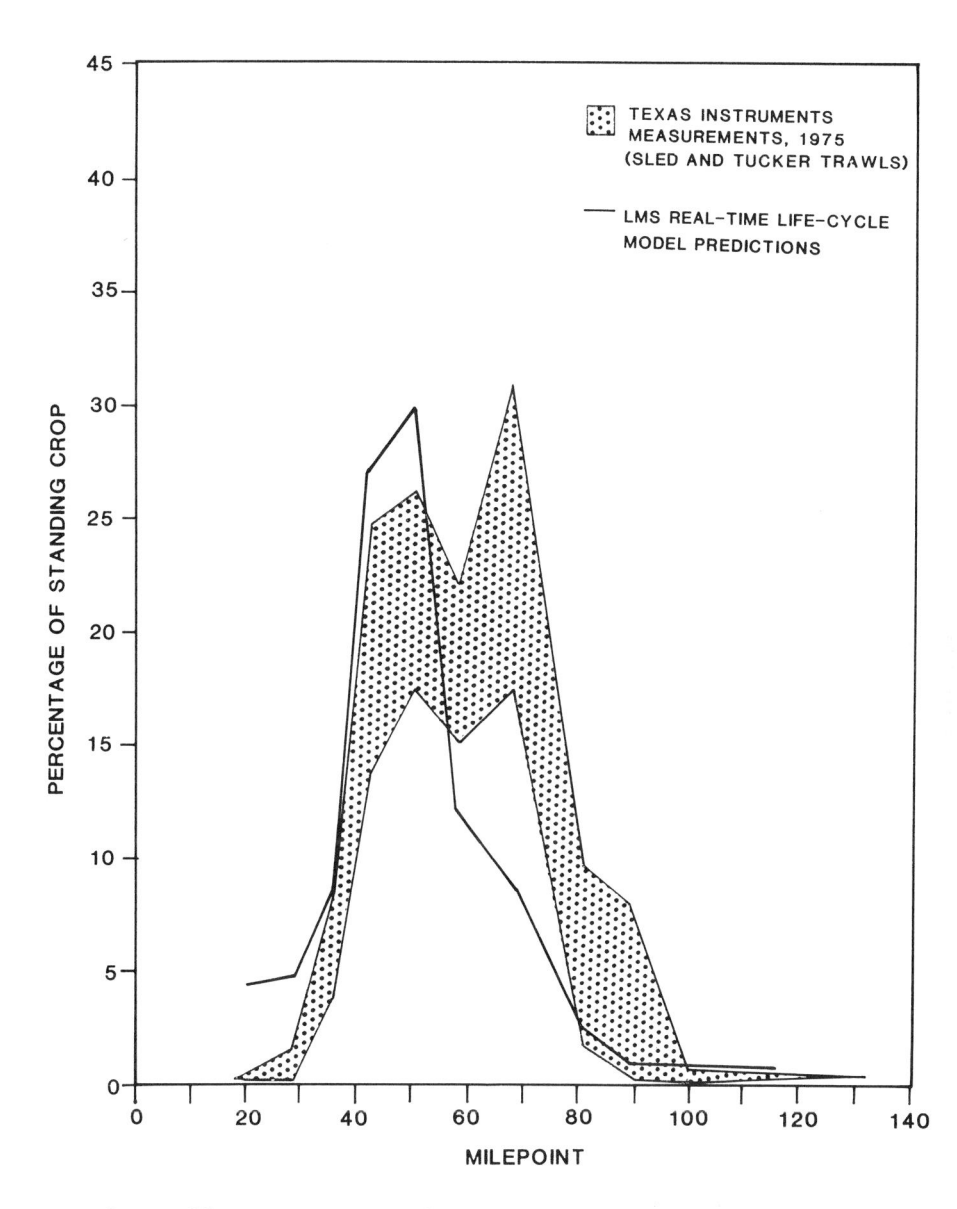

Figure 25. *Comparison of real-time life cycle model predictions with field measurements of the spatial distribution of striped bass post yolk-sac larvae 1–7 June 1975.*

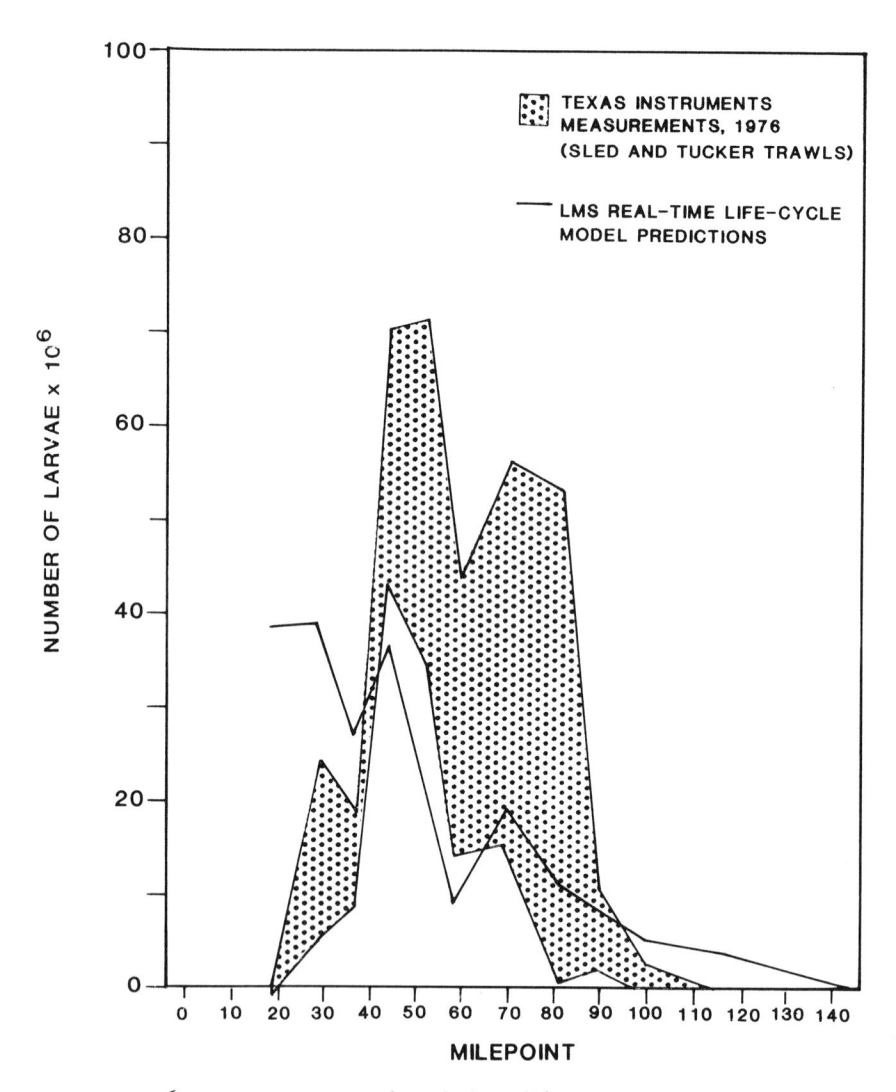

Figure 26. *Comparison of real-time life-cycle model predictions with field measurements (bimodal egg distribution) of the spatial distribution of striped bass post-yolk-sac larvae 13–19 June 1976.*

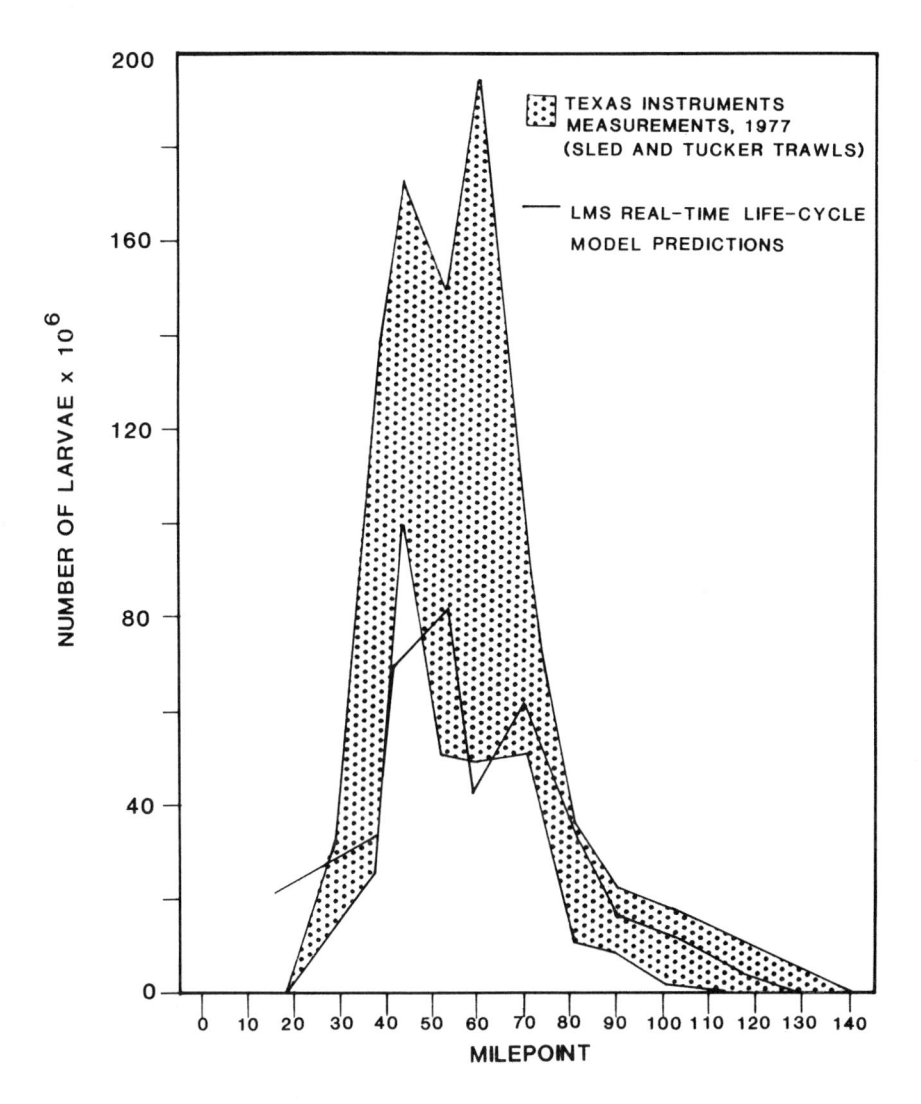

Figure 27. *Comparison of real-time life-cycle model predictions with field measurements of the spatial distribution of striped bass post-yolk-sac larvae during 5 June–11 June 1977.*

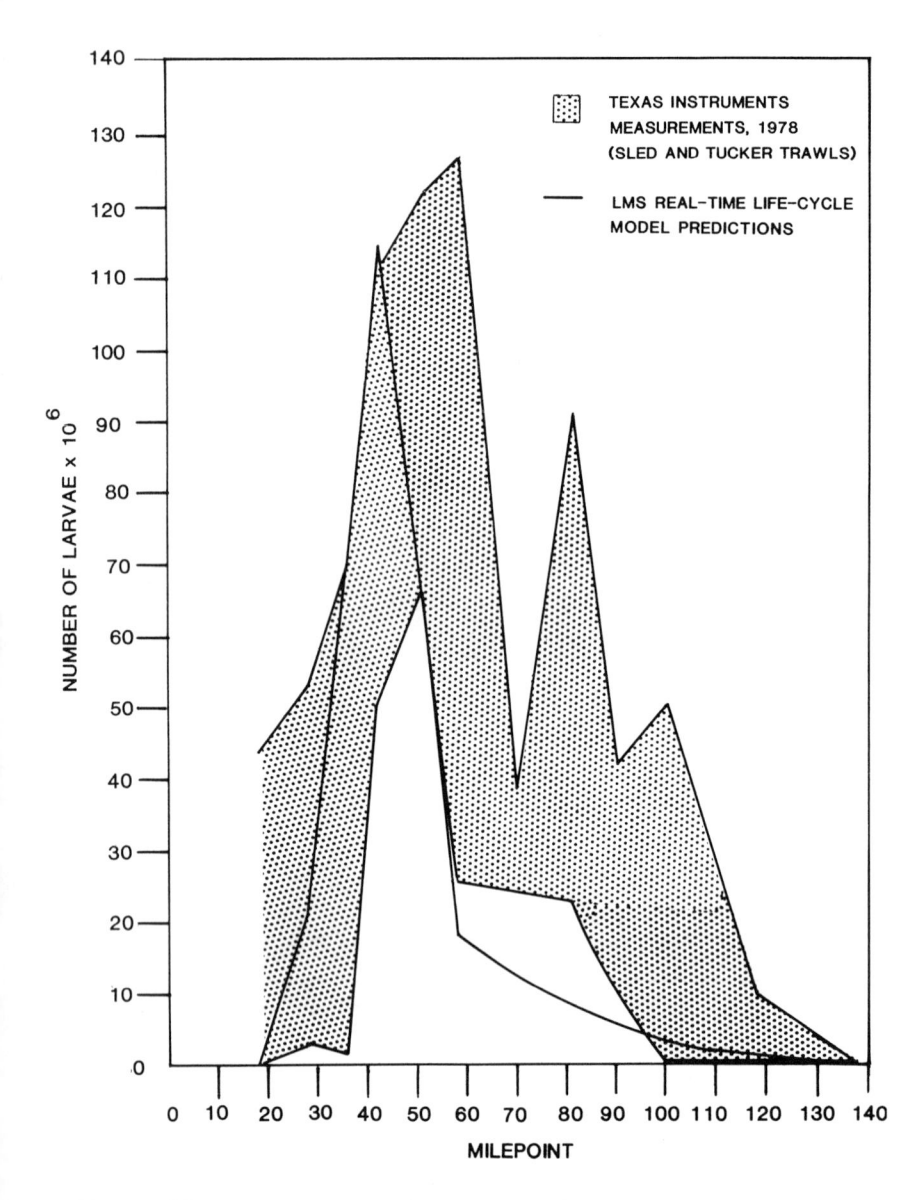

Figure 28. *Comparison of real-time life-cycle model predictions with field measurements of the distribution of striped bass post yolk-sac during 4–10 June 1978.*

158

4. The early life stages of striped bass and other species may have the ability to avoid the net downstream transport in the Hudson. At present, the precise mechanisms for this avoidance are unknown but may include coupling of vertical migration with tidal flows.

DEFINITION OF RIVER MILE INDEX

The available field data on striped bass and white perch provide further insights into patterns of larval movement. One of the most challenging aspects of this investigation is the development of a means of analyzing the data on spatial distributions in a way that will permit statistical analyses.

It is helpful to have a single index of the epicenter of the distribution. For the early life stages of fishes, temporal variations should be included in such an index. The River Mile Index (RMI) described here does this by giving highest weighting to the distributions during periods of peak abundance.

In computing the RMI, one begins by determining the temporally weighted fraction of the standing crop found in each sampling region. This fraction (D_j) is computed as follows:

$$D_j = W_i \cdot S_{i,j} \qquad (2)$$

where:

W_i = fraction of total annual production of a given life stage present during week i.

$S_{i,j}$ = fraction of the life-stage standing crop for week i which was found in region j.

The RMI is then computed as follows:

$$\text{RMI} = \sum_{j=1}^{N} MP_j \times D_j \qquad (3)$$

where:

MP_j = river mile midpoint of sampling region j.

N = total number of sampling regions.

Table 36. *Values of the River Mile Index (RMI) for striped bass and white perch larval stages during 1974–1980*

YEAR	LIFE-STAGE YOLK-SAC	POST YOLK-SAC	RMI DIFFERENCE YS-PYS
STRIPED BASS			
1974	64	59	5
1975	56	53	3
1976	57	51	6
1977	57	56	1
1978	57	53	4
1979	58	43	15
1980	68	55	13
WHITE PERCH			
1974	73	68	5
1975	67	69	−2
1976	82	82	0
1977	81	79	2
1978	80	77	3
1979	87	76	11
1980	96	82	14

Table 36 shows the RMI values for striped bass and white perch larvae in 1974–1980 and the differences between the indices for the larval stages in each year. This difference can be used as a measure of net downstream movement during the early life-history stages. Since the freshwater flow provides a measure of net downstream transport in the Hudson, one might expect that the RMI differences in Table 36 would be correlated with freshwater flow during the period of larval abundance. However, this expected correlation does not obtain. In fact, in 1980, when the flows were lowest, the RMI differences were second highest.

INFLUENCE OF SALT FRONT ON RMI

Studies on the Hudson (TI 1980a) and Potomac (Polgar et al. 1976) have suggested that the location of the salt front can influence the location of peak abundance of early life stages of striped bass. It has been well-established, for instance, that the striped bass generally spawn above the salt front (McFadden and Lawler 1977). There is also some evidence that natural mortality may be higher below the salt front (Polgar et al. 1976). Furthermore, since the location of the salt front is directly influenced by freshwater flow, its influence may also be reflected in any correlation between the RMI and salt front location.

The location of the salt front can be determined using an empirical relationship derived by Abood (1977). The salt front location, L, expressed as milepoint, is computed as follows:

$$L = 135(Q_f(LH)/1000)^{-0.38} \text{ for } Q_f(LH) \leq 27{,}000 \text{ cfs} \quad (4)$$
$$1948(Q_f(LH)/1000)^{-1.19} \text{ for } Q_f(LH) > 27{,}000 \text{ cfs}$$

where $Q_f(LH)$ is the freshwater flow in the lower Hudson in cfs. This value is determined from the freshwater flow at Green Island using regression presented by LMS (1981a) for the months of May, June, and July (Table 37).

To provide an average salt front location for comparison with the RMI, the monthly values were combined into a single value based on a temporal weighting scheme. The scheme is analogous to the approach used in computing the RMI, namely:

$$SFL = \sum_{i=1}^{3} SF_i \times FP_i$$

where

SFL = milepoint of average salt front location.

SF_i = milepoint of the salt front location during month

i (May = 1, June = 2, July = 3).

FP_i = fraction of the total life-stage production

found in month i.

Table 37. *Location of salt front in the Hudson during May, June, and July of 1974–1980 computed from equation (4)*

YEAR	SALT-FRONT LOCATION (RIVER MILE)		
	MAY	JUNE	JULY
1974	31	54	49
1975	37	47	58
1976	21	44	44
1977	42	58	65
1978	39	52	70
1979	38	55	70
1980	52	59	68

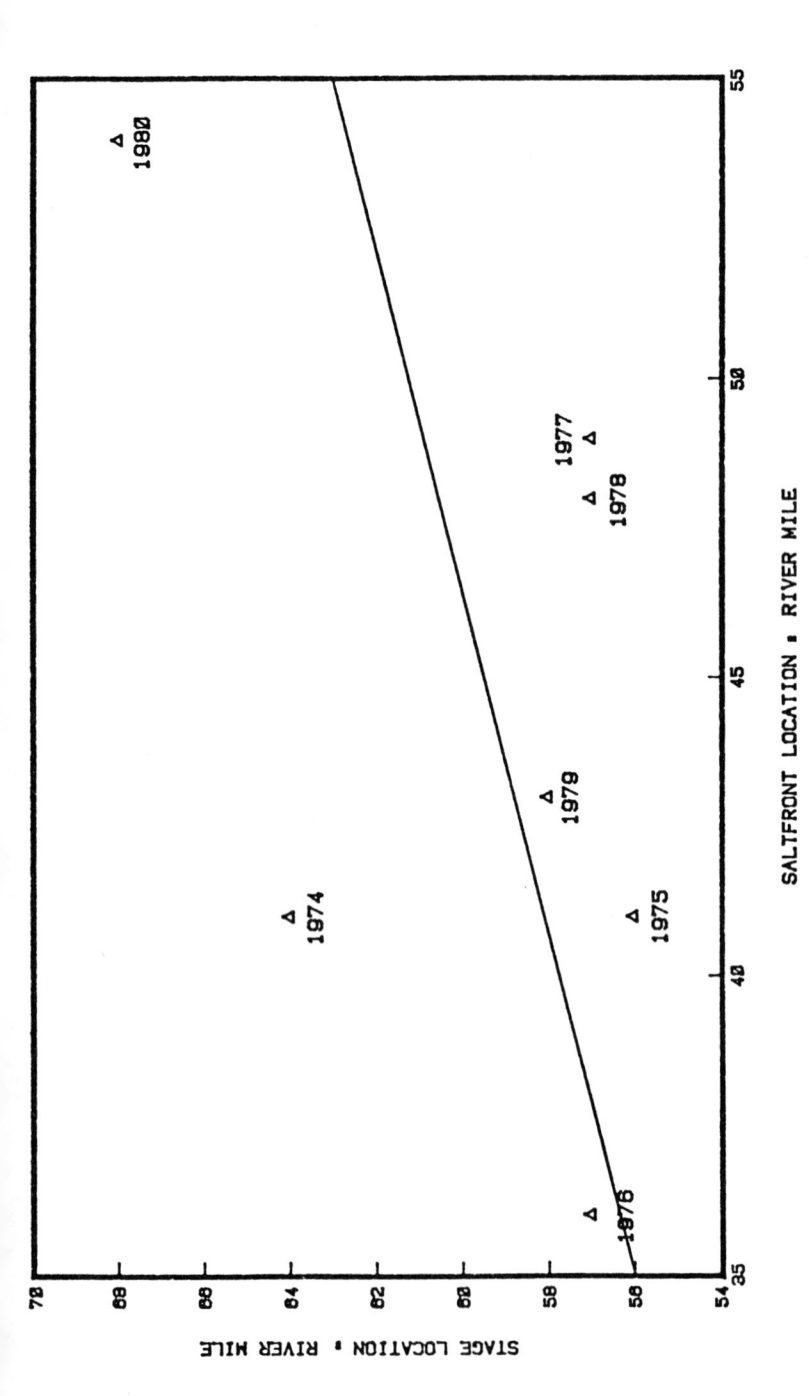

Figure 29. *Influence of salt-front location on RMI for striped bass yolk-sac larvae.*

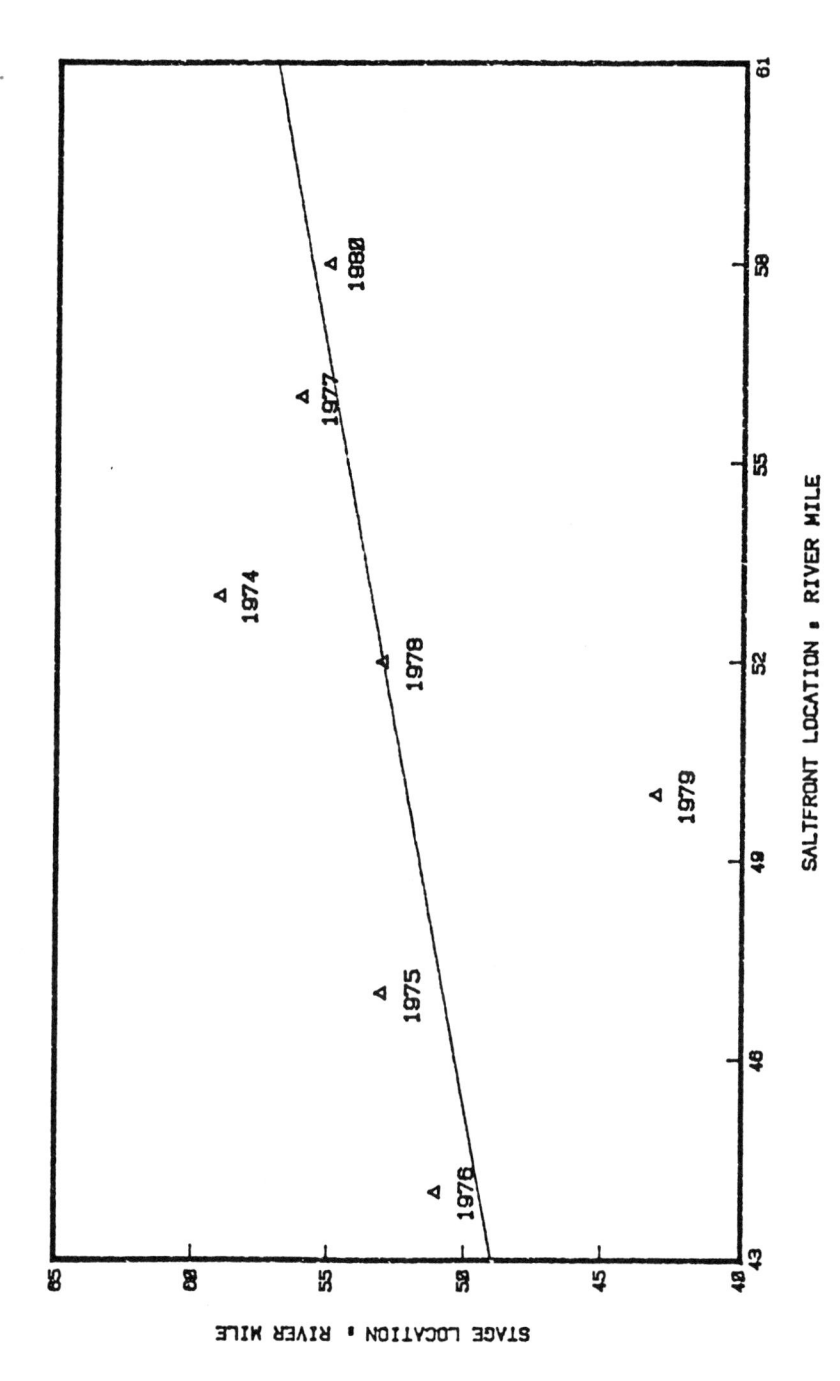

Figure 30. Influence of salt-front location on RMI for striped bass post yolk-sac larvae.

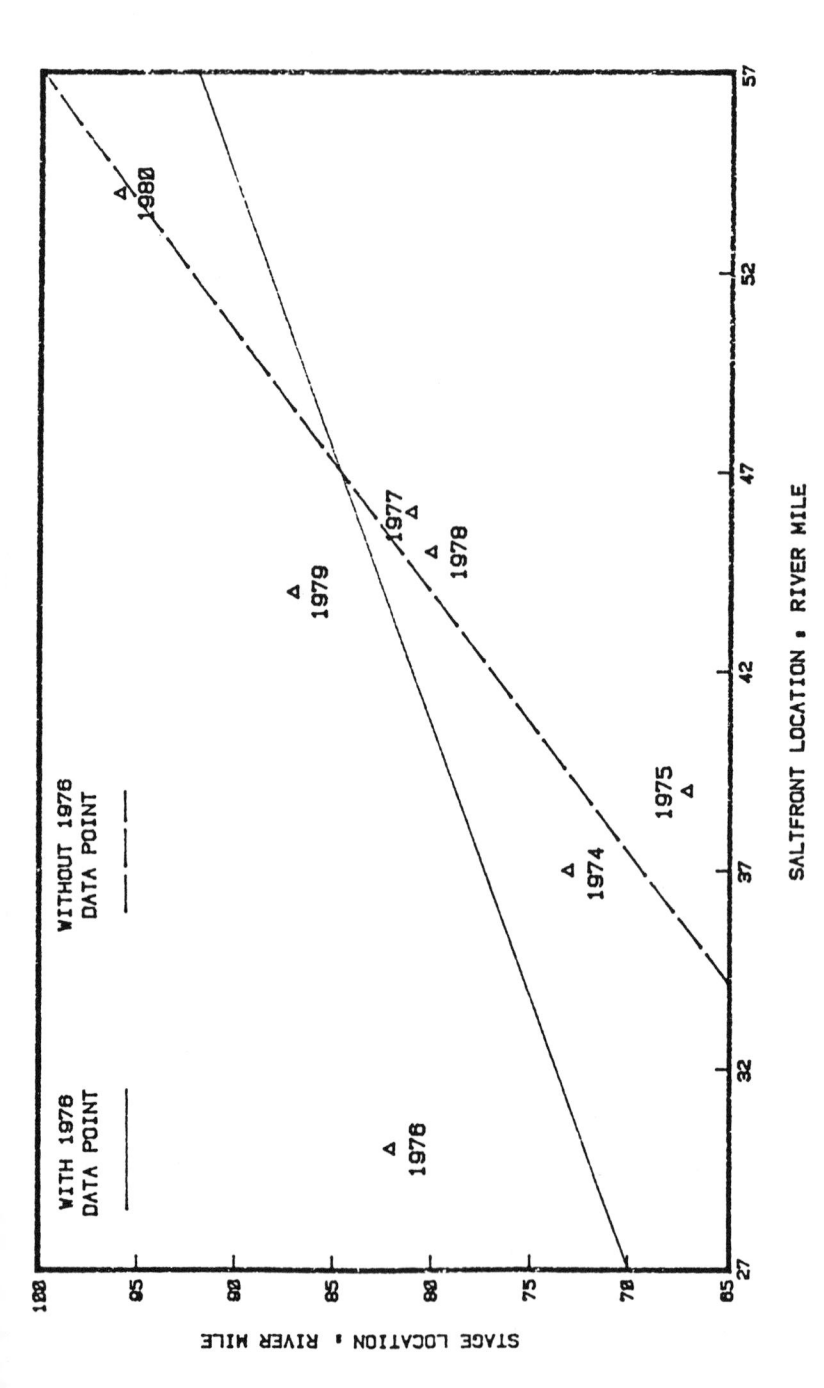

Figure 31. Influence of salt-front location on RMI for white perch yolk-sac larvae.

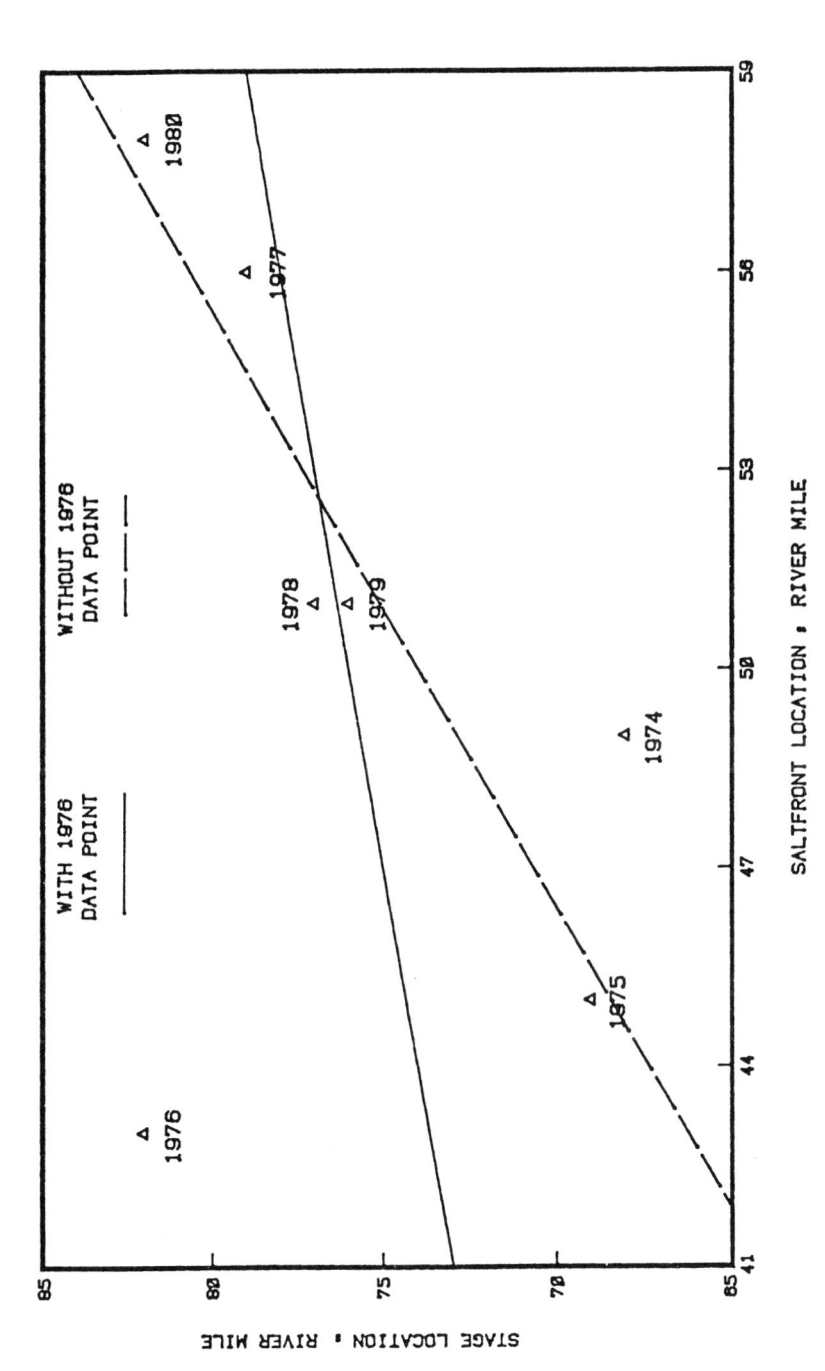

Figure 32. *Influence of salt-front location on RMI for white perch post yolk-sac larvae.*

Figures 29 and 30 compare the RMI and SFL values for striped bass larvae. In general, the trend is as expected. The RMI, the measure of the epicenter of life-stage abundance is found farther upstream in years when the salt front is also farther upstream. However, the correlations are rather weak with correlation coefficients of only 0.48 and 0.44 for yolk-sac and post yolk-sac stages, respectively. As shown in Figures 31 and 32, white perch larval stages show a similar pattern. When all data points are included, the correlations are again relatively weak with r values of 0.6 and 0.36, respectively. However, as shown by the dashed line in these figures, when the 1976 data point is eliminated, the r value is approximately 0.9, and the correlation is significant at the 5% level. As explained in LMS 1981f, there were unusual temperature patterns during the 1976 spawning season, resulting in two spawning peaks for both white perch and striped bass. There are also indications that the larvae from the first peak had unusually high mortality. This may explain the anomalous nature of the 1976 data point for white perch.

Figures 29 through 32 also reveal that with the exception of striped bass post yolk-sac larvae (Figure 30) the RMI values consistently lie above the salt front and that the RMI values for striped bass are consistently less than those for corresponding white perch life stages. These two observations suggest that, as expected, both species spawn above the salt front and the white perch tend to spawn upstream of the striped bass.

CORRELATION BETWEEN DOWNSTREAM MOVEMENT OF STRIPED BASS AND WHITE PERCH LIFE STAGES

While the differences between yolk-sac and post yolk-sac RMI values do not appear to be correlated with freshwater flow, there is a strong correlation between the downstream movements of striped bass and white perch. For instance, it is evident that the river mile differences are comparable for the years 1974–1980 and that the values for striped bass are slightly less than those for white perch. These general observations are confirmed by the regression shown in Figure 33. The correlation coefficient is 0.86 and is significant at the .05 level. This suggests that in a given year, the distance in river miles between the temporally weighted peaks in the distribution of

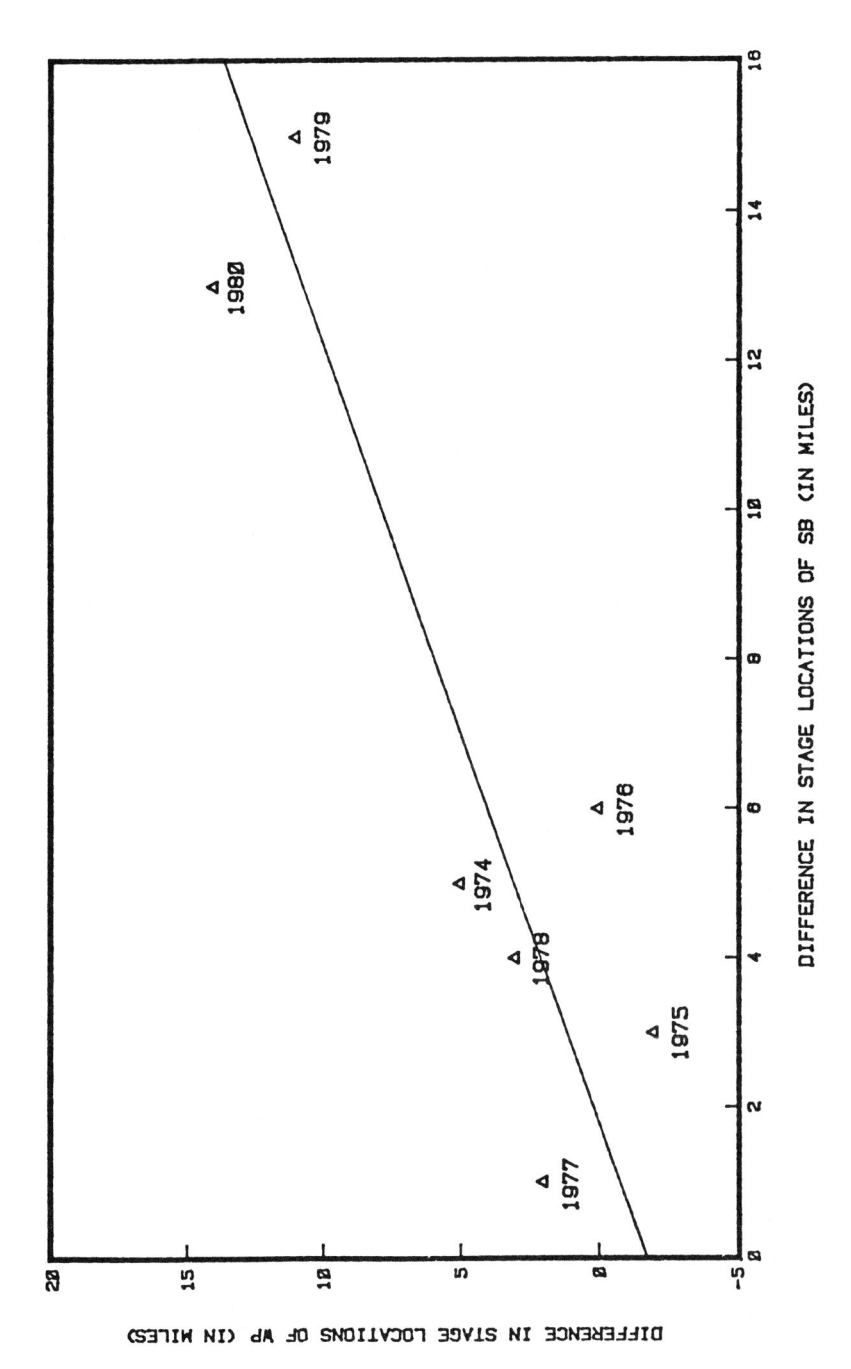

Figure 33. *Difference between yolk-sac and post yolk-sac stage locations for white perch versus striped bass.*

yolk-sac and post yolk-sac stages of the two species is highly correlated. There are several possible reasons for this correlation, including the following:

1. Certain characteristics of the river hydrodynamics may influence the two species in the same manner.
2. The availability of food which is critical to survival of the post yolk-sac and older stages may influence the net downstream movement of the two species in the same manner.

At present, the precise reason for the similarity in movement patterns of the two species is not known, but its documentation is an example of the types of analyses that can be made using the extensive data base that has resulted from the 316(b) studies.

CONCLUSIONS

Observed patterns of larval movement suggest that larvae are transported downstream at rates considerably less than those computed for theoretical flow values. This is true even when the simulation incorporates the observed diurnal migration of striped bass larvae. Model calibrations and observed differences in a temporally averaged index of the spatial distribution of yolk-sac and post yolk-sac larvae of striped bass reveal that the effective hydrodynamic transport of the larvae is approximately one-tenth the theoretical value.

Striped bass and white perch both spawn above the salt front, and white perch spawn upstream from striped bass. Both species spawn farther upstream in years when low freshwater flows allow the salt front to move upstream. Downstream movements of the two species are highly correlated.

The extensive long-term data base that has resulted from the power plant studies on the Hudson River makes possible statistical analyses of patterns of fish behavior. The biggest challenge in using this information is to find innovative ways to analyze the data so as to provide new insights into the factors that affect distribution and abundance of key fish species in the Hudson.

ACKNOWLEDGEMENTS

The modeling and field studies discussed in this paper were originally funded by ConEd, NYPA, ORU, and CHGE.

Part III
Sturgeons

6. Contribution to the Biology of Shortnose Sturgeon in the Hudson River Estuary

Thomas B. Hoff, Ronald J. Klauda, and John R. Young

INTRODUCTION

The shortnose sturgeon (*Acipenser brevirostrum*), the smallest of five North American sturgeon species, is distinguished from the Atlantic sturgeon (*Acipenser oxyrhynchus*), the only other Atlantic Coast estuarine sturgeon, by blackish viscera, wider mouth, absence of a fontanelle, almost complete absence of postdorsal shields, and preanal shields arranged in a single row. The growth rates of the two species are similar in early life; their different adult sizes are due to faster growth of the Atlantic sturgeon after it emigrates from the estuaries to the ocean (Vladykov and Greeley 1963). The shortnose sturgeon attains adult proportions when about 2 feet in total length, whereas Atlantic sturgeon up to 4 feet long may retain juvenile characteristics.

Shortnose sturgeon are restricted to the east coast of North America from the Saint John River, New Brunswick, to Indian River, Florida. Distribution of the species has historically centered in Atlantic seaboard rivers, including the Saint John, Connecticut, Hudson, Delaware, Potomac, and Saint Johns (Florida); reports of shortnose, however, have always been uncommon because of their similarity to the Atlantic sturgeon and the lack of incentive for fishermen and scientists to record their distribution and abundance (Shortnose Sturgeon Recovery Team 1981). The shortnose sturgeon is less migratory than the Atlantic sturgeon and spends most of its life in large tidal rivers, but it has been taken in brackish and salt water (Scott and Crossman, 1973). All captures at sea have been within a few miles of land (Dadswell 1981). A landlocked population occurs in the Holyoke Pool, Connecticut River (Taubert 1980, Taubert and Dadswell 1980).

The shortnose sturgeon was listed on the original United States endangered species list in 1967 (32 Federal Register 4001, March 11, 1967), and it is classified as rare in Canada (Gorham and McAllister

1974). In early 1977, evaluation of the status of shortnose sturgeon in the United States was assigned to a panel of experts, four from the United States and one from Canada, with the establishment of a shortnose-sturgeon recovery team. This team's draft Recovery Plan report was submitted to NMFS in January 1981 with this major conclusion: "Take no action at this time to change the endangered species status of the shortnose sturgeon; however, determination of regional changes in the endangered species status of the shortnose sturgeon may be possible, especially in the Northeast."

The sturgeons have traditionally been a desirable commercial species in most northeast estuaries. For example, in 1811 the shortnose sturgeon was "more sought after... than the larger common species..." (Atlantic sturgeon) by commercial fishermen of the Delaware River (LeSueur 1818). By the late 1880s, the species was called "not uncommon in the Hudson..." (Ryder 1888). Then, during the 1900s, the abundance of shortnose sturgeon apparently declined. Extremely slow growth and late maturation are likely causes for the species' apparent inability to withstand heavy fishing exploitation. Prior to 1937, a few fishermen set gill nets in the Hudson River for shortnose sturgeon (pinksters and roundnosers, in the fishermen's vernacular) just before the yearly run of American shad; but as the sturgeon declined, so did fishing pressure.

While the flesh is considered delicious and the eggs desirable for caviar, shortnose sturgeon are not currently exploited either commercially or by sport fishermen in the United States because of their endangered status. Some, however, may be marketed inadvertently with Atlantic sturgeon. The shortnose may be more abundant now than in the recent past (Scott and Crossman 1973, McCleave et al. 1977, Heidt and Gilbert 1978, Dadswell 1975 and other references in Hoff 1979). But efforts to develop a commercial fishery for shortnose sturgeon at present, even if they are sufficiently abundant, would be hampered by its long residence time in estuarine waters and resultant accumulation of high concentrations of heavy metals and other toxicants (Dadswell 1976).

The recent evaluation of shortnose sturgeon by the Shortnose Sturgeon Recovery Team has generated renewed interest in the biology of this species along the Atlantic Coast of North America, including the Hudson River. Research activity focused on the general life history of the shortnose sturgeon in the Hudson River was relatively intensive during the 1930s. Greeley (1937) reported that the shortnose sturgeon is a permanent resident of freshwater areas and noted that "there is no evidence that it was ever exceedingly numerous." Life-history and distribution information for shortnose sturgeon was presented by Greeley (1935 and 1937). Curran and Ries (1937) examined ninety-five shortnose sturgeon ranging in length from 17.7 to 34.8 inches (450–884 mm) during their 1936 investigations and commented that the commercial importance of the species had recently increased. They also stated that the diet of sturgeon in the lower Hudson included insects, crustaceans, molluscs, and annelids in order of decreasing importance. Hudson River shortnose sturgeon mature at 50–60 cm TL or 4–8 years of age (Greeley 1937).

Research on sturgeons in the Hudson estuary was limited from the 1940s through 1960s. Recently Dovel (1976–1981a) conducted studies specifically designed to collect information on Atlantic and shortnose sturgeon in the estuary. Much of what is presently known about shortnose sturgeon comes from Dovel's work, but the information is in the "gray" literature and not widely available. An important finding of his studies was that spawning occurs during the last two weeks of April and the first two weeks of May (Dovel 1979b) in the upper reaches of the estuary (between Coxsackie and Troy, km 200–245).

Additional recent information on shortnose sturgeon biology was gathered during the 1970s by several environmental consulting groups conducting research focused primarily on striped bass. The objective of this paper is to present an analysis and summary of these data.

MATERIALS AND METHODS

Study Area

The Hudson River estuary (Figure 34) is the tidal part of the Hudson River. The estuary stretches 246 km (154 miles) from the Battery to the Federal dam at Troy. The narrow, salt-intruded channel was shaped by glaciers which covered the region during the Pleistocene (Shuberth 1968). The estuary has extreme depths of about 53 m, but is mostly much shallower, with depths of 10–20 m being most common. The greatest depths and narrowest widths occur where the river penetrates the Highlands near West Point. There the cross-sectional area is small, hence, the currents are strong. In areas of less rugged terrain, the river is wider and shallower, with a greater cross-sectional area and weaker currents. The overall gradient is low (1.5 m in 246 km) and the channel bottom is approximately at sea level near Albany.

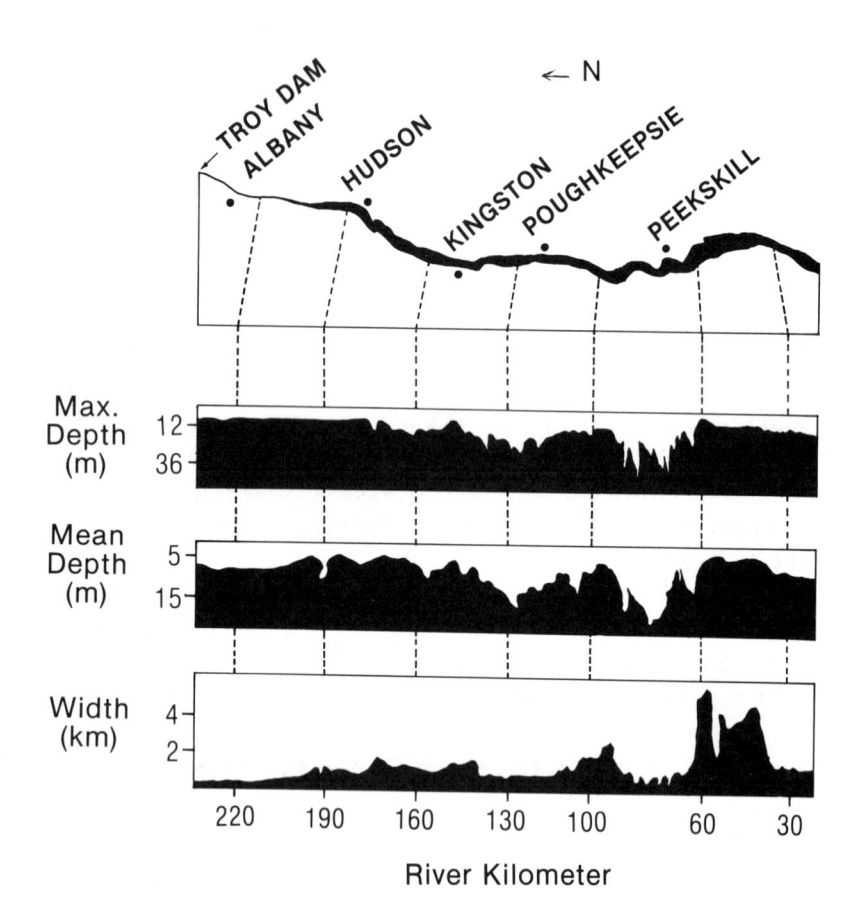

Figure 34. *Major morphometric characteristics of the Hudson River estuary.*

Water temperatures in the estuary generally reflect seasonal air temperatures and range from 0°C to approximately 28°C. Freshwater discharge into the estuary is low in the summer and midwinter, and is partially controlled by the Troy dam and other flood-control and water supply reservoirs in the 34,644 km^2 drainage basin (Klauda et al. 1980). Peak discharges occur in November–December and March–April. The influence of tidal flow in the estuary is much greater than that of freshwater flow (Darmer 1969).

Salinities greater than 1.0 ppt rarely occur above Cornwall (km 91). Salinity is usually greater than 1 or 2 ppt below the Tappan Zee Bridge (km 43). Based on a ratio between estuary volume and mean freshwater flow (Simpson et al. 1973), the Hudson estuary can be described as relatively unstable (ratio = 0.4) and susceptible to relatively rapid salinity changes (Cooper et al. 1979). Other East Coast estuaries have higher stability ratios: Delaware Bay (0.6), Potomac River (0.7), and Chesapeake Bay (1.2).

Sampling Programs

Fisheries research in the Hudson estuary has been continuous since 1965 in conjunction with power plant siting, licensing, and impact assessment studies. The major emphasis of these utility funded studies (collectively referred to as the Hudson River Ecological Study) was to determine the effects of power plant operations on selected species (e.g., striped bass). Although the research programs were not designed to emphasize sturgeons, many specimens of both species were collected. Sampling methods are summarized by Klauda et al. in the initial paper in this volume; a brief summary of the gear that frequently collected sturgeon is presented here.

Ichthyoplankton surveys, with a 1.0-m^2 epibenthic sled and 1.0-m^2 Tucker trawl on bottom in channel and shoal areas collected larval, young, and adult sturgeons throughout the estuary. All sturgeons younger than yearlings were recorded as "*Acipenser* spp." Older juveniles through adults were also collected in a bottom trawl survey between the George Washington and Newburgh-Beacon Bridges (km 19–98).

In 1976, an adult striped bass stock assessment program was initiated which sampled the estuary from the Tappan Zee Bridge (km 43) to the Newburgh-Beacon Bridge March through June (McLaren et al., this volume). Anchored gill nets (91 m long) of various mesh sizes (10.2, 12.7, 15.2, and 17.8 cm stretch mesh) were used. In addition, the catches of commercial fishermen who incidentally captured shortnose sturgeon were sampled. During the spring and fall of 1979 and 1980, a large mesh 15.2-m headrope otter trawl (10-cm wings with either 4.4 or 8.9 stretch in the cod end) was used to sample striped bass and incidentally collected numerous sturgeons.

Impingement studies at electrical generating plants on the Hudson estuary have been conducted since 1969. Detailed information on daily impingement sampling at the Indian Point power plant was reported by TI (1976b) and Raytheon (1971). Impingement data have been recorded since 1971 (or on-line date) at four other generating plants (Bowline Point, Lovett, Roseton, and Danskammer Point). The sampling schemes and methods used over the years at each power plant were described in nearfield reports for the Bowline Point and Roseton plants (CHGE 1977, ORU 1977).

RESULTS

Catch

Five hundred and eighty-eight yearling and older shortnose sturgeon were collected from 1969 through 1980 by the four major environmental consulting firms (TI, LMS, EA, and Raytheon) sponsored by the utilities to study the Hudson River estuary (Table 38). Nearly half (281) were caught since 1976 in gill nets set for adult striped bass by TI. The increase in shortnose sturgeon catches by bottom trawls beginning in 1979 was also associated with TI's adult striped bass stock assessment program and corresponded with the use of the large bottom trawl. This large trawl can easily sample the channel bottom areas in the deepest regions of the estuary that are generally inaccessible by gill nets and collect sturgeon with little or no visible evidence of capture or handling-related stress.

Carlson and McCann (1968), summarizing the work done by Northeast Biologists for the utilities from 1965 through 1968, do not report any catches of shortnose sturgeon. Raytheon (1971) collected five shortnose sturgeon from June 1969 to June 1970. No shortnose were collected in 1971 owing to limited sampling. Incidental catches of shortnose sturgeon were uniformly low from 1972 to 1975, but generally increased from 1976 to 1980. The adult striped bass stock assessment program was initiated in 1976, and catches of shortnose sturgeon increased concomitantly with the increasing sampling experience and collecting efficiency.

More than half (297) of the shortnose sturgeon captures between 1969 and 1980 occurred during June (Figure 35). The TI adult striped bass program continued from mid-March through June from 1976 through 1980 with nearly equal monthly effort. The preponderance of June catches may indicate a movement of shortnose sturgeon into the striped bass sampling area (Tappan Zee to Newburgh-Beacon Bridges) during May and June from upstream areas of the estuary. These data support Dovel's (1978a) hypothesis of a downstream migration soon after spawning. This hypothesis was also supported by recaptures of shortnose which had been tagged and released near

Table 38. *Number of yearling and older shortnose sturgeon collected in various gear used in the Hudson River ecological study from 1969 through 1980*

GEAR	1969	1970	1971**	1972	1973	1974	1975	1976	1977	1978	1979	1980	TOTAL
Bottom Trawls	2	3	0	5	8	5	9	9	7	16	109	65	238
Gill Nets	0	0	0	1	2	0	3	41	70	92	16	56	281
Ichthyoplankton Nets	0	0	0	0	0	7	0	3	6	0	0	1	17
Impingement	0	0	0	8	5	6	1	2	8	5	4	1	40
Other*	0	0	0	2	1	0	0	0	3	4	2	0	12
TOTAL	2	3	0	16	16	18	13	55	94	117	131	123	588

*Includes gear used by sampled commercial fishermen, and surface trawls.
**Only limited sampling in 1971.

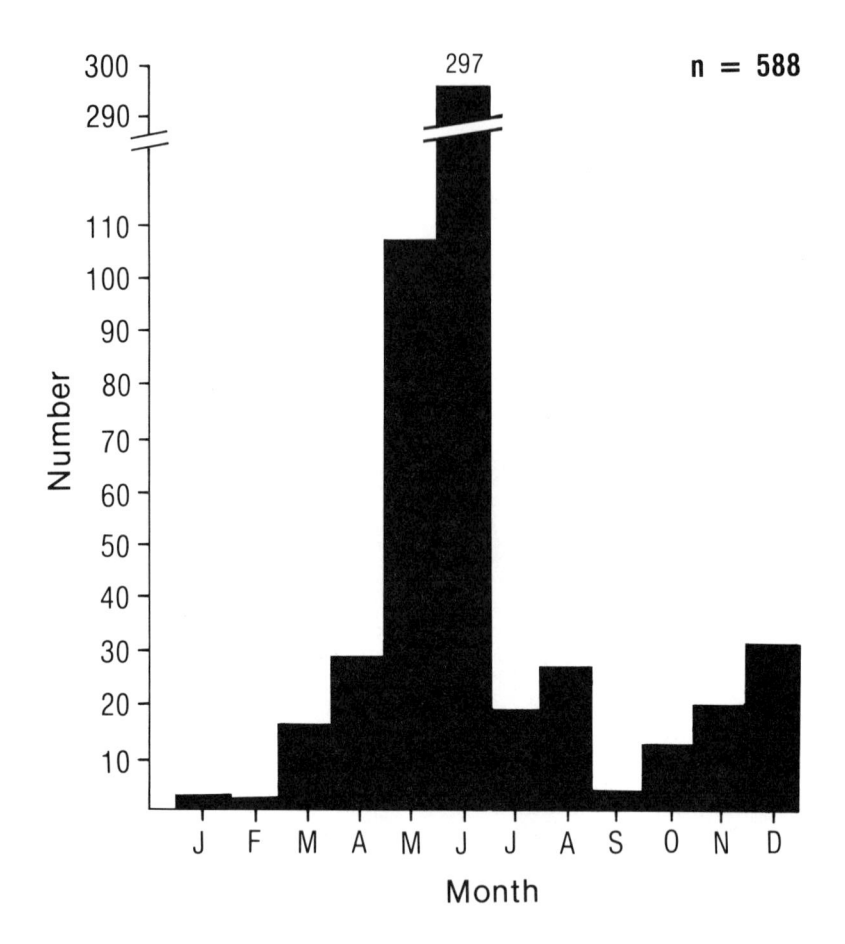

Figure 35. *Monthly totals of yearling and older shortnose sturgeon collected in the Hudson River Ecological Study, 1969–1980.*

the spawning area. On 8 June 1979, TI recaptured an adult at km 67 which had been tagged at km 134 on 7 May 1979 (Dovel 1979d).

The sampling regions where most shortnose sturgeon were collected (Figure 36) probably reflect the distribution of sampling effort for adult striped bass. Nearly equal numbers of shortnose were caught in Tappan Zee (km 38–53) and Croton-Haverstraw (km 54–61) sampling regions. None were collected above Saugerties (km 166).

Length and Weight

Since the shortnose sturgeon is an endangered species, efforts were made to minimize stress associated with handling, but length

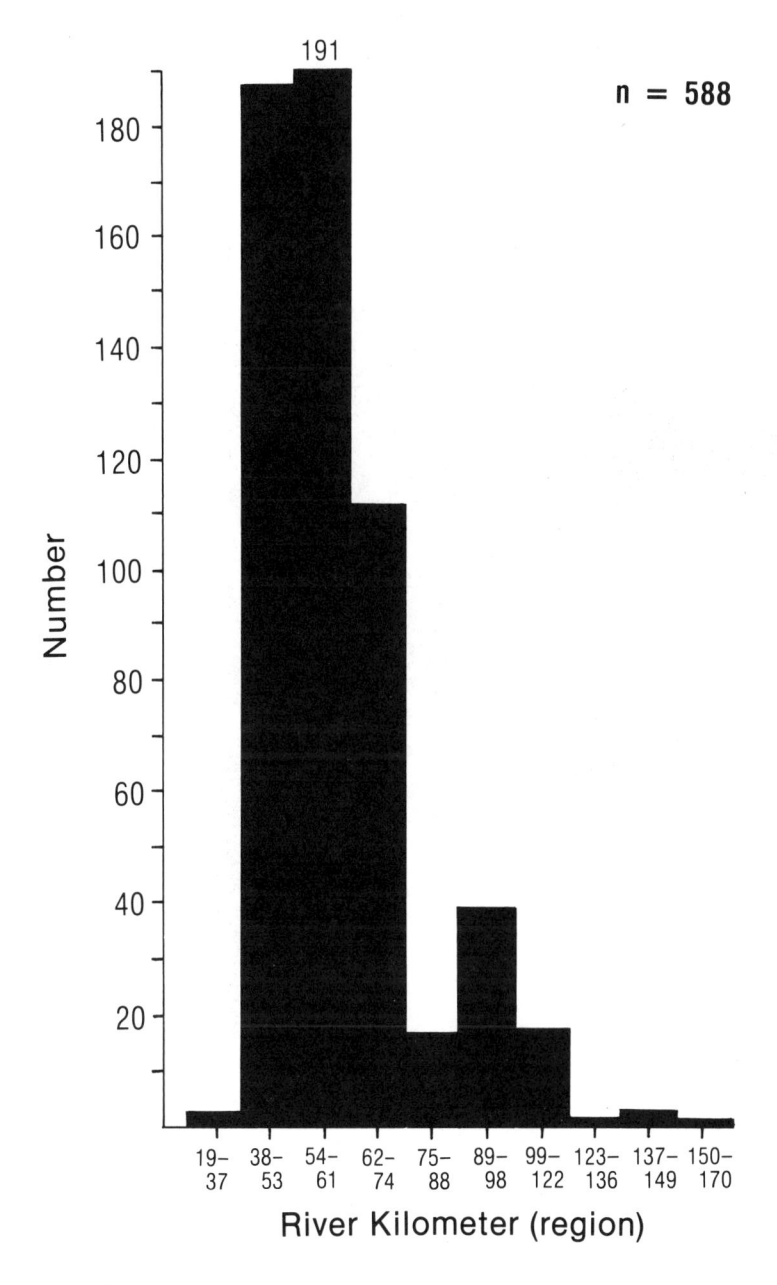

Figure 36. *Regional distribution of yearling and older shortnose sturgeon collected in the Hudson River Ecological Study from 1969 through 1980.*

179

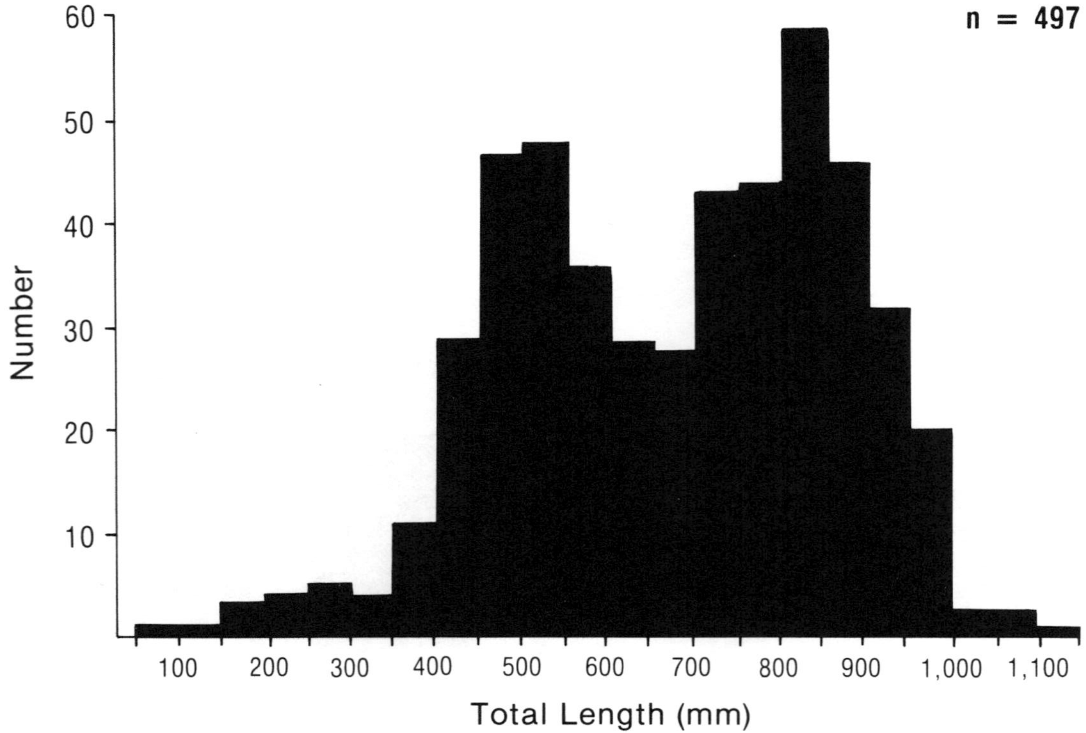

Figure 37. Length frequency distribution of shortnose sturgeon collected in the Hudson River Ecological Study from 1969 through 1980.

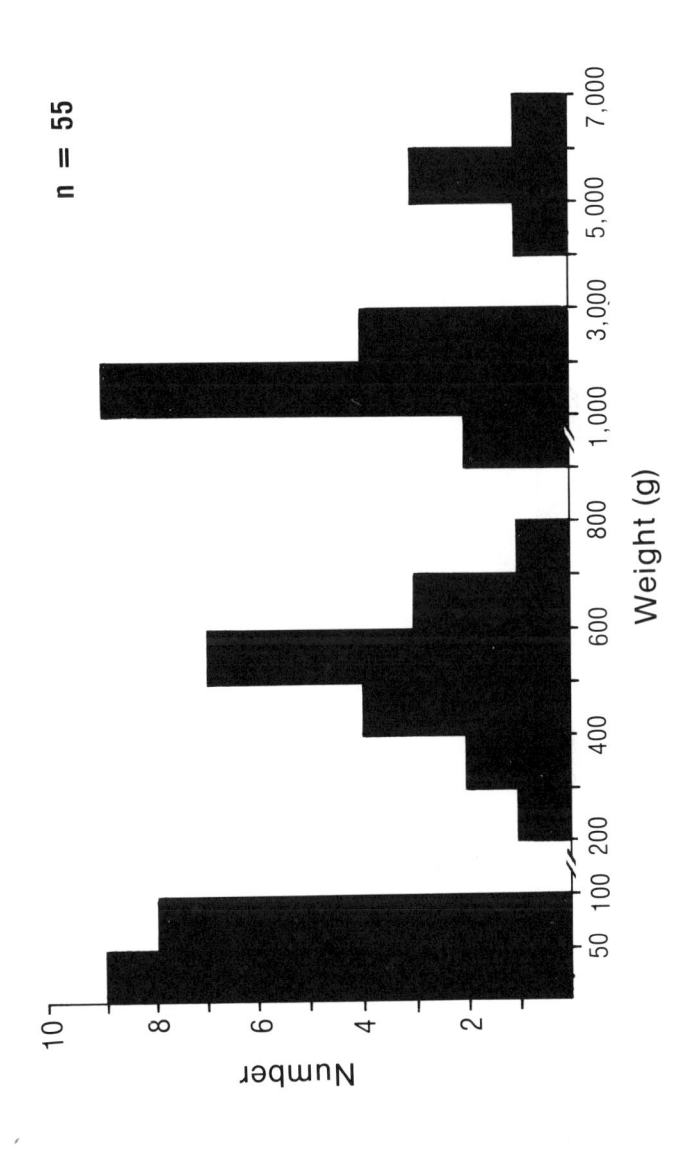

Figure 38. *Weight frequency distribution of shortnose sturgeon collected in the Hudson River Ecological Study from 1969 through 1980.*

Table 39. *Meristic and morphometric measurements* from seven shortnose sturgeon incidentally caught by Texas Instruments in gill nets during mid- to late June 1978 (fish died in the nets)*

INDIVIDUAL	1	2	3	4	5	6	7	MEAN	STANDARD DEVIATION	RANGE
Collection Date	6/19/78	6/20/78	6/21/78	6/23/78	6/28/78	6/29/78	6/30/78	—	—	—
Collection River Mile	30	27	30	29	30	30	27	—	—	—
Collection Sample Number	531436	531438	531373	531466	531501	531508	531517	—	—	—
Fork Length (mm)	810	589†	860	810	809	627	648	761	89	589–860
Total Length (fresh)(mm)	935	683†	980	925	905	700	726	862	117	683–935
Total Length (mm)	918	—†	952	910	910	681	715	848	107	715–952
Head Length (mm)	175	117	166	154	150	127	120	144	21	117–175
Snout Length (mm)	64	40	155	57	52	43	42	65	38	40–155

										Mean	SD	Range
Postorbital Length (mm)	99	66	104	92	94	78	68	86	14	66–104
Interorbital Length (mm)	73	41	66	56	61	49	49	56	10	41–66
Mouth Width (mm)	47	36	49	27	40	34	36	38	7	27–49
Total Weight (fresh)(g)	5600	2300†	6700	4350	5650	2550	2650	4583	1559	2300–6700
Total Weight (g)	5400	–†	6350	4160	5450	2300	2450	4352	1536	2300–6350
Gonad Weight (g)	139.8	82.1	481.4	65.9	134.1	57.5	94.9	151	138	57.5–481
Anal Ray Count	20	15	18	17	–†	16	17	17	2	15–20
Number of Gill Rakers on First Brarchial Arch	28	24	27	28	30	18	19	25	4	18–30
Sex	F	F	F	F	M	F	F	—	—	—

*All measurements and counts were from frozen fish except where denoted by (fresh).
†Not included in calculations of mean and standard deviation because some measurements for these individuals were not taken.

data were routinely recorded on most of the sturgeon collected in gill nets and trawls. Shortnose sturgeon collected from 1969 to 1980 ranged from 98 mm to 1,100 mm total length (TL). The length-frequency distribution (Figure 37) is indicative more of sampling gear selection than of the population structure and is certainly biased toward larger specimens. TI captured dozens of individuals over 900 mm TL with several fish larger than 1,000 mm. The largest shortnose caught by TI was caught in a 17.8-cm gill net at km 58 on 19 May 1980. This 1,100-mm individual was released in good condition, and sex and weight were not recorded. The previous size record in the scientific literature was a 36.7-inch (932-mm) specimen (Koski et al. 1971).

Greeley (1937) reported that shortnose in the Hudson estuary mature at between 500 and 600 mm TL. Dadswell (1979) reported that 50% maturity is reached at 550 mm (Fork Length, FL) for males ι and 700 mm FL for females in the Saint John River, New Brunswick. If 550 mm TL is used as a lower size limit for mature fish, nearly 70% (344) of the shortnose collected in the Hudson between 1969 and 1980 were possibly mature.

Weight data on shortnose sturgeon are limited. Weights for fifty-five individuals, however, ranged from 3 g to 6,700 g (Figure 38). The length-weight relationship is Log $W = -13.7 + 3.24$ Log L ($R^2 = 0.976$).

Meristic and Morphometric Characteristics

In mid- to late June 1978, seven large shortnose sturgeon incidentally caught by TI in gill nets died before they could be released. High water temperature may have been the cause of death. To salvage as much data as possible from these adult specimens, they were frozen and later thawed, dissected, and measured (Table 39). The meristic and mophometric features of these seven individuals were very similar to data presented by Dadswell (1981) for shortnose sturgeons from the Saint John, Kennebec-Sheepscot, and earlier Hudson River studies (Table 40). Six of the seven were females. These specimens have been deposited in the American Museum of Natural History (Hoff and Klauda 1979b).

Larvae and Young-of-the-Year

Few young-of-the-year (larvae and juvenile) shortnose or Atlantic sturgeons have been collected in the Hudson River estuary. From 1973 to 1979, forty-nine larvae (twenty-four yolk-sac, twenty-five post

Table 40. *Comparative morphometric and meristic data for adult Acipenser brevirostrum. TL = total length, MW = mouth width (inside lips), SL = snout length, IOW = interorbital width, POL = postorbital length, HL = head length (table modified from Dadswell 1981)*

CHARACTER	SAINT JOHN (Gorham and McAllister 1974)	MEAN FOR RIVER SYSTEM KENNEBEC-SHEEPSCOT (Squiers and Smith 1978) (Fried and McCleave 1973)	HUDSON (Valdykov and Greeley 1963)	HUDSON (This Study)
MW/SL	60.1	70.2	—	58
MW/IOW	76.8	68.6	74	68
SL/HL	44.0	38.6	35	45
SL/POL	—	73.7	70	76
POL/HL	—	56.7	55	60
IOW/HL	—	35.9	37	39
HL/FL	—	19.9	22	19
TL/FL	12	11	11	11
Gill rakers	27.6	26.2	25.5	25
Anal rays	20.8	—	—	17
Dorsal scutes	10.2	9.8	10	—
Ventral scutes	8.5	8.4	8	—
Lateral scutes	—	26.9	28	—

Table 41. *Number of unidentified sturgeon* (Acipenser spp.) *larvae and juveniles collected in the Hudson River ecological study, 1973 through 1979*

YEAR	YOLK-SAC LARVAE	POST YOLK-SAC LARVAE	JUVENILES
1973	6	1	17
1974	9	9	1
1975	7	11	5
1976	0	3	1
1977	0	1	0
1978	2	0	1
1979	0	0	1
Total	24	25	26

yolk-sac) and twenty-six juveniles were collected (Table 41). All were taken near the river bottom at depths of 3 m to 23 m in freshwater areas between km 70 and 190 in May through mid-July (Hoff et al. 1977b).

Shortnose and Atlantic sturgeons are difficult to distinguish during early life stages; therefore, it is not known how many of these larvae and juveniles were shortnose sturgeon. It is noted, however, that adult Atlantic sturgeon outnumber adult shortnose sturgeon by 8.2 to 1 (Table 42).

The paucity of collections of sturgeon eggs, larvae, and young-of-the-year reflect not only their relatively low abundance but also the difficulty in collecting sturgeon early life-history stages and a distribution of sampling effort designed to estimate abundance of striped bass. The eggs are demersal and adhesive and thus are not easily sampled. Larval stages also appear to be demersal, based on sampling depths where the few specimens were caught (Hoff et al. 1977b). Dadswell (1975) also reported that young shortnose sturgeon "are difficult to catch because they live in deep channels, where strong currents make sampling nearly impossible."

DISCUSSION

These incidental catches contribute to the growing body of data which suggests that a viable shortnose sturgeon population exists in the Hudson River estuary. The number of yearling and older individuals collected each year steadily increased since 1975, but this increase is at least partly due to increased use of a 15.2-m bottom trawl and

Table 42. *Annual ratio of yearling and older Atlantic sturgeon to shortnose sturgeon collected during the Hudson River ecological study, 1972 through 1979*

YEAR	NUMBER OF ATLANTIC STURGEON	NUMBER OF SHORTNOSE STURGEON	RATIO
1972	70	16	4.4:1
1973	211	16	13.2:1
1974	334	18	18.6:1
1975	353	13	27.2:1
1976	339	55	6.2:1
1977	567	94	6.0:1
1978	292	117	2.5:1
1979	1,590	131	12.1:1
Total	3,756	460	8.2:1

multiple mesh anchor gill nets (TI 1981b). The large trawl, added to TI's sampling program in 1979, was particularly effective and collected 70% of the shortnose sturgeon caught that year. Only one sturgeon died in the gear; the rest were in good condition and were released alive. These incidental catches also demonstrate that the shortnose sturgeon population contains a number of large individuals, another positive sign.

A five-year study of Hudson River sturgeons began in 1975. The results of this study are scattered in a series of reports by Dovel, and the preliminary findings suggest that the Hudson estuary has one of the largest populations of shortnose sturgeon in the United States (Dadswell 1981, Shortnose Sturgeon Recovery Team 1981). The population of subadults and adults numbers at least ten thousand to thirteen thousand individuals, and conceivably as many as thirty thousand, if only part of the population returns to the spawning grounds each year (Dovel 1980a and 1981a). Dadswell (1979) compared the findings of Dovel (1979a, 1979f) and Greeley (1937), and concluded that the shortnose sturgeon population in the Hudson estuary appears to have been stable during that forty-year period.

The shortnose sturgeon is not a numerically dominant fish in the Hudson estuary, but because of its endangered status (justified or not), its sympatry with Atlantic sturgeon, and its use of the entire estuary during its life cycle, it should be included in any management plan for the estuary. In their 1981 draft report, the Shortnose Sturgeon Recovery Team urged that research on shortnose sturgeon in the Hudson be continued; but because a relatively extensive data base exists for the Hudson population, they recommended that other Atlantic Coast populations receive more attention in the near future. Therefore, the rapid accumulation of data on Hudson River sturgeons which occurred in the 1970s is not likely to be matched in the 1980s.

Monitoring the expanding commercial fishery for Atlantic sturgeon in the Hudson may prove to be one source of useful data on shortnose sturgeon, at least in the lower estuary. Reported landings of "common" sturgeon have steadily increased in recent years from only 70 kg in 1974 to almost 6,500 kg in 1980 (data from National Marine Fisheries Service survey records). More than 90% of the sturgeon were taken in stake and anchor gill nets in Rockland and Westchester Counties. The rest were taken upriver near the spawning grounds (Ulster County) in drift gill nets. Any plan to expand the commercial fishery for Atlantic sturgeon should be cognizant of shortnose sturgeon. Dadswell (1981) suggested that the two species may compete for space in some areas where their distributions overlap. TI collected yearlings and older age groups of both species in gill nets and bottom trawls in Croton and Haverstraw Bays in areas where most Atlantic sturgeon are taken in the commercial fishery (Hoff et al. 1977, TI 1980e, 1981b).

Plans for navigational dredging or industrial development in the upper Hudson estuary should be scrutinized carefully, because shortnose sturgeon apparently spawn their demersal eggs during April and May between Coxsackie (km 200) and Troy (km 245) (Dovel 1979b). The larvae are also bottom oriented (Pekovitch 1979) and apparently disperse gradually downstream during the summer. Dovel (1979b, 1980c, 1981) suggested that the entire estuary from Haverstraw Bay to Coeymans (km 213) represents the nursery area for larval and juvenile shortnose sturgeon.

The 1970s were a productive decade for fisheries research in the Hudson River estuary, especially for striped bass, white perch, and Atlantic tomcod. But information on Atlantic and shortnose sturgeon also increased significantly. We agree with the conclusion of the Shortnose Sturgeon Recovery Team that the data base on shortnose sturgeon is extensive compared to other United States coastal populations. But many basic aspects of shortnose sturgeon biology in the Hudson are still unknown, especially the role of environmental factors, including pollutants, on survival. Like several other fish species in the estuary, shortnose sturgeon are contaminated with PCBs (the highest concentration in the brain tissue) and also appear to suffer a high incidence of fin rot (Dovel 1980b, 1980c, 1981). Regular monitoring of shortnose sturgeon could yield valuable data on the population's status and also information of trends in levels of environmental contaminents.

ACKNOWLEDGEMENTS

Financial support for this research was provided by ConEd, CHGE, ORU, and the New York Power Authority. Dr. James G. Hoff, Dr. Kenneth L. Marcellus, and Mr. William Dovel reviewed the manuscript and offered many helpful comments. We thank the TI field crews for their work in data collection, Michele Predmore for secretarial assitance, and Ingrid Farnam for her artistic expertise. Finally, we thank the members of the Shortnose Sturgeon Recovery Team for sharing many ideas, as well as draft documents, with us and for their concern for this unique species.

Part IV
River Herrings

7. Distributions and Movements of the Early Life Stages of Three Species of Alosa in the Hudson River, with Comments on Mechanisms to Reduce Interspecific Competition

Robert E. Schmidt, Ronald J. Klauda, and John M. Bartels

INTRODUCTION

American shad, alewife, and blueback herring are anadromous species that support major sport and commercial fisheries along the northeast Atlantic coast. The general life histories of the species have been summarized by Bigelow and Schroeder (1953), Scott and Crossman (1973), and others. Descriptions of the biology of these fishes also exist for specific geographic areas; for example, Chittenden (1969), Hildebrand and Schroeder (1928), Marcy (1969, 1976b), Kissil (1974), Loesch and Lund (1977), and Leggett (1976).

Spawning of *Alosa* spp. in Atlantic coastal rivers is spatially and temporally segregated. American shad spawn in the mainstream of rivers in relatively shallow water with moderate current (Marcy 1972, Leggett 1976). Alewives prefer sluggish, very shallow water (30 cm deep) in large rivers or small tributaries (Jones et al. 1978). Blueback herrings also spawn in large rivers and tributaries but prefer swift currents and hard substrata, and spawn later in the summer (Loesch and Lund, 1977).

The Hudson River estuary contains large populations of these species but there is little published information on their biology. Because of its importance in the commercial fishery of the Hudson River, the American shad has received considerable attention (Talbot 1954, Burdick 1954, Medeiros 1974, Klauda et al. 1976) but these studies were focused almost exclusively on the adults. These studies and limited information on the early life-history stages (Stira and Smith

1976) indicate that the general biology of the American shad in the Hudson is similar to that of other populations.

Studies of other northeast populations of alewives and blueback herring have generally dealt with populations in drainage systems much smaller than the 35,000 km^2 Hudson River (TI 1976c). Because small drainage systems have different environmental regimes, the existing literature on these two species is not necessarily applicable to the Hudson River system.

From 1973 through 1980, TI extensively sampled the tidal part of the Hudson River. An important part of these studies was the distribution, in space and time, of early life stages of selected fish species, including the species of *Alosa*. The objectives of this paper are to summarize a portion (1976–1979) of these extensive data; describe the distributions and movements of eggs, larvae, and juveniles; and examine the evidence for resource partitioning mechanisms which may reduce interspecific competition and promote niche segregation among the three alosine species.

MATERIALS AND METHODS

The analyses in this paper are based primarily on surveys designed to collect early life stages of fishes, conducted by TI in 1976–1979. Routine physicochemical data (e.g., water temperature) were collected concurrently with all biological samples.

Study Area

The Hudson River extends about 500 km from its origin in the Adirondack Mountains to the Atlantic Ocean at New York City. The river is tidal and unimpounded 243 km below the Federal Lock and Dam at Troy, New York.

Salinity intrusions commonly reach km 100 during the summer months. Descriptions of the physical and chemical characteristics of the estuary can be found in Abood (1974, 1977) and Darmer (1969).

The study area, from the George Washington Bridge (km 19) to the Troy Dam (km 243) was divided into twelve sampling regions which, for the purposes of this paper, were grouped into three sampling zones (Figure 39). Although the boundaries of these zones are somewhat arbitrary, they roughly divide the estuary into three distinct physiographic areas. The lower zone extends from the George Washington Bridge to about km 61 and is relatively wide and shallow. The middle zone (km 61–122) is narrow and contains the deepest parts

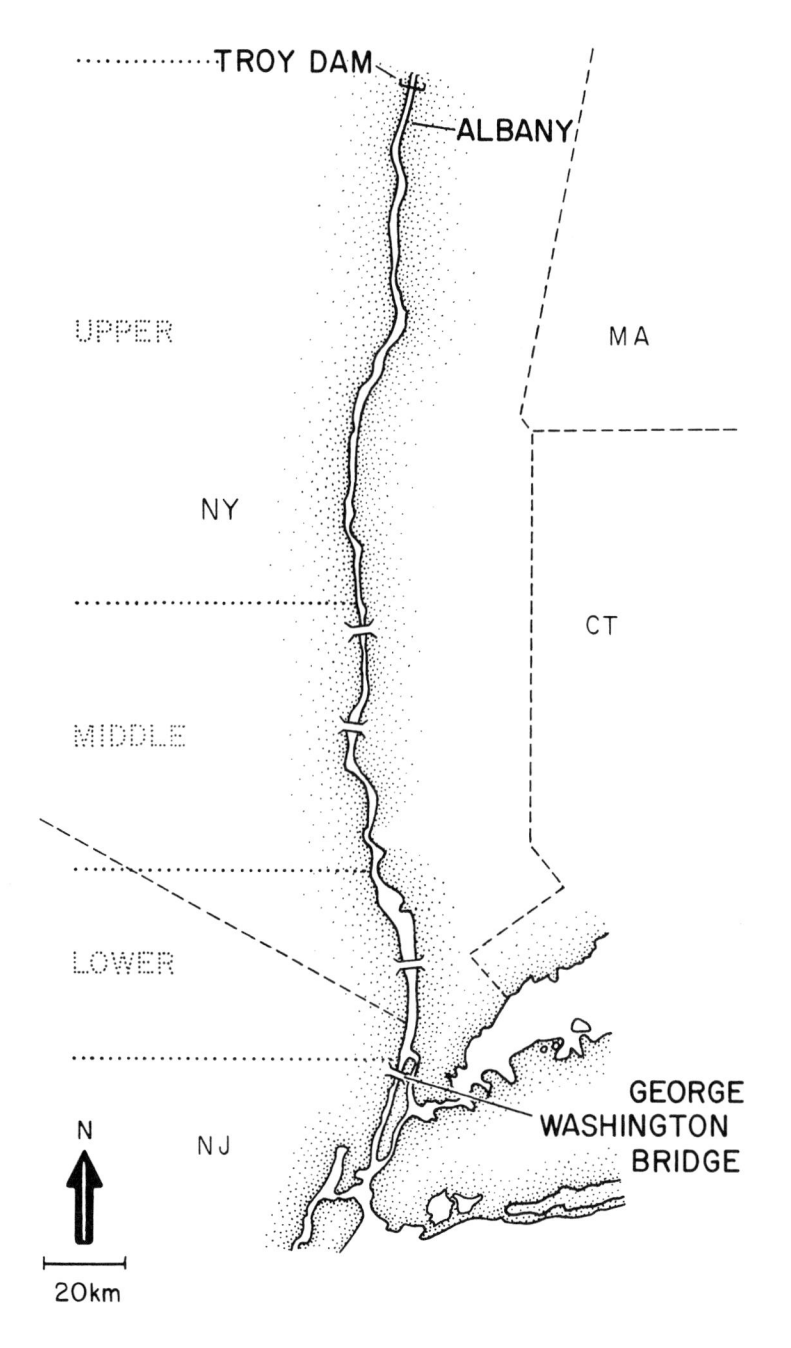

Figure 39. *Study area in the Hudson River estuary, New York, showing the three sampling zones and the political boundaries of the contiguous states.*

of the estuary. The upper zone (km 123–243) is moderately wide but shallower than the middle zone. The salt front extends into the middle zone during low flows but never reaches the upper zone.

Eggs and Larvae

American shad eggs and larvae are readily distinguishable from the other species of *Alosa* (Mansueti and Hardy 1967). Distinguishing alewife from blueback herring larvae is possible for some sizes (Chambers et al. 1976) but is time-consuming and costly; hence, we designated all alewives and blueback herring eggs, larvae, and juveniles less than 4 cm total length as 'river herring.' The survey methods, gear, and sample allocations were consistent within and across years from 1976 through 1979. The surveys began in mid-April and two hundred samples were collected weekly during the day until June, and at night thereafter through early August. A stratified random design keyed to the observed distributions of striped bass (the focal species) eggs and larvae in prior years was used to allocate samples to the twelve sampling regions.

Near-bottom samples in waters 2 m deep were collected with a 1.0-m^2 epibenthic sled (505-μm mesh); midwater and near surface samples were collected with a 1.0-m^2 Tucker trawl (505-μm mesh). Tow speeds for the sled and trawl were 1.0 m and 0.9 m per sec, respectively, and tow duration for both gears was 5 minutes. The mouth openings, net areas, net porosities, tow speeds, and tow durations yielded sustained filtration efficiencies greater than 85% (Tranter and Smith 1968, TI 1977a). Calibrated digital flow meters were suspended within the nets to record volume sampled. Calibrated electronic flowmeters were mounted on the towing cable just above the gear to record towing speed. Double-trip release mechanisms permitted opening and closing the nets at specific depths.

Descriptions of the distributions were based on weekly estimates of the densities of three developmental stages: eggs, yolk-sac larvae, and post yolk-sac larvae (postlarvae). Larvae were considered to be in the yolk-sac stage until the digestive system was fully developed, regardless of the degree of yolk retention. The post yolk-sac stage included those with a complete digestive system but without a full complement of fin rays. Once the fin rays were fully developed the young were defined as juveniles.

The mean weekly densities of each life stage in each of the twelve sampling regions were plotted as isopleths. Data for 1976–1979 were averaged, since the variation among years was negligible. The values plotted as isopleths were arbitrarily selected to represent the range of densities (in orders of magnitude) and to emphasize the spatial-temporal distributions of the highest densities.

Juveniles

The distribution of juveniles was monitored with two surveys conducted throughout the study area. One survey used a 30-m seine (0.48 cm in the bunt) and sampled one hundred beaches biweekly during the day from early June through mid-December 1976–1979. A stratified random design keyed to juvenile striped bass distributions in previous years was used to allocate samples in 1976–1978. In 1979, the samples were randomly allocated to each of the twelve regions on the basis of shore zone (0–3 m deep) area. The seine was set perpendicular to the shoreline with one end on shore. The offshore end was hauled to shore in a semicircle against the prevailing current. This procedure was used during all years and assured a relatively constant area being swept during each haul (TI 1981b). Distributions were calculated on the basis of mean biweekly catch per haul in each region.

The second survey used a 3.6-m (headrope) otter-type bottom trawl (0.6-cm mesh cod-end liner). This survey (known as the "try trawl" survey) sampled offshore areas 1.5–6.0 m deep. Approximately one hundred randomly selected sites were sampled biweekly during the day from April through mid-December in 1979 only. Samples were allocated to each of the twelve regions in proportion to the shoal volume (area less than 6 m deep). Tows were made at 1.5 m per sec against the prevaling current for approximately 10 minutes. Trawl data are presented only for juvenile alewife, because catches of juvenile American shad and blueback herring were generally very low and erratic.

A subsample of juveniles collected during each seine and bottom trawl survey was preserved in 10% formalin and returned to the laboratory for length measurements. Total length measurements were taken after the fish had been in formalin for at least two days, since laboratory studies showed that no significant shrinkage occurred after this period.

RESULTS

Eggs

In general, the distribution of eggs reflected species specific differences in spawning times, although some river herring eggs were

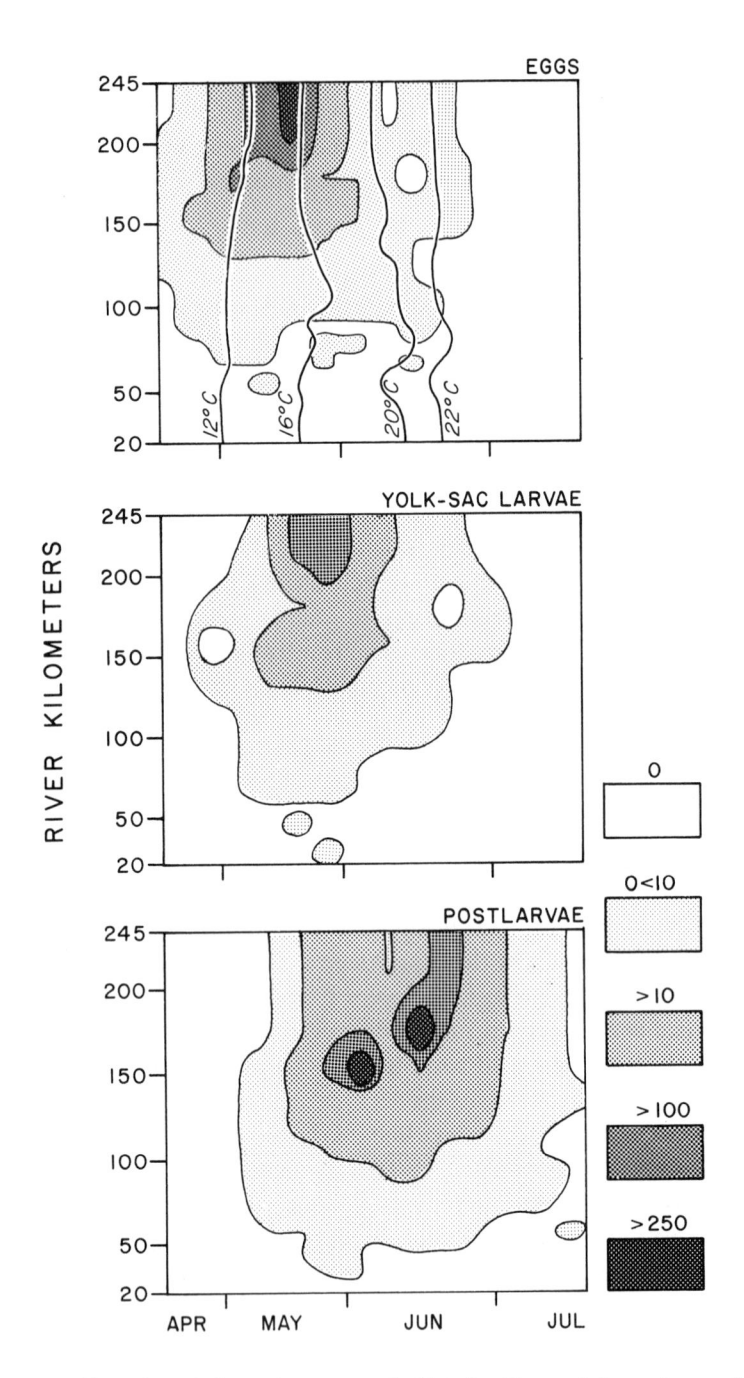

Figure 40. *Spatial and temporal distribution of American shad eggs and larvae in the mainstream Hudson River estuary, 1976–1979. Boundaries are isopleths of mean densities (1000 m^{-3}) for each week and sampling region, all years combined.*

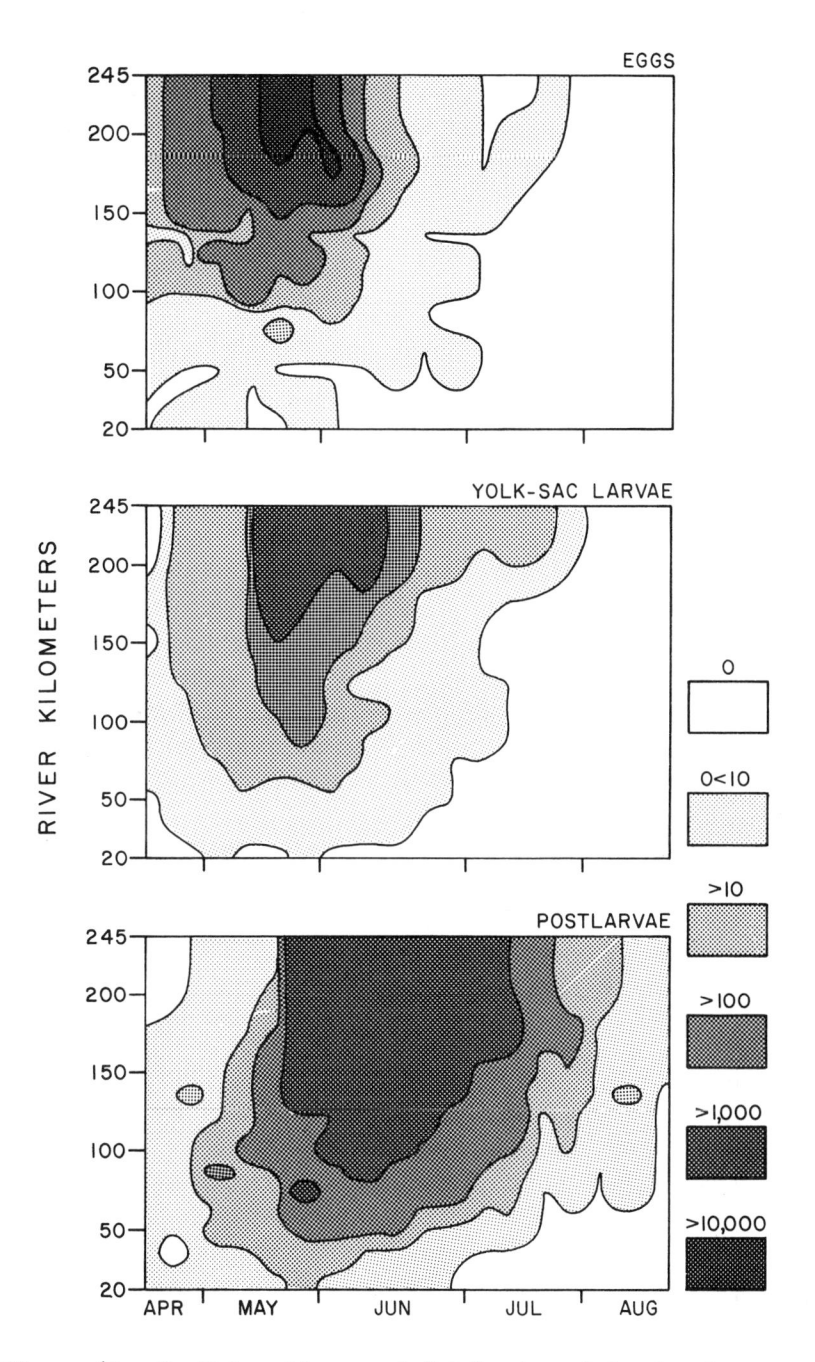

Figure 41. *Spatial and temporal distribution of river herring eggs and larvae in the mainstream Hudson River estuary, 1976–1979. Boundaries are isopleths of mean densities (1000 m^{-3}) for each week and sampling region, all years combined.*

taken with American shad eggs. Peak densities of shad eggs (up to 1,047 eggs per 1000 m^3) occurred in mid-May, while peak densities of river herring eggs (to 63,376 eggs per 1000 m^3) occurred in mid- to late May. River herring eggs persisted later in the summer than American shad eggs (Figures 40 and 41).

The majority of all *Alosa* eggs were taken in the upper zone of the estuary and few were collected below the salt front. The location of the salt front (defined by the 0.3 mS/cm isopleth) varied between km 29 and km 103 during the study, depending on season and freshwater flow (TI 1981b). Dovel (1981b) collected few *Alosa* eggs below Stony Point, the approximate upstream boundary of our lower zone. The most intense spawning activity was within 70 km of the Troy Dam and Lock at km 243. The dam is the upper limit for American shad migration, but both alewife and blueback herring negotiate the locks and spawn in the lower Mohawk River (Gann and Brandt, personal communication 1981).

Marcy (1972) stated that the development of shad eggs was prolonged and mortality concomitantly increased when water temperatures were below 16°C. Since most Hudson River shad eggs were produced at temperatures below this, it appears that shad in the Hudson River spawn in conditions that are suboptimal for survival of their eggs and larvae.

We did not attempt to characterize the environmental conditions during river herring spawning, since we did not distinguish between the eggs and larvae of the two species. The majority of river herring eggs were presumably alewives prior to mid-May and blueback herring after mid-May, based on the occurrence of adult blueback herring in the estuary (TI 1981b).

Larvae

The distribution of American shad and river herring larvae showed progressive downstream dispersal from areas of highest egg densities. The yolk-sac larvae and postlarvae of all three species are planktonic (Jones et al. 1978) and therefore vulnerable to dispersal by passive transport mechanisms. The net flow in the upper and middle zones is seaward (Abood 1977); thus, the distributions plotted in Figures 40 and 41 probably resulted in passive downstream drift of the early life stages.

Juveniles

AMERICAN SHAD Juvenile American shad first appeared in seine collections in early June and reached peak abundance in the upper

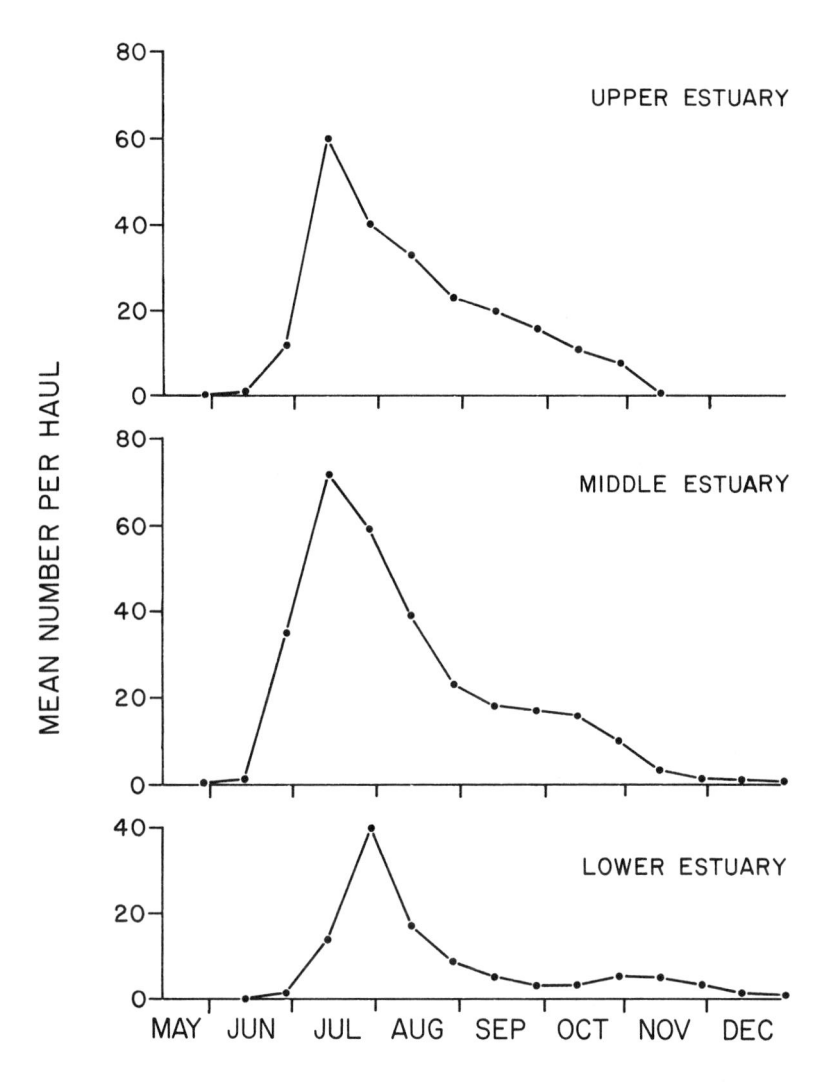

Figure 42. *Mean catch per haul of juvenile American shad in seines from the three zones of the mainstream Hudson River estuary, 1976–1979.*

and middle zones of the estuary by early July. In the lower zone, peak abundance occurred about two weeks later (Figure 42). This pattern probably resulted from downstream transport of juveniles rather than from recruitment of larvae into the shore zone of the lower zone, because few larvae were present earlier in the lower zone. Marcy (1976b) noted that downstream movement of juvenile American shad in the Connecticut River did not begin until early August, when water temperatures were highest or beginning to decline. In the Hudson, maximum water temperatures usually occur about mid-August (TI 1981b). Concomitant with their downstream movement, modal lengths of juvenile American shad in the Hudson during July were longer in the middle and lower zones than in the upper zone (Figure 43).

From August until late October, juvenile American shad in the upper zone were about the same size or slightly larger than juveniles collected further downstream. This change in length frequency distribution could result from the larger juveniles moving downstream through the lower zone and out of the study area at the same time there was a general downstream movement of all sizes of the remaining juvenile American shad, beginning in August when the water temperatures start to decline.

The majority of juvenile American shad left our study area by late October, when water temperatures approached 15°C (Figure 42). Leggett and Whitney (1972) stated that temperatures near 15°C stimulated juvenile American shad to leave Atlantic coastal rivers. Most juveniles left the Hudson by mid-November, before water temperatures became detrimental at 4°–6°C (Chittenden 1972). Milstein (1981) reported that American shad juveniles overwinter off the mouth of the Mullica River, New Jersey. This group probably includes juveniles from the Hudson stock that overwinter in nearby coastal waters.

ALEWIFE Juvenile alewives appeared in seine collections from late June to early July, about a month later than juvenile American shad (Figure 44). In the upper and middle zones of the estuary, catches of juveniles declined soon after the beginning of July, while offshore bottom trawl catches increased. Apparently, the juvenile alewives remained in the shallows, or near mouths of the tributary streams where they were spawned, for several weeks, then moved offshore as they grew.

Peak seine catches of juvenile alewives in the middle and lower zones occurred two and four weeks later, respectively. Although some of the juveniles collected in the lower and middle zones were prob-

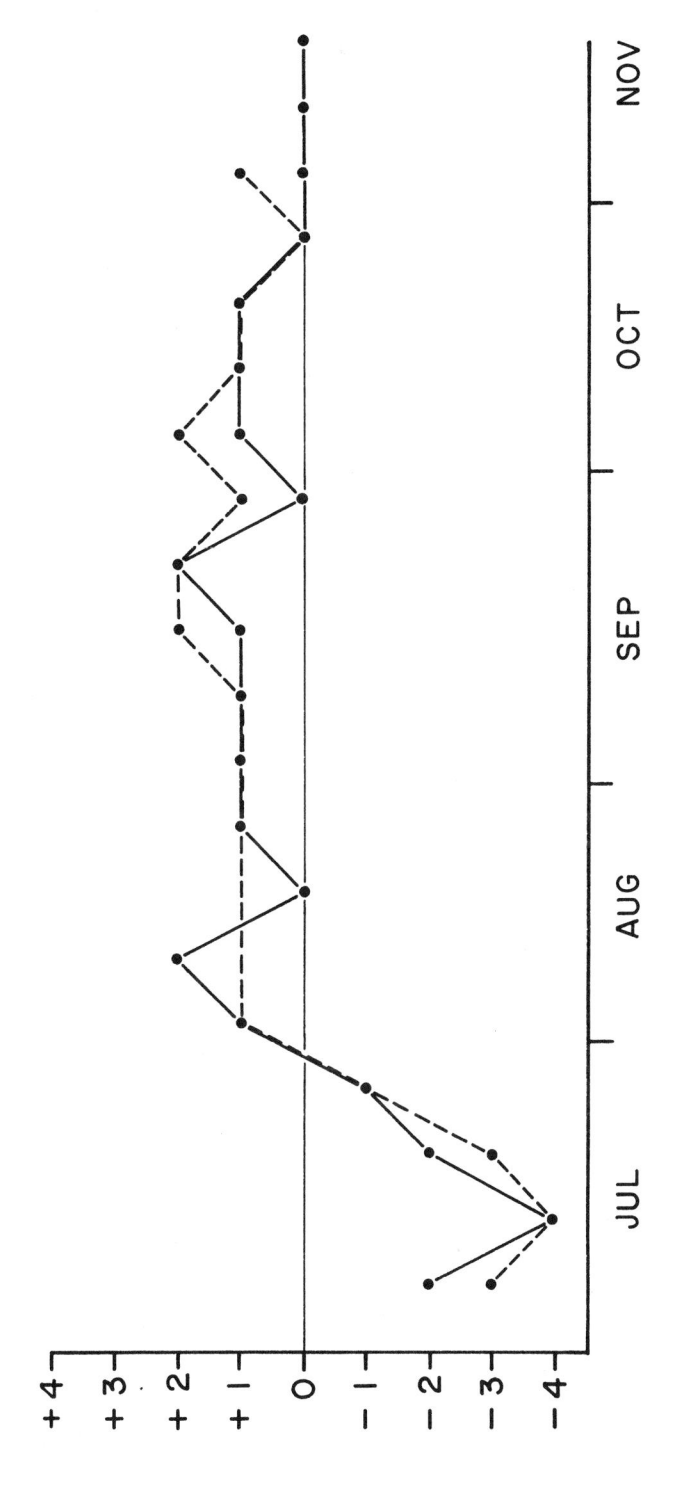

Figure 43. Differences in modes of 5 mm length classes between the juvenile American shad collected by all surveys in the upper zone of the estuary and those in the middle zone (——) and between the upper zone shad and those in the lower zone (---), Hudson River estuary, 1979. A positive sign (+) indicates that American shad in the upper zone have a larger modal length and a negative sign (−) indicates the converse.

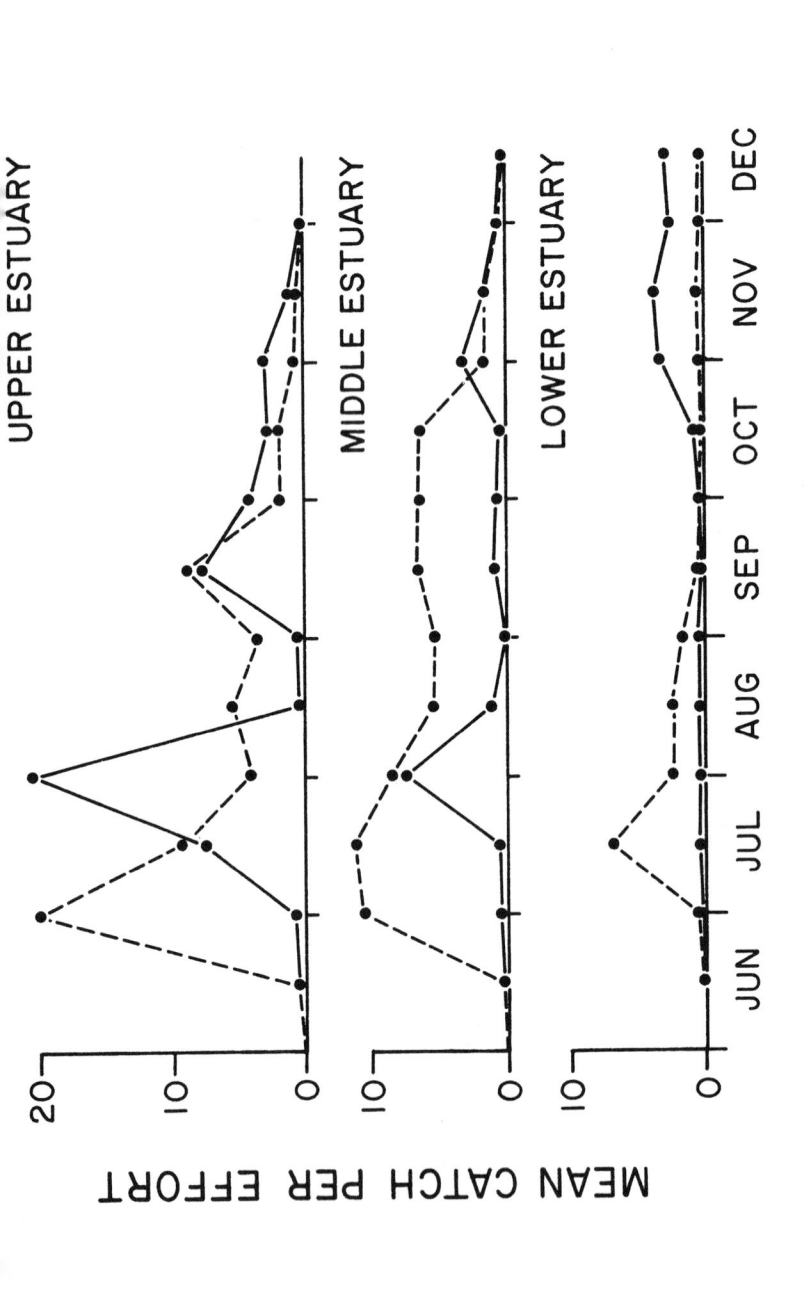

Figure 44. Mean catch per haul of juvenile alewives in seines, 1976–1979 (----) and mean catch per tow in bottom trawls, 1979 (———), Hudson River estuary.

ably spawned there in the mainstream or in the tributaries, the fact that the catches declined in the upper zones as they were increasing downstream suggests that the juveniles were moving downstream. Richkus (1975) reported that downstream movements of juvenile alewives in the Annaquatucket River, Rhode Island, were correlated with sudden temperature drops, increases in precipitation, and increases in flow. None of these factors was obviously related to the midsummer downstream movements of juvenile alewives in the Hudson in 1976–1979.

Not only did juvenile American shad and alewives begin downstream movements relatively early in the summer, but their patterns of length frequencies among the three zones of the study area were similar. In late July and early August, the modal lengths of juvenile alewives collected in the middle and lower zones were longer than the modal lengths in the upper zone (Figure 45). A bimodal length frequency was evident in the juvenile alewife data from mid-July through the end of August (Figure 46). This bimodality corresponded to a bimodal coefficient of maturity (and presumably an extended disjunctive spawning period) reported for Hudson River alewives by LMS (1975f).

After August, the juvenile alewife length frequency distribution was unimodal, and size differences among those of the upper zone versus those in the middle and lower zones became less pronounced (Figure 45). Juvenile abundance also gradually declined in the study area after August (Figure 44). Like juvenile American shad, larger juvenile alewives may be leaving the study area when most of the remaining juveniles are just beginning to move downstream from the upper zone of the estuary. Most juvenile alewives left the upper and middle zones before the water temperatures reached 7°C in late November (TI 1976a). Some juveniles remained in the lower zone until sampling ended in mid-December (Figure 44), and some may remain in the estuary all winter, as Pacheco (1973) observed in Chesapeake Bay.

BLUEBACK HERRING The distribution and movements of blueback herring differ from those of the alewife and American shad. Recruitment to the juvenile stage begins later, in early July, because they spawn later than the other two species, and peak catches in the shore area occurred in early to mid-August (Figure 47). During this time, seine catches of juvenile bluebacks were an order of magnitude larger than catches of juveniles of the other two species (Figures 42, 44, and 47). Catchability of blueback herring in surface and shore-

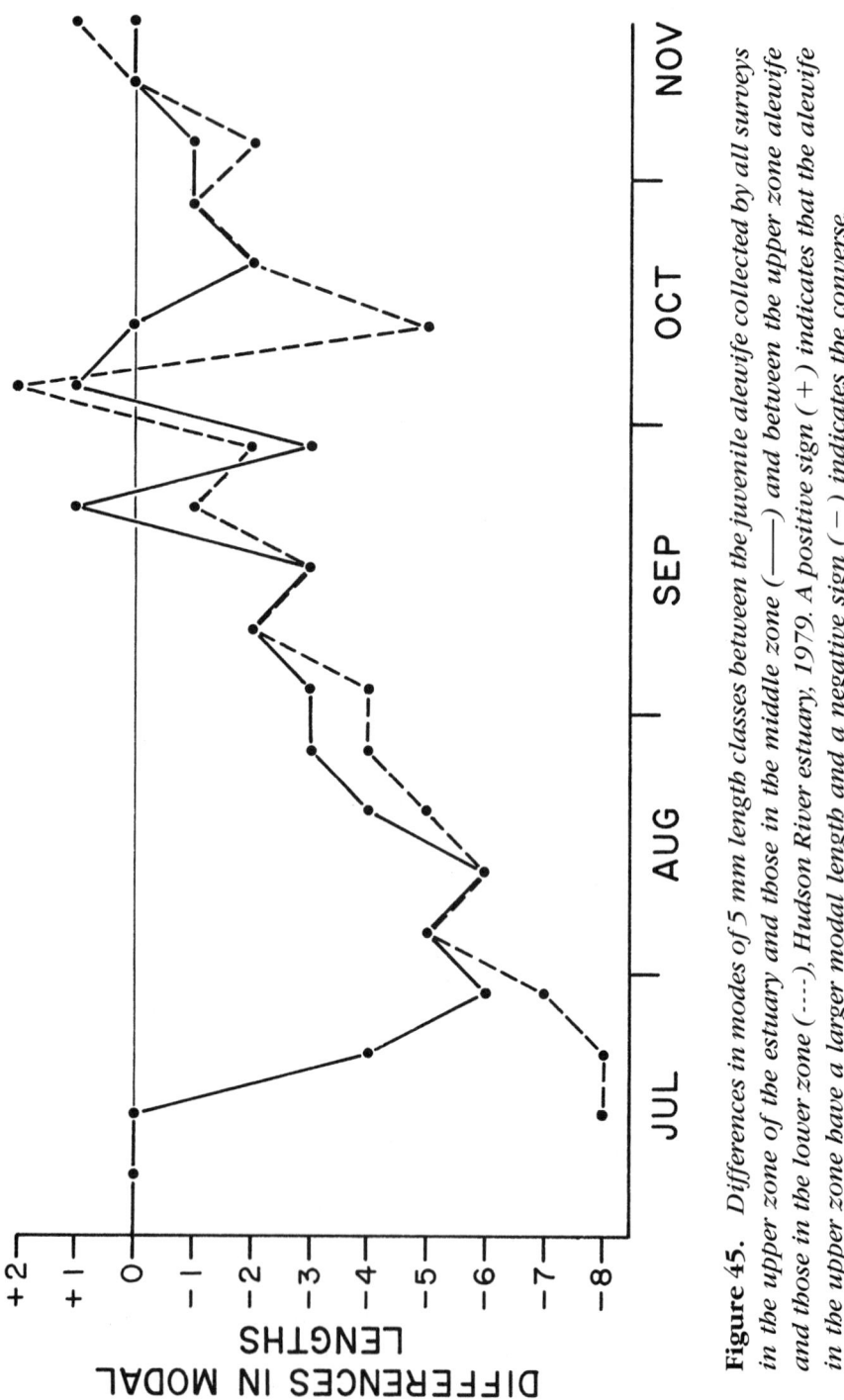

Figure 45. Differences in modes of 5 mm length classes between the juvenile alewife collected by all surveys in the upper zone of the estuary and those in the middle zone (——) and between the upper zone alewife and those in the lower zone (---), Hudson River estuary, 1979. A positive sign (+) indicates that the alewife in the upper zone have a larger modal length and a negative sign (−) indicates the converse.

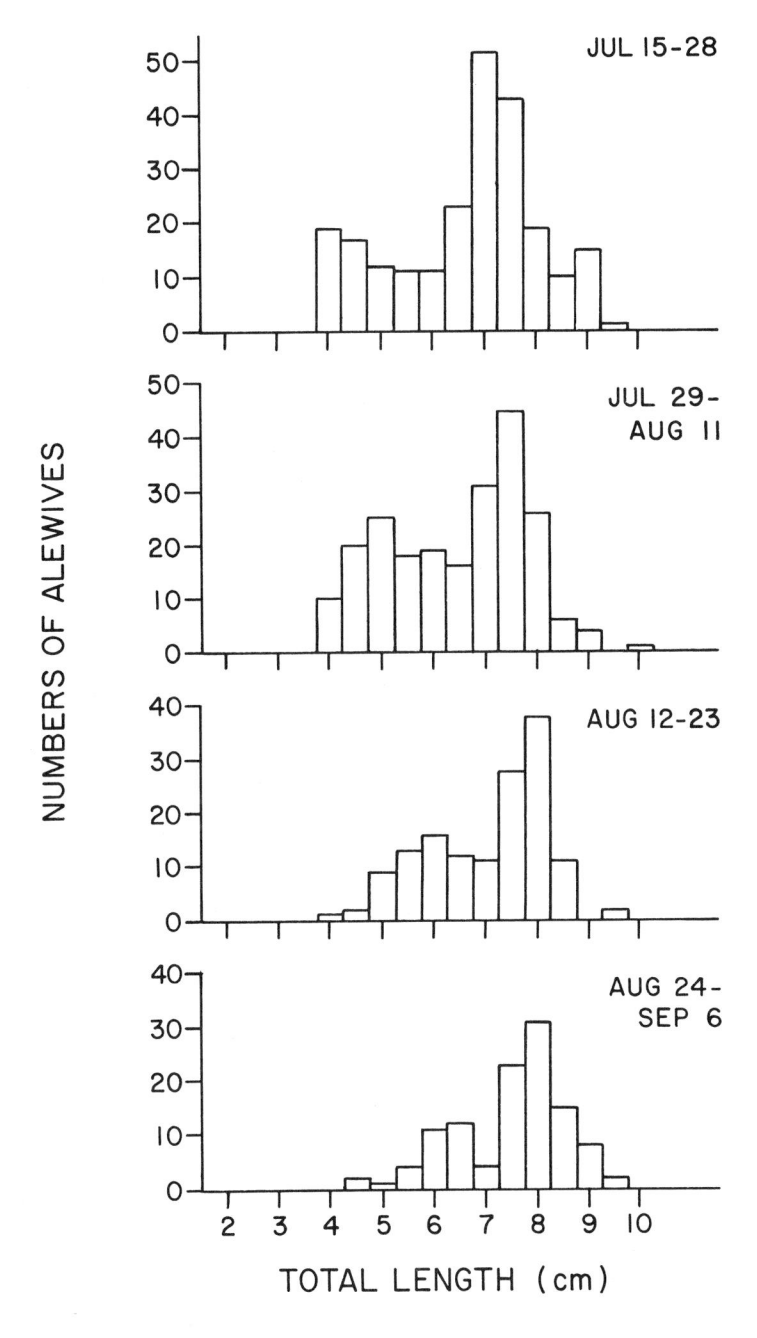

Figure 46. *Length frequency distribution for juvenile alewife collected in all surveys in the Hudson River estuary, 1979. Data from all samples in each biweekly period from the entire study area were combined.*

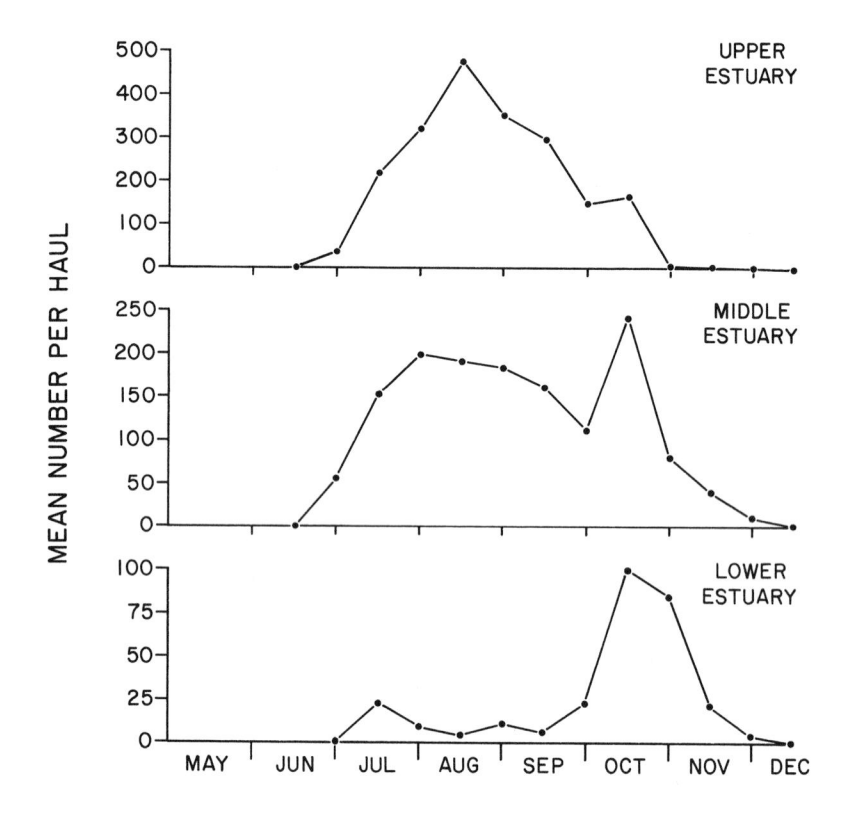

Figure 47. *Mean catch per haul of juvenile blueback herring in seines in the mainstream Hudson River estuary, 1976–1979.*

zone gear is greater than that of alewives in the daytime (Loesch et al. 1982); nevertheless, we feel that the larger catches of juvenile blueback herring in seines reflect their true abundance in the Hudson estuary. Blueback herring dominate impingement collections at most Hudson River power plants and far outnumber alewives and American shad (e.g., TI 1980d).

The apparently low abundance of juvenile blueback herring in the lower zone through the end of September suggests that, unlike the other two species of *Alosa*, juvenile blueback herring remain in the vicinity of their natal areas throughout the summer. The increase in catches in the middle and lower zones at the same time that the catches drop in the upper zone (Figure 47) suggests a rapid downstream migration during October. By the end of November, virtually all juvenile blueback herring have left the study area. The few that

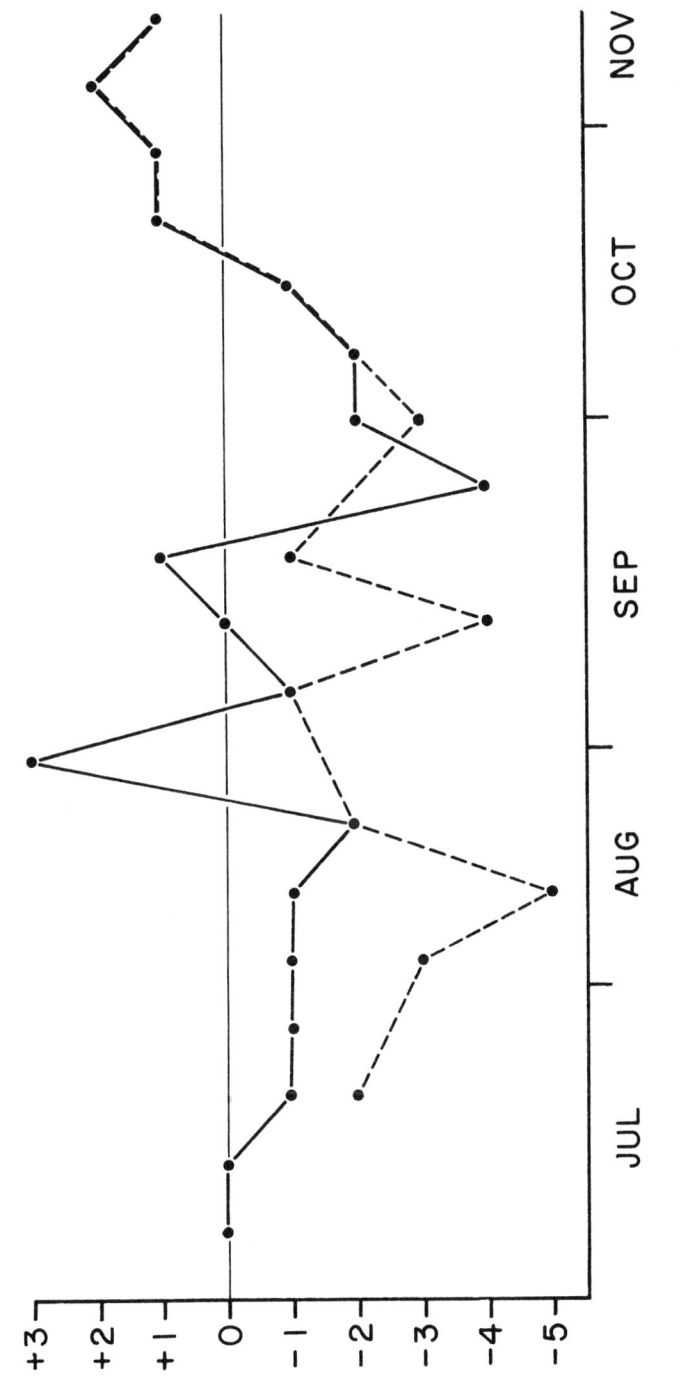

Figure 48. Differences in modes of 5 mm length classes between the juvenile blueback herring collected by all surveys in the upper zone of the estuary and those in the middle zone (——) and between the upper zone bluebacks and those in the lower zone (----), Hudson River; 1979. A positive sign (+) indicates that the blueback herring in the upper zone have a larger modal length and a negative sign (−) indicates the converse.

were caught in December may have been individuals that would have overwintered in the lower zone of the estuary, a phenomenon that Davis and Cheek (1973) observed in Virginia waters.

We have little evidence of clear zonal size differences in juvenile blueback herring in the Hudson. Those collected in the lower zone were consistently longer than those in the upper zone, but the extreme fluctuations of modal length differences between the middle and upper zones did not reveal any patterns (Figure 48).

DISCUSSION

Eggs

Spawning of American shad at suboptimal temperatures (below 16°C) has also been reported in other populations. Marcy (1972) reported water temperatures similar to those we observed in the Hudson during spawning of the American shad in the Connecticut River, and Carscadden (1975) reported consistent shad spawning at temperatures below 16°C in one tributary of the St. John River, New Brunswick. To paraphrase Ulanowicz and Polgar (1980), there is no reason to expect that anadromous fishes lay their eggs in optimal water quality conditions. The effects of suboptimal conditions on eggs of American shad in the Hudson River are intriguing but unknown.

Since American shad spawn entirely within the mainstream of the estuary, our estimates of egg densities should reflect the true distribution. This is not necessarily true for alewife and blueback herring eggs, however. In addition to the portions of the populations of these two species that spawn above the Troy Dam, alewife spawning runs occur in most tributaries from the Croton River (km 56) north (Gann and Brandt, personal communication, 1981). Blueback herring also probably spawn in or near these tributaries. Therefore, river herring eggs collected in the mainstream Hudson represent an unknown portion of the total river herring eggs deposited in the drainage system.

The eggs and larvae of four other clupeids may also be included in our river herring data: Atlantic menhaden (*Brevoortia tyrannus*), hickory shad (*Alosa mediocris*), Gizzard shad (*Dorosoma cepedianum*), and Atlantic herring (*Clupea harengus*). Atlantic herring and Atlantic menhaden spawn at sea (Hildebrand 1963), but some eggs and larvae could be transported into the lower portions of our study area by tidal currents. Dovel (1981b) collected small numbers of the eggs of both species in the lower Hudson. Hickory shad and gizzard shad are present in the Hudson River estuary (Smith 1976,

Dew 1973), but they were collected infrequently during the intensive sampling efforts of TI in 1976–1979 (e.g., one hickory shad and thirteen gizzard shad in 1979). Early life stages of Atlantic menhaden, Atlantic herring, hickory shad, and gizzard shad were present but uncommon in the estuary and should therefore have contributed little to our estimates of river herring egg and larval densities.

Larvae

The passive drift hypothesis offered to explain the distribution of *Alosa* larvae relative to eggs is further supported by temporal flow patterns in the Hudson estuary. Freshwater flow is generally greatest in May (Figure 49). Therefore, if early hatched larvae are passive, they should drift farther downstream than larvae that hatch later in the spawning period. This was indeed the case (Figures 40 and 41) and supports the notion that freshwater flow is the major influence on the distribution of larval *Alosa* in the Hudson; however, other factors, such as spatial variations in mortality rates for eggs and larvae, could also influence distribution patterns. Recruitment of river herring larvae from the tributaries could also account for the distribution patterns observed in our data.

Juveniles

The length-frequency patterns in Figures 43 and 45 suggest that the larger individuals of alewives and shad move downstream first. Chittenden (1969) and Marcy (1976c) reported this phenomenon for juvenile American shad in the Delaware and Connecticut Rivers, respectively. Loesch et al. (1982) reported early downstream movements of large juvenile alewives in Virginia.

Length frequency data for blueback herring (Figure 48) do not suggest size-related movements in the Hudson, but Loesch (1969) reported that blueback herring in the Connecticut River did segregate by size prior to emigration.

An alternative explanation for the observed differences in length frequencies is that juvenile alewives and shad grew faster in the lower zone than in the upstream zones. We cannot falsify either hypothesis with our data, but examination of daily growth rings on the otoliths in the future should provide definitive data on growth rates. This technique has been used successfully to age clupeids by Taubert and Tranquilli (1980) and Jacobs and Crecco (1980).

The bimodal length frequency distribution observed for juvenile alewives in the Hudson (Figure 46) has also been observed in other

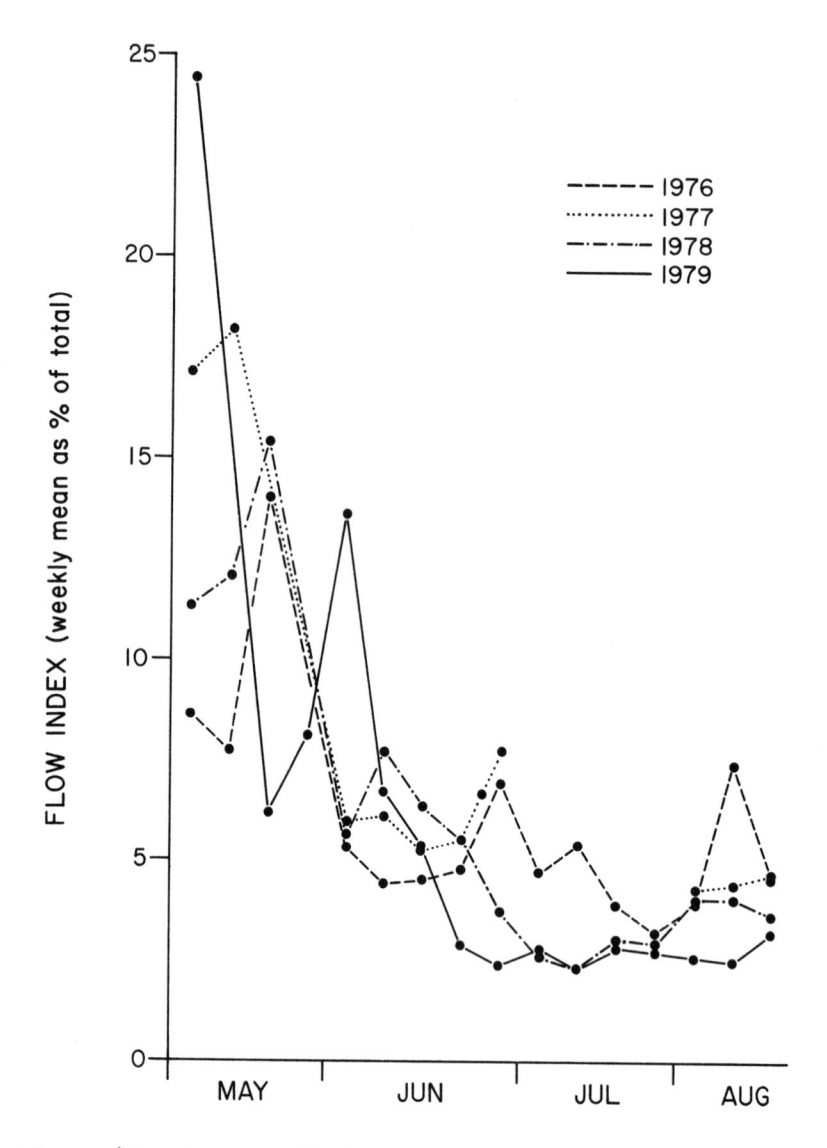

Figure 49. *Patterns of freshwater flow in the Hudson River estuary, 1976–1979.*

Data are weekly mean flows in any given year (calculated from daily measurements by USGS at Green Island gauging station near Troy Dam) divided by the total of all weekly means for the period May through the third week in August of that year and expressed as a percentage. Data for July 1977 were unavailable.

populations. Messiah (1977) reported a bimodal maturity index for alewives in the St. John River, New Brunswick. Richkus (1975) observed a bimodal length frequency distribution in juveniles from a Rhode Island drainage. Neither author attached much significance to these observations, and we have no way to determine its cause or whether bimodality is common for this population. Bimodality in young-of-the-year cohorts of fishes has been attributed to an extended spawning period and the advantage that larger individuals gain in capturing food (DeAngelis and Coutant 1982). If a bimodal or extended spawning period occurs regularly in Hudson River alewife populations, the juveniles could be segregating by size with the larger individuals from the early spawning moving farther downstream than the smaller juveniles from the later spawning. Other researchers (Shelton et al. 1979, Timmons et al. 1980) have presented evidence that shortages of food of particular sizes can lead to bimodality in largemouth bass cohorts when smaller individuals cannot obtain sufficient food and their growth rates are retarded. At present, however, the extended spawning period hypothesis appears to be the most likely explanation for the bimodality in the length distribution of juvenile alewives in the Hudson.

Evidence of Mechanisms of Niche Segregation Among Juveniles of Alosa *Species*

Larvae and juveniles of *Alosa* spp. are often a large component if not the dominant fishes of estuaries along the northeastern Atlantic coast. Marcy (1976b) noted that river herring eggs and larvae were the dominant species in ichthyoplankton collections from the Connecticut River. Species of *Alosa* are also dominant fishes in the Hudson River estuary. Blueback herring, shad, and alewife juveniles were the first, third, and fourth most abundant species collected in seines in the Hudson during 1973 (TI 1976a). LMS (1975f) reported that alewife and blueback herring comprised 40% of all fishes impinged at the Albany Steam Electric Generating Station (km 232).

The juveniles of the three species discussed in this paper are morphologically so similar that accurate identification requires a trained individual. Because of their abundance, and their similar morphology and life history, there is great potential for competition among these three species. However, the patterns of distribution and movements described here appear to reduce interspecific competition. Interspecific competition may also include differences in diel activity patterns and feeding habits.

DISTRIBUTION AND MOVEMENTS Spatial segregation of the species of *Alosa* begins when early summer downstream movements remove individuals of shad and alewives from the upper zone. Later, the more abundant blueback herring dominate the upstream area. However, all three species were present in samples from throughout the river and were often collected simultaneously if not in the same seine haul. Thus, other more finely tuned mechanisms must also be important in resource partitioning among these three species.

DIEL ACTIVITY All three *Alosa* species are known to exhibit diel vertical migrations from near the bottom during the day to the surface at night in the Mattaponi River, Virginia (Loesch et al. 1982). Bonomo and Daly (1981) reported similar vertical movements for blueback herring in the Hudson River estuary. Besides vertical movements, diel inshore-offshore movements that segregate juvenile alewives from the other two species have been observed in the Hudson. Juvenile alewives are most abundant in inshore areas at night, while juvenile American shad and blueback herring are most abundant near shore during the day (McFadden et al. 1978, Dey and Baumann 1978). Loesch et al. (1982) caught very few alewives in offshore areas at night in Virginia waters, but they did not sample near shore. Juvenile American shad exhibit peak activity periods and schooling behavior during the day (Katz 1978) when they are near shore. Competition between juvenile alewives and the other two *Alosa* presumably would be minimized by differences in diel activity patterns.

FEEDING HABITS Juvenile American shad and blueback herring both occur in shallow nearshore areas during the day but their feeding habits differ. American shad juveniles feed on Chironomidae larvae, Formicidae (Hymenoptera), and Cladocera (Massman 1963, Hirschfield et al. 1966, Levesque and Reed 1972, Marcy 1976b). Blueback herring juveniles tend to be more planktivorous and feed on Copepoda, Cladocera, and larval Diptera (Hirschfield et al. 1966, Burbidge 1974). Domermuth and Reed (1980) stated that juvenile American shad fed more on terrestrial insects at the surface, less on Copepoda, and on different cladoceran families than do blueback herring juveniles. Massman (1963), Levesque and Reed (1972), and Davis and Cheek (1966) also commented on the importance of terrestrial insects in the diet of juvenile American shad. Grabe and Schmidt (1978) conducted a short-term study in the Hudson and found that during the afternoon juvenile American shad fed primarily at the air-water interface on terrestrial insects, mostly Formicidae. During the same period, juvenile blueback herring fed mostly on Cladocera. Grabe and

Schmidt suggested that the period of greatest interspecific competition for food occurred in early morning, when all three species of *Alosa* fed on epiphytic animals on the macrophyte, *Myriophyllum spicatum* (see also Menzie 1980). Levesque and Reed (1972) and Marcy (1976c) also suggested epiphytic foraging for juvenile American shad, but Domermuth and Reed (1980) hypothesized that Chironomidae were taken directly from the water column by juvenile American shad and blueback herring. Because of these species differences in feeding habits, direct competition between American shad and blueback herring is likely to be minimal in the Hudson River estuary.

It appears that there are several mechanisms that reduce competition between juveniles of the three species of *Alosa* in the Hudson River estuary. Differences in distribution, diel activity patterns, and feeding habits are evident. These mechanisms have been implicated in reducing competition among other species of fishes (Keast 1965, 1970, Baumann and Kitchell 1974, Weaver 1975, McEachren et al. 1976, Werner and Hall 1976), and it is reasonable to assume that they are also important for reducing competition among the species of *Alosa* in the Hudson River. We hope that this paper will stimulate further investigations.

ACKNOWLEDGEMENTS

We are grateful to the men and women who labored countless hours, day and night, to collect, process, and compile the data used in this paper. Special thanks are extended to former colleagues with whom we engaged in many stimulating discussions about the distribution of fishes in the Hudson River estuary, especially Howard C. Barker, C. Allen Beebe, K. Perry Campbell, William P. Dey, Lynn E. Foster, Donald J. Grosse, Kathryn Farinacci Klauda, James T. McFadden, Michael I. Riner, Roy T. Rowland, Roger A. Rulifson, Barry A. Smith, Robert J. Stira, and Michael A. Tabery. Joseph G. Loesch and William A. Richkus provided constructive criticisms of the manuscript. Kathy Schmidt prepared the illustrations. Funding for this study was provided to TI by ConEd, PASNY, ORU, and CHGE.

Part V
Tomcod

8. Life History of Atlantic Tomcod, Microgadus tomcod, in the Hudson River Estuary, with Emphasis on Spatio-Temporal Distribution and Movements

Ronald J. Klauda, Richard E. Moos, and Robert E. Schmidt

INTRODUCTION

The Atlantic tomcod, *Microgadus tomcod* (Walbaum), is a small gadid fish that ranges along the eastern coast of North America from southern Labrador to Virginia (Scott and Crossman 1973). Spawning typically occurs during winter in shallow, low-salinity areas of estuaries and near the mouths of tributary streams (Booth 1967). After an extended incubation period of twenty-four to sixty days (Scott and Crossman 1973, LMS 1980f), the newly hatched larvae drift or move downriver to the lower portions of estuaries (Scott and Crossman 1973, Hardy and Hudson 1975, Peterson et al. 1980).

The Hudson River estuary is probably the southernmost major spawning area for Atlantic tomcod (Dew and Hecht 1976; Grabe 1978). No spawning has been documented in the Connecticut River (Marcy 1976c) or Long Island Sound (Richards 1959), and the status of Atlantic tomcod in New Jersey is uncertain (Miller 1972), although limited spawning may occur in the Raritan river and/or Raritan Bay (IA 1977). In the Delaware River estuary, no Atlantic tomcod catches were reported in shore-zone samples from 1958 through 1960 by de Sylva et al. (1962), and no Atlantic tomcod have been caught in recent seine and bottom-trawl surveys in Delaware Bay (personal communication with Ronald W. Smith, Delaware Division of Fish and Wildlife). Massman (1962) reported catching only one Atlantic tomcod during four years of sampling in Chesapeake Bay.

The Atlantic tomcod is sought after by sport and commercial fishermen throughout its range, yet published information on the life history of this species is limited, presumably because sampling is

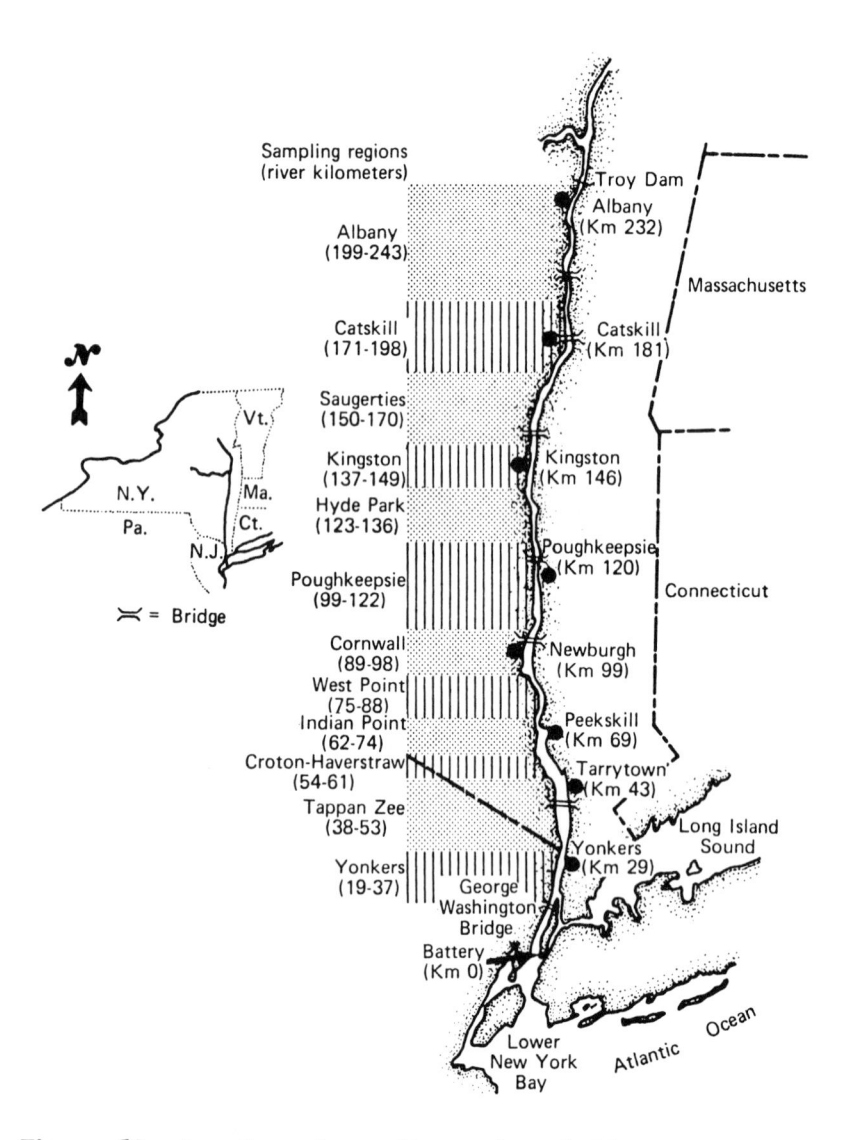

Figure 50. *Location of sampling regions (with river kilometer boundaries) in Hudson River estuary.*

difficult during the winter spawning and early nursery periods. Texas Instruments studied the biology of Atlantic tomcod in the Hudson River estuary from 1975 through 1980 as part of a multidisciplinary project investigating the impact of power plants on several fish species. This paper summarizes recent data on the life history of Atlantic tomcod adults, larvae, and juveniles in the Hudson River estuary with emphasis on spatio-temporal distributions and movements.

STUDY AREA

The Hudson River extends about 500 km from its origin in the Adirondack Mountains to New York City. The river is unimpounded and tidal for 246 km to the Federal Dam and Lock at Troy, New York, and typically estuarine to about Poughkeepsie (Figure 50). Physicochemical descriptions of the system were presented by Fenneman (1938), Busby (1966), Giese and Barr (1967), Texas Instruments (1976c), Abood (1974, 1977), and Klauda et al. (1980).

We sampled the Atlantic tomcod population between the George Washington Bridge and the Troy Dam, km 19–243, year-round from 1973 through 1980. The study area was divided into twelve sampling regions. Each region was subdivided into shoal, bottom, and channel strata (Figure 51). The estuary is connected through Upper and Lower

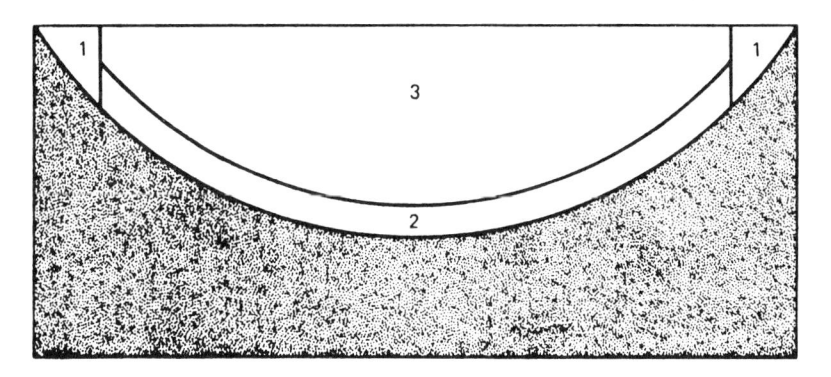

1 Shoal [depths less than 6 m]
2 Bottom [bottom (3 m) in depths greater than 6 m]
3 Channel [above bottom (3 m) in depths greater than 6 m]

Figure 51. *Diagrammatic cross-section of Hudson River estuary showing strata sampled by Tucker trawl and epibenthic sled.*

New York Bays to the Atlantic Ocean through the Harlem and East Rivers to Long Island Sound; hence, a portion of the estuary lies outside the study area.

MATERIALS AND METHODS

Several field surveys were designed to collect Atlantic tomcod throughout most of their life cycle. The primary data for this paper were collected from March 1975 through March 1980. The principal sampling gear (box traps, Tucker trawl, and epibenthic sled) and sampling designs were described by Texas Instruments (1980a, 1980e, 1981b). Data from other studies conducted in the Lower Bay and Long Island Sound were included to provide a more complete description of the distribution and movements of Hudson River Atlantic tomcod.

SPAWNING ADULTS

The distribution and movement of adults from early December through mid-March were monitored with a box-trap survey. About twenty box traps (0.9 m × 9.9 m × 1.8 m, 0.95-cm mesh, no wings or leads) were set at from fourteen to eighteen sites in the shoals (areas less than 6 m deep) from about km 22 to km 134 (Figure 52). The unbaited traps were set adjacent to docks, piers, and bulkheads; hence, the allocation of sampling effort was influenced by availability and accessibility of these structures. Ice and unpredictable winter weather prevented the use of boats to set traps and precluded the use of a more randomly or uniformly distributed sampling effort. Box traps were fished around the clock but checked and emptied only once per twenty-four hours, Monday through Friday. Catches recorded on Tuesday through Friday represented a series of approximately 24-hour efforts. Catches recorded on Monday represented approximately 72-hour efforts. Duration of fishing effort (in hours) was recorded for each box trap catch and used in catch-per-unit-effort calculations.

Adult tomcod were counted, sexed, marked, measured, and released. Generally, all fish from catches of less than twenty individuals were marked with carlin tags (custom-made for Texas Instruments by Floy Tag Co.). When more than two hundred tomcod were caught, about one-half were tagged and about one-half were fin-clipped to reduce handling time and improve post-marking survival. Tagged fish were measured to the nearest millimeter (total length). Finclipped

Figure 52. *Typical distribution of box-trap sampling sites used to collect adult Atlantic tomcod during winter spawning period in Hudson River estuary.*

fish were counted by 25-mm length groups from less than 125 mm to greater than or equal to 276 mm total length. Movements of individuals were evaluated from recoveries of marked fish taken during project field surveys, by fishermen, and from power plant intake screens.

The box-trap survey also supplied data on several population parameters. A 24-hour catch of adults from each trap set at seven standard sites (km 43–90) was returned fresh to the laboratory each

Table 43. *Criteria used in visual classification of Atlantic tomcod gonads for state of maturity.*

CLASSIFICATION	DEFINITION
Immature	A specimen that is either male or female but is too young to spawn (subadult). Gonads which have not developed are transparent or pinkish.
Developing	Applicable to subripe fish just prior to the spawning season which may or may not spawn that season. Testes and ovaries are opaque and reddish to reddish-white (ovaries may appear orange); eggs are visible to the naked eye, granular, and whitish to orange-reddish.
Ripe	Adult in spawning condition and having well developed gonads, but no milt or eggs extruded upon application of pressure to abdomen. Will spawn in current season.
Ripe and Running	Adult prepared to spawn immediately; expulsion of eggs or milt from body with little provocation.
Partially Spent	Sexual products partially discharged; gonads somewhat flaccid rather than firm as a developing gonad; genital aperture usually inflamed, with some hemorrhaging present.
Spent	Applied to adult specimens at completion of spawning activity. The sexual products have been discharged, the genital aperture is usually inflamed and some hemorrhaging is present. The gonads have the appearance of deflated sacs; the ovaries usually contain a few eggs (in a state of reabsorption), and the testes contain some residual sperm; the ovarian wall is leathery.

week from December through February. Fish were sorted into 25-mm length groups and up to twenty individuals were randomly subsampled from each length group for determination of total length, weight, sex, and state of maturity. Gonads were classified into six maturity categories (Table 43) based on visual examination. Sex ratios were calculated for all tomcod subsampled for each date. Sex ratios were tested by chi square ($\alpha = 0.05$) against the null hypothesis that the ratio was 1 male to 1 female.

EARLY LIFE STAGES

The distributions of early life stages during late winter, spring, and early summer were monitored from early March through early

April of each year (depending on weather and ice conditions). A total of one hundred samples was collected biweekly during daytime between km 22–122 in the shoal, bottom, and channel strata (Figure 51). Sampling effort within regions and strata was allocated by a stratified random design keyed to the observed distributions of tomcod early life stages in prior years.

In mid-April, the ichthyoplankton survey was extended upriver to km 224. Two hundred samples were then collected weekly during daytime until early June and at night thereafter. A stratified random design was also used for this expanded ichthyoplankton survey, but sample allocations were keyed to the observed distributions of striped bass eggs and larvae in prior years (striped bass was the focal species of the impact assessment studies). Since most samples were allocated to the Indian Point and West Point regions near the middle of the study area (Figure 50), catches of Atlantic tomcod in the extended survey may have underestimated their relative abundance in the most downstream regions of the study area (Yonkers and Tappan Zee).

Samples were collected with a 1.0-m epibenthic sled (505-µm mesh) in shoal and bottom strata, and with a 1.0-m Tucker trawl (505-µm mesh) in the channel stratum. Tow speeds for the sled and trawl were 0.9 and 1.0 m per sec, respectively. The standard tow duration for both types of gear was five minutes. Calibrated digital flowmeters suspended within the nets recorded sample volume and calibrated electronic flow meters mounted on the towing cable just above the gear recorded towing speed. Double-trip release mechanisms permitted opening and closing of the nets at specific depths.

The distributions of yolk-sac larvae, post yolk-sac larvae (postlarvae), and juveniles were described from weekly estimates of densities (number/1000 m) in each sampling region. Few Atlantic tomcod eggs were collected in the ichthyoplankton survey because the demersal eggs are spawned mostly during December and January, a time when we were unable to sample with plankton gear. Larvae were considered to be in the yolk-sac stage from hatching through development of a complete and functional digestive system. The stage from initial development of the digestive system (regardless of degree of yolk and/or oil retention) to possessing a full complement of fin rays was defined as the postlarval stage. Transition from postlarvae to the early juvenile stage was defined as complete when the postlarvae had a full complement of fin rays. Transition from the early juvenile to juvenile stage was arbitrarily defined to occur in early to mid-August and not based on developmental or growth-related characteristics.

The regional density (d_r) of each life stage was calculated by summing the strata (Figure 51) densities (d_k) within each region as follows:

$$d_r = \frac{\sum V_k d_k}{\sum V_k}$$

where V_k = volume of water (m^3) in the kth stratum within the rth sampling region.

Average densities of each life stage within each stratum were multiplied by the volume of water within that stratum to estimate regional standing crops. The standing crops in each region were summed to obtain weekly standing crop estimates for the entire study area. Estimates of standing crop were not adjusted for gear efficiency and presumably underestimated true population size of the larger, more motile, early life stages.

Freshwater flows during the late winter-early spring are reportedly an important influence on the distribution of Atlantic tomcod eggs and larvae. Dew and Hecht (1976) suggested that relatively high freshwater flows during March and April in the Hudson River estuary resulted in a fairly rapid downstream movement (about 104 km) by tomcod larvae after hatching in February and March. Peterson et al. (1980) suggested that the early life history of Atlantic tomcod in southern New Brunswick, Canada, was adapted to the hydrodynamics of the spawning areas and that more eggs drifted downstream during relatively high flow years. Therefore, we calculated an annual index of freshwater flow potential for displacing tomcod larvae downriver in the Hudson from the average daily flow measured at the United States Geological Survey gauging station at Green Island (km 243) during February through April, multiplied by a weighting factor which took into account rapid increases (surges or pulses) of flow. This weighting factor was the average daily flow for the week of highest flow divided by the average daily flow for the week of lowest flow during the three-month period from February through April.

The extent of downriver displacement of the postlarval population in each year was represented by the spatial distribution index for the Yonkers region (km 19–37), the most downstream sampling region, for the period from the last week in March through about mid-June. The spatial distribution index is the regional standing crop estimate divided by the sum of all regional standing crops for the life stage. Since only five pairs of data points for weighted flow and downriver displacement were available, the data were plotted and examined visually.

JUVENILES

The distribution of juveniles during late summer and fall was monitored with two surveys conducted primarily in the downstream half of the study area. One survey used an otter-type bottom trawl equipped with 0.8-m × 1.2-m doors, a 7.8-m headrope, and a fine mesh (1.27 cm) cover over the cod end. This trawl survey sampled from thirty-two to thirty-eight stations between km 38 and km 122 during daytime on alternate weeks from April through November. The trawl was towed against the current at 1.3 m per sec for five minutes while maintaining a 3 to 1 cable length to depth ratio. The bottom trawl sampled primarily in areas deeper than 6 m.

The second juvenile survey used a 1.0-m epibenthic sled equipped with a 3-mm mesh net and a conical fyke attached to the cod end. In 1975–1978, one hundred samples were taken at night on alternate weeks from mid-August through December in the shoal and bottom strata from Yonkers to Poughkeepsie, km 22–122 (Figure 50). In 1979, this survey was extended upriver to include all strata from the entire study area (km 19–243), a 1.0-m Tucker trawl (3-mm mesh net) was added, sampling began earlier (in mid-July), and two hundred samples were collected on alternate weeks. Tow speed and duration for both gears were 1.5 m per sec and five minutes, respectively. A digital flow meter measured sample volume. Tow speed was measured by an electronic flow meter.

Descriptions of the distribution of juveniles during late summer and fall were based on alternate week estimates of regional catch-per-unit-effort (CPUE) from the bottom trawl and regional density from the epibenthic sled and Tucker trawl. Bottom trawl CPUE was calculated from total catch divided by total number of tows. Regional densities were calculated as described above for the early life stages.

RESULTS AND DISCUSSION

Sexual Maturity

Hudson River Atlantic tomcod become sexually mature during the fall of their first year of life at approximately eleven to twelve months. The age composition of the adult population during five winter spawning seasons (1975—76 through 1979–80) averaged 96.7% yearlings and 3.3% age II individuals (TI 1981b). Age III adults are rare in the population and were collected only during the 1976–77 season, when they comprised only 0.02% of the spawners (TI 1980e).

Atlantic tomcod are relatively short-lived throughout their distribution range (Cox 1921, Bigelow and Schroeder 1953, Booth 1967), but longevity seems to be most substantially reduced in the Hudson estuary population. Age III and older individuals are more common in populations farther north (Legendre and Lagueux 1948; Howe 1971, Roy et al. 1975; Fleming et al. unpublished manuscript).

Spawning Time

Sampling for Atlantic tomcod eggs in the Hudson was not attempted due to winter weather conditions; hence, spawning time was determined indirectly from collections of adults in box traps and female

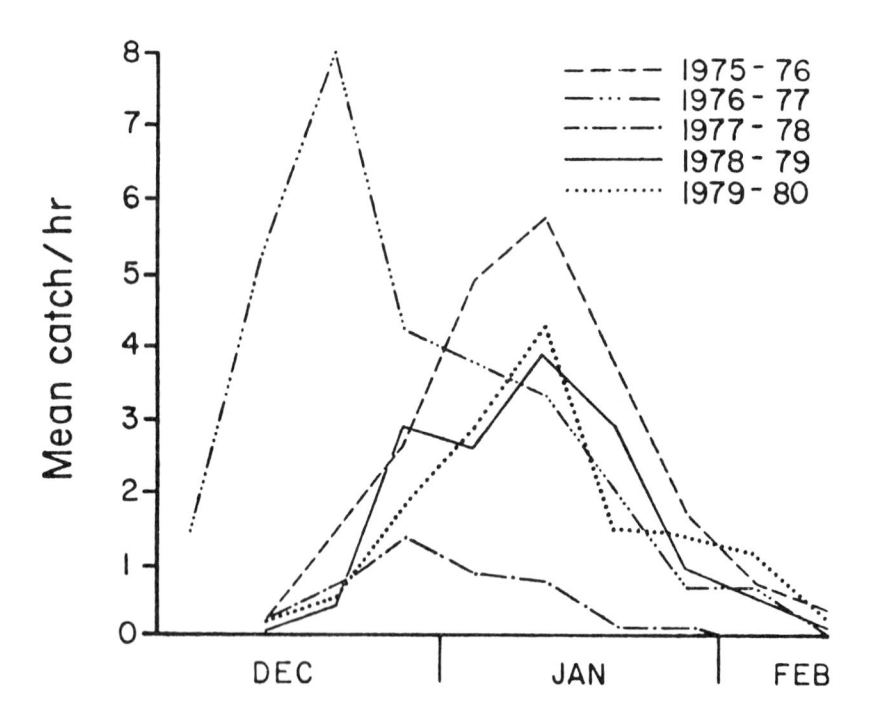

Figure 53. *Catch rates of adult Atlantic tomcod in box traps during five spawning seasons, Hudson River estuary.*

maturity. Adult catches were highest between mid-December and mid-January during five spawning seasons (Figure 53). The earliest peak in adult tomcod catch occurred in mid-December during the 1976–77 season. During 1977–78, peak catches occurred in late December and early January. Catches peaked around mid-January during the remaining three seasons. In all five seasons, catches of adult tomcod in box traps declined sharply after late January. Spawning time was also estimated indirectly from the percent of prespawning females (ripe plus running ripe) collected in box traps (Figure 54). Information on female maturity during the winters of 1975–76 through 1978–79 was based on the examination of 3625 females (Table 44). During all five spawning seasons, the highest percentage of prespawners (nearly 100%) occurred in early December, followed by a decline in the percentage of prespawners beginning in mid-December through the end of January (Figure 54). The February data were inconsistent due to the small number of female Atlantic tomcod collected. Based on these trends in adult catches and the percentage of prespawning females, we deduced that the spawning period extends from mid-December through the end of January, with some annual variation.

For unexplained reasons, spawning appeared to begin earlier in 1976–77 than during the other four seasons. November–December water temperatures were lower in 1976–77 (Figure 55). We suggest that gonadal maturation in Atlantic tomcod may be cued, in part, by water temperature; that is, lower than normal late fall water temperatures may accelerate the onset of spawning. But this hypothesis remains to be tested.

Sex-specific movements of adults to and from the spawning area were inferred from sex-ratio data. The sex ratio of the spawning population during early December and again during February was often significantly different ($\alpha = 0.05$) from a 1 to 1 ratio (Figure 56). Males were predominant in seventy-five of the seventy-eight samples that had sex ratios significantly different from 1 to 1. From late December through mid-January, however, the sex ratio was more frequently 1 to 1. Sex ratios of juvenile Atlantic tomcod in the summer prior to maturation were almost always 1 to 1 (TI 1981b). Based on these data, we conclude that male and female tomcod segregate during gonadal maturation in the late fall and exhibit different movement patterns during the spawning period.

The observed temporal variation in winter sex ratios can be explained by a delayed arrival and earlier departure from the spawning area by females, with males occupying the spawning area for several

Table 44. Total number of adult Atlantic tomcod (males and females) collected by box traps in the Hudson River estuary during spawning period and number of females examined for state of maturity.

MONTH	1975–76 TOTAL	1975–76 FEMALE	1976–77 TOTAL	1976–77 FEMALE	1977–78 TOTAL	1977–78 FEMALE	1978–79 TOTAL	1978–79 FEMALE
December	2991	103	25,806	599	2511	359	2425	305
January	4405	378	5,476	595	1572	333	4886	823
February	733	56	579	40	29	5	713	29

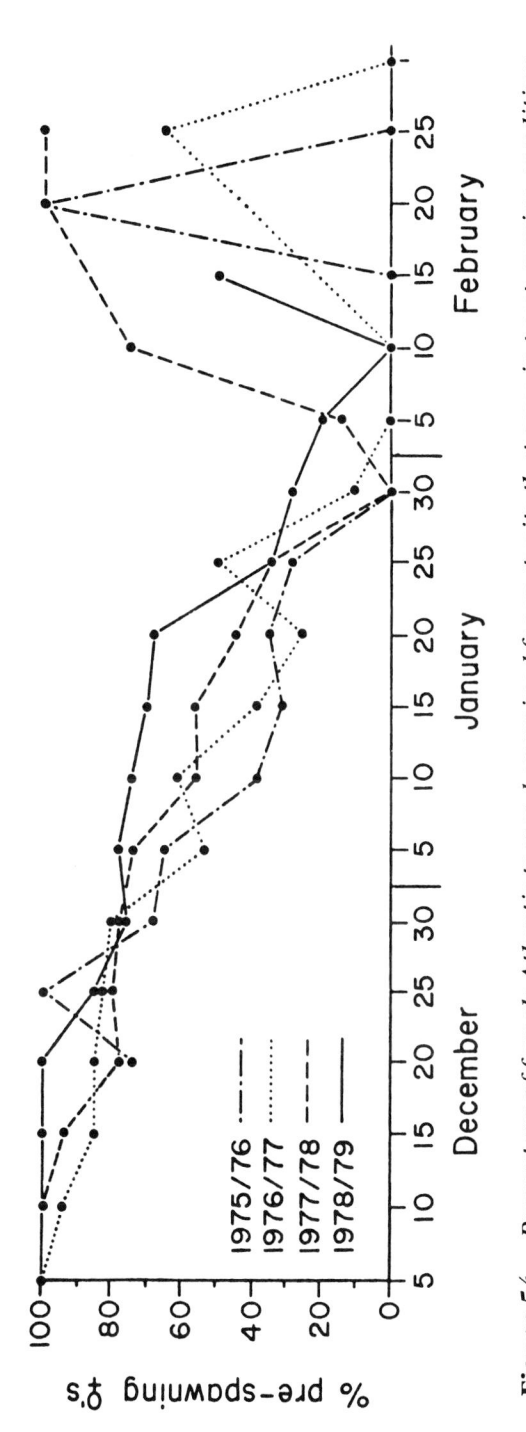

Figure 54. *Percentage of female Atlantic tomcod examined for maturity that were in prespawning condition (ripe and running ripe) during winter spawning season, Hudson River estuary. Each data point represents a five-day period prior to and inclusive of date plotted on x-axis.*

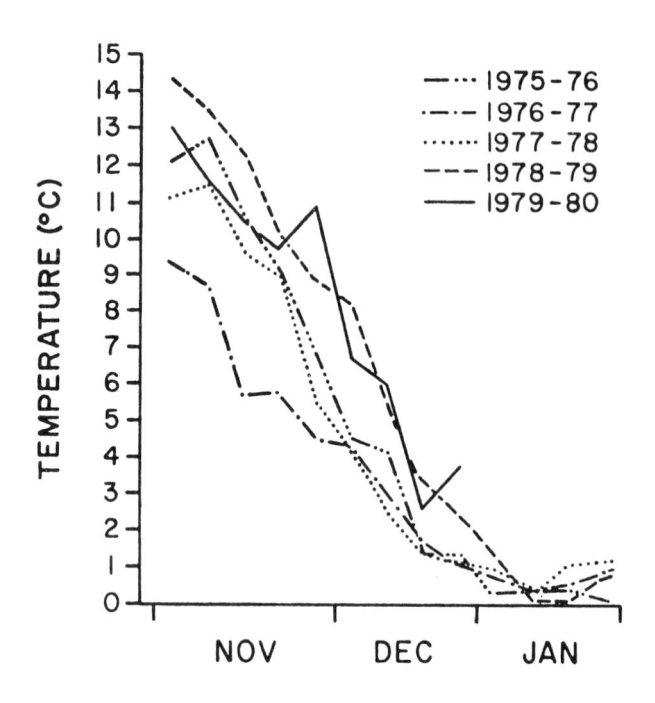

Figure 55. *Late fall and winter water temperatures in West Point region (km 75–88), Hudson River estuary.*

months. Females apparently do not arrive in the spawning area until they are ready to spawn and then return to the lower estuary, or die, soon after spawning. When the sex ratio favors males in early December (Figure 56), those few females present are mostly in prespawning condition (Figure 54). When the sex ratio is about 1 to 1, both prespawning and postspawning (partly spent and spent) females are present. In February, the sex ratio again favored males, suggesting that

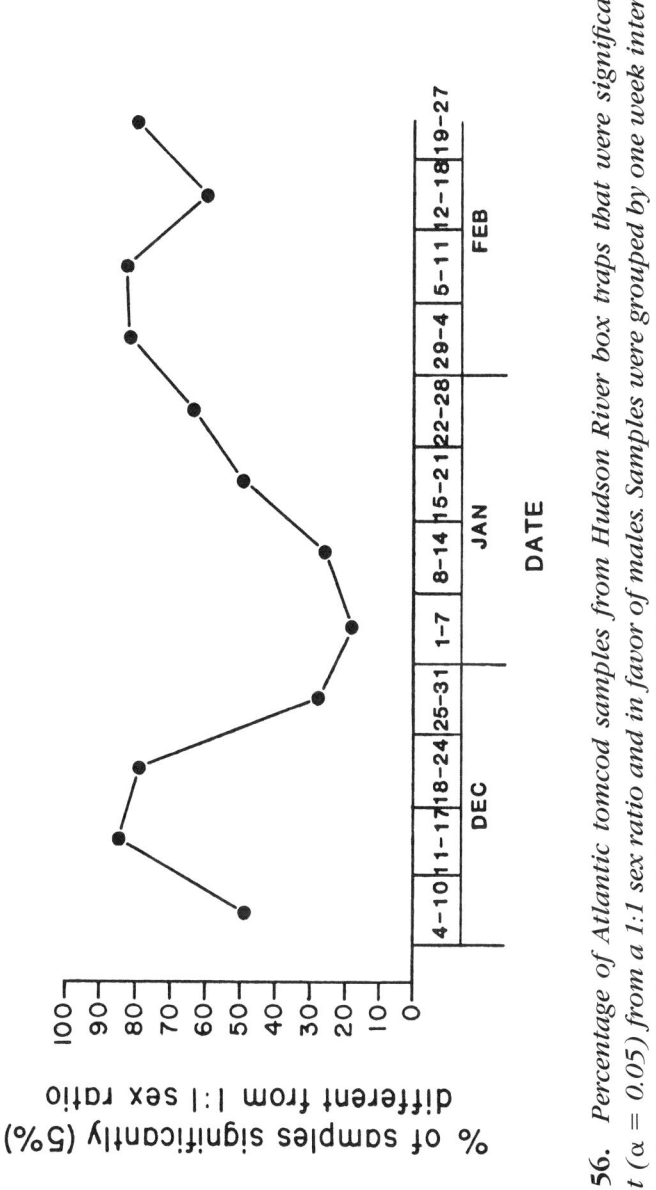

Figure 56. Percentage of Atlantic tomcod samples from Hudson River box traps that were significantly different ($\alpha = 0.05$) from a 1:1 sex ratio and in favor of males. Samples were grouped by one week intervals corresponding to sampling schedule and data for 1975–79 combined.

most of the females have vacated the spawning area and moved offshore, emigrated from the study area, or died.

Spawning Location

Hardy and Hudson (1975) reported that Atlantic tomcod eggs hatch in thirty-six to forty-two days at a mean temperature of 3.4°C. Since tomcod spawning ceases by the end of January in the Hudson River, and winter conditions typically precluded sampling until March, few eggs were collected in our ichthyoplankton survey. In 1976 and 1977, sampling began in February and a few eggs were collected between km 75 and km 122. The location of spawning was therefore determined indirectly from the distribution of adult catches in box traps.

During five spawning seasons, adults were most abundant between Tappan Zee and Poughkeepsie (Figure 57), with substantial catches generally occurring in the West Point region (km 76–89). The overall catch trends suggest that the West Point sampling region is located near the center of the Atlantic tomcod spawning area in the Hudson River estuary, with some annual variation. During the 1975–76 and 1976–77 spawning seasons, the highest catches were generally in the West Point region and upriver to Poughkeepsie (km 99–122). During the three most recent seasons, adult tomcod were most abundant in the West Point region and in regions further downriver, such as Tappan Zee (km 39–53) and Croton-Haverstraw (km 55–61). Whether annual shifts in the distribution of peak spawning activity is associated with shifts in salt-front position, variation in freshwater flow, or differences in the timing of spawning migrations with early migrants moving farther upstream remains to be tested.

Tagging and recapture data were also examined for information on adult tomcod movements during the spawning season. During five spawning seasons, a total of 50,048 adults were tagged and released (range of 7,700–25,000 per year). The recapture rate was less than 5% in any year. The majority of recaptures were taken in the same river mile in which the fish were tagged and released (defined as no movement). Of those individuals that moved, more moved downstream than upstream from the point of release within any month or year (Table 45). Prior to spawning, mature Atlantic tomcod may move upstream through the deeper channel areas and are unavailable for tagging until they move into shallow areas to spawn. Therefore, it is likely that most of the adults were tagged just prior to spawning or shortly thereafter, and any short-term movements (few days to few

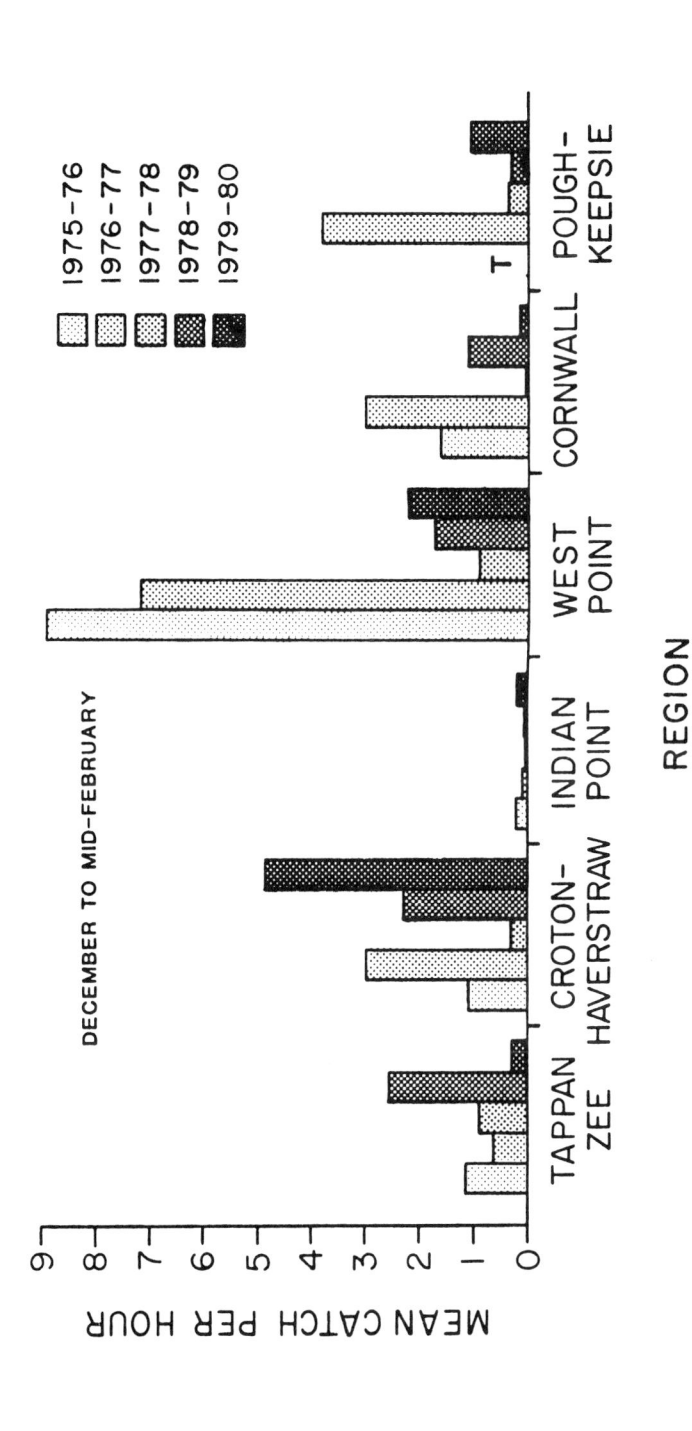

Figure 57. Catch rates of adult Atlantic tomcod in box traps during five spawning seasons, Hudson River estuary (T = trace, <0.005).

Table 45. *Number of tagged adult Atlantic tomcod recaptured more than 1.6 river kilometers from point of release and within three–four months after release.*

YEAR	NUMBER RECAPTURED	
	DOWNSTREAM	UPSTREAM
1975–76	58	17
1976–77	44	36
1977–78	45	20
1978–79	77	24

weeks) would expectedly be limited or reflect a postspawning downstream migration. The tagging data offered few insights on the movements of adult tomcod during the spawning season.

Recovery of tagged individuals by sport fishermen from outside of the study area (Figure 58) indicate that some Hudson River Atlantic tomcod move into New York Harbor and Long Island Sound after spawning. Although the number of tags returned from adjacent coastal areas is small (twenty), these returns indicate that any spawning populations of Atlantic tomcod in New Jersey and Long Island Sound are not necessarily distinct spatially from the Hudson River population.

Spawning Behavior

Atlantic tomcod often spawn close to shore in estuaries and freshwater tributaries (Nichols and Breder 1926, Curran and Ries 1937, Greeley 1937, Booth 1967, Howe 1971) where they should be readily observable. However, because they spawn during the winter and usually under ice cover, little is known about their spawning behavior.

Howe (1971) recounted a single observation of spawning activity during early December in Muddy Creek, Massachusetts.

> "Many tomcod had swarmed on the windward side of the creek in emergent cord grass (*Spartina alterniflora*) and under mats of floating detritus. Fish, approximately 5 to 13 inches in length, were observed twisting and rolling together in groups of from three to seven. One specimen, caught by hand, had running milt. Water temperature was 36.5°F (2.5°C) (Coates, personal communication)."

During laboratory experiments with Atlantic tomcod collected from the Hudson River estuary in January and February 1978, we observed courtship (prespawning) behavior in mixed groups of one or two females with three or four males. Ripe adults captured in box

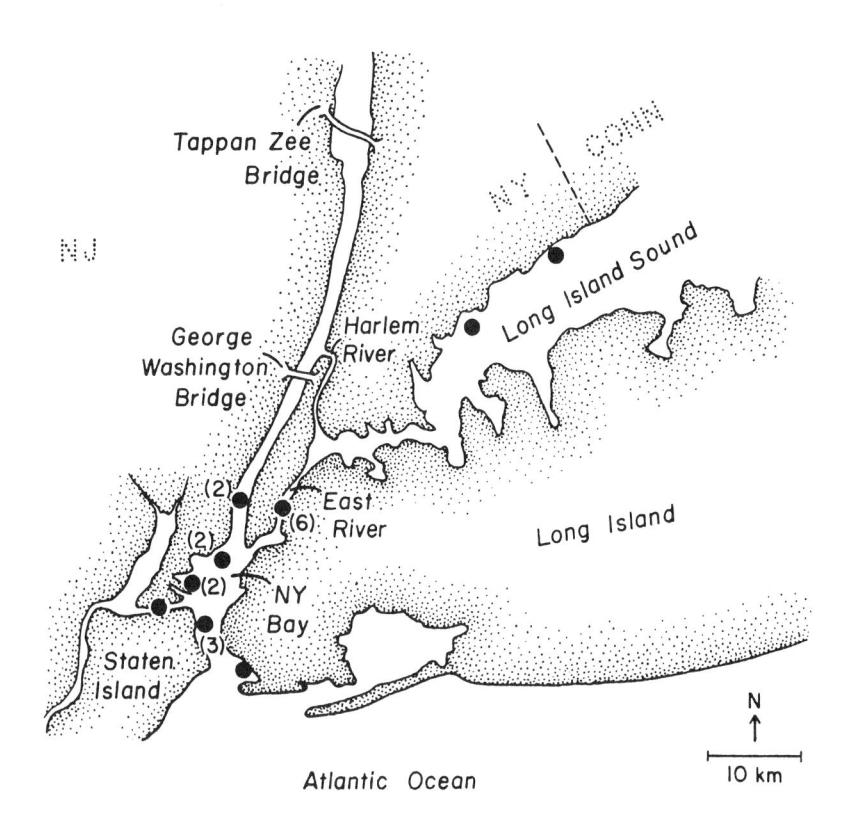

Figure 58. *Areas contiguous to mouth of Hudson River estuary showing recapture locations of Atlantic tomcod outside of the study area, 1975–79. Number in parentheses indicates number of multiple recaptures from same general locality.*

traps were transported to the laboratory and housed in 100 L aquaria receiving a three L/min flow of filtered ambient (2–3°C) river water. Overhead fluorescent lights were on from 0800–1630 hours. The aquaria were covered with stryrofoam sheets (2.5 cm thick) to reduce illumination. The fish were neither tagged nor marked and, except for periodic checks to determine if spawning had occurred, were undisturbed for approximately one week (Watson, MS).

Courtship behavior was observed and filmed during this undisturbed period when the lights were on. Even though several visibly ripe fish were present in each aquarium, the observed courtship display sequences involved only one male (usually the largest in the aquarium) paired with one female. In some aquaria, all fish eventually participated in courtship behavior, but group displays (e.g., several males with one female) were not observed. A male initiated courtship by approaching a female and nudging her flank with the lateral side of his head or snout (Figure 59). If the female remained passive and did not swim away, he would roll over and attempt to maneuver under her into a position with their ventral surfaces aligned, touching and facing in the same direction (Figure 59D). The male used his pelvic fins to embrace the female and gain close alignment. The courtship behavior of *Gadus morhua* and *Lota lota* (Brawn 1961, Fabricus 1954) includes a similar ventral surface alignment and embrace. Several courtship sequences were observed, but all were terminated by the female tomcod prior to discharge of eggs and sperm. Actual spawning was not observed, although six females did spawn naturally in the aquaria during the experiments, but only at night (Watson, unpublished manuscript). Two females spawned all their eggs during one night, while four females were spawning during two successive nights. Four of the six natural spawnings had fertilization rates greater than 50%.

Distribution of Early Life Stages

EGGS Although few Atlantic tomcod eggs were collected during our Hudson River studies, major characteristics of the eggs are briefly reviewed for completeness. Atlantic tomcod eggs are spherical and about 1.5 mm in diameter. They were once thought to possess a distinct oil globule (Bigelow and Schroeder 1953; Scott and Crossman 1973), but Booth (1967) stated that no single large oil globule exists, although many small globules are often noticeable (with magnification) and clustered near one side of the yolk (personal communication, H. Peterson).

There is general agreement in the literature that Atlantic tomcod eggs are heavy, sink to the bottom, and settle into the substratum over which they are spawned. Watson (unpublished manuscript) measured the settling rates of three ten-egg samples (thirty to thirty-four days post-fertilization) spawned naturally from three Hudson River females housed in laboratory aquaria. The settling rate through 25 cm of fresh water (2°C, total hardness about 300 mg/l) in modified McDonald jars (Bonn et al. 1976) was 1.4 cm per sec. The specific gravity of Atlantic tomcod eggs incubated in freshwater remains constant at 1.030, according to Peterson et al. (1980). For comparison, the semibuoyant eggs of striped bass from the Sacramento-San Joaquin

A) SINGLE MALE APPROACHING RIPE FEMALE

B) MALE ATTEMPTING TO ALIGN HIMSELF UNDER FEMALE

C) MALE ALMOST ALIGNED WITH FEMALE SO VENTRAL SURFACES COINCIDE (MALE USES PELVIC FINS TO ACHIEVE ALIGNMENT)

D) MALE ALIGNED WITH FEMALE WITH VENTRAL SURFACES COINCIDING

Figure 59. *Courtship (prespawning) behavior of Atlantic tomcod observed in laboratory aquaria.*

estuary settle at a rate of 0.24 cm per sec in freshwater (California Fish and Game 1974) and have a mean specific gravity of 1.0005 (Albrecht 1964).

There is less agreement as to whether Atlantic tomcod eggs are adhesive. Bigelow and Schroeder (1953), Scott and Crossman (1973), Ryder (1887), and Pearcy and Richards (1962) stated that the eggs are adhesive or mentioned the collection of egg clusters. Booth (1967) essentially agreed when he reported that eggs manually stripped from female fish may not be quite ripe and were less adhesive than those spawned under natural conditions. Peterson et al. (1980) observed that the eggs were weakly adhesive immediately after being stripped from the female and fertilized, but the adhesiveness disappeared when the eggs were separated. The opinion that Atlantic tomcod eggs are not adhesive is also scattered through the literature back to the late 1800s (Mather 1886, 1887, 1889, 1900; Nichols and Breder 1926; Breder 1948). Recently, Watson (1978) reported that tomcod eggs that were spawned naturally in aquaria or manually stripped when the females were running ripe were not adhesive. Eggs were adhesive if they were stripped before the females were fully ripe. Our observations agree with other first-hand observations that naturally spawned Atlantic tomcod eggs do not appear to be adhesive.

Adult tomcod consume their own eggs (Cox 1921, TI 1976, Nittel 1976, Grabe 1980). This source of egg mortality would tend to offset the advantage of a non-broadcast mode of egg discharge suggested by their courtship behavior, but the eggs are small, heavy, and deposited in shallow areas, so a large portion of the spawn probably sinks into the bottom where the probability of being eaten by the parents is reduced. Egg survival in Atlantic tomcod should therefore be relatively high based on observations of courtship and spawning behavior, the demersal characteristics of the eggs, and the relatively narrow range of temperature and flow fluctuations to which the eggs are exposed during the winter spawning season.

YOLK-SAC LARVAE Although Atlantic tomcod yolk-sac larvae were collected during all five years of our study, only in 1976 and 1977 did the sampling schedule seem likely to encompass the period of peak abundance and provide useful information on spatio-temporal distributions. Peak densities occurred in mid-March in both years, about the time the ichthyoplankton survey began. Densities decreased rapidly after mid-March to less than ten yolk-sac larvae/1000 m^3 by early April (Table 46).

Table 46. Regional density (number/1000 m^3) of Atlantic tomcod yolk-sac larvae, Hudson River estuary.

WEEK	YEAR	YONKERS	TAPPAN ZEE	CROTON-HAVERSTRAW	INDIAN POINT	WEST POINT	CORNWALL	POUGHKEEPSIE
Feb. 23–27	1976	101.6	77.3	65.1	77.8	51.2	9.9	3.4
Feb. 21–25	1977	7.1	15.6	16.6	37.7	66.4	52.8	47.6
Mar. 1–5	1976	12.8	76.7	107.2	123.4	126.0	83.1	61.0
Mar. 1–5	1977				No Samples Collected			
Mar. 8–11	1976	113.3	349.2	619.5	308.1	190.7	99.8	34.0
Mar. 7–11	1977	54.3	375.8	738.2	455.1	1,627.8	3,119.7	0
Mar. 22–25	1976	157.6	199.7	151.3	120.3	21.1	8.7	4.4
Mar. 21–26	1977	19.1	13.3	15.2	24.9	51.9	29.2	14.7
Apr. 5–8	1976	0.1	0	0	0	0	0	0
Apr. 4–7	1977	0	1.2	1.3	1.7	0.2	0.3	0

Sampling region

Yolk-sac larvae were distributed farther downstream in 1976 than in 1977 during periods of peak densities (Table 46). Several explanations for these annual differences are plausible, including annual variation in spawning location, early life-stage transport, regional variation in egg and larval survival, or a combination of these factors. Although the catches of adults in box traps did not suggest that spawning locations differed between the 1975–76 and 1976–77 seasons, freshwater flow rates at Green Island Dam were two to four times greater during January–February 1976 than during these same months in 1977. Annual differences in downstream transport rates of eggs and larvae is a plausible hypothesis explaining annual variation in yolk-sac larval distribution in the Hudson River estuary. Peterson et al. (1980) suggested that the drift of Atlantic tomcod eggs and larvae in a Canadian stream was directly associated with stream discharge. We have not evaluated an alternative hypothesis that regional variation in egg and larval survival influenced the differences in the distribution of yolk-sac larvae observed in the two seasons.

POST YOLK-SAC LARVAE Atlantic tomcod post yolk-sac larvae (postlarvae) were usually present in low densities when sampling began in late February or early March. Densities increased through March to maxima in early to mid-April (Figure 60). Transformation to the early juvenile life stage occurred from late April through mid-May, and very few postlarvae were collected after mid-May.

A few postlarvae were collected as far upriver as the Albany region (km 199–243), but high postlarval densities were almost never observed upriver of the apparent center of spawning activity near West Point. Postlarval densities were generally highest downstream in the Yonkers and Tappan Zee region (km 19–53). Larvae have also been collected further downriver in the vicinity of lower Manhattan (km 1–5) from late March through May (Dew and Hecht 1976, LMS 1980f). Thus, the lower boundary of our study area (Geroge Washington Bridge, km 19) was above the downriver limit of distribution of Atlantic tomcod larvae spawned in the Hudson River estuary.

There appeared to be a gradual displacement of larvae from the spawning area into the downstream regions (km 19–53) by early May. Other studies have shown that the downriver shift appears to accelerate during the transformation from the yolk-sac to postlarval stage (Cox 1921, Howe 1971). The extent of displacement of the postlarval population in the Hudson River estuary varied among the five years of our study (Figure 60). Several factors could explain these annual

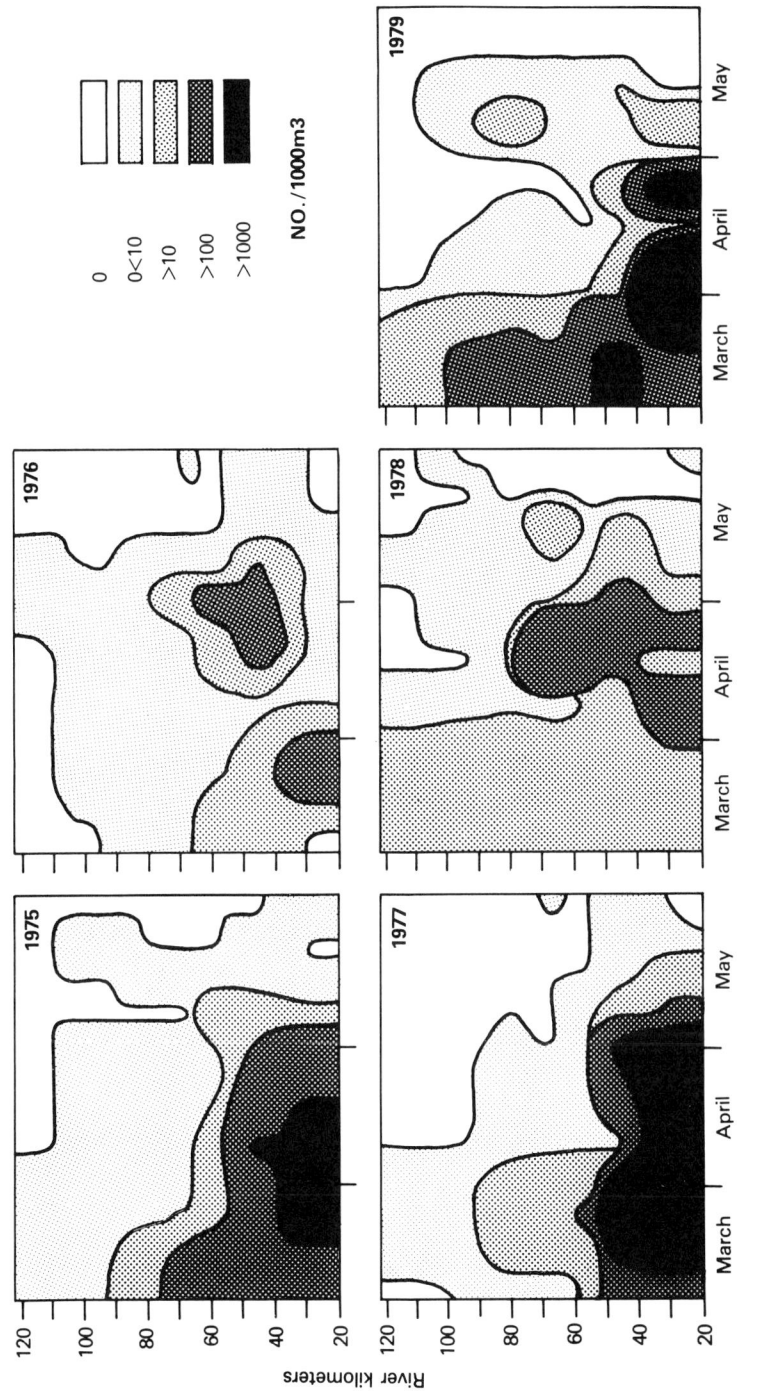

Figure 60. Density isopleths of Atlantic tomcod postlarvae collected by Tucker trawl and epibenthic sled, Hudson River estuary.

variations, but the most plausible explanations are shifts in the location of major spawning activity and differences in freshwater flow during February–April.

Catches of mature tomcod in box traps and data on stage of maturity in females showed that the major spawning area shifted annually either upriver or downriver around the West Point region (Figure 57). However, these spawning location shifts did not closely match annual variations in the distribution of postlarvae. For example, based on data from the adults, spawning activity in 1976–77 was apparently concentrated farther upstream than in other years (between West Point and Poughkeepsie, km 76–122), yet the resultant postlarvae were concentrated in the most downriver regions of the study area (Figure 60).

Variation in freshwater flow is a more plausible explanation for yearly differences in distribution of postlarvae. A typical annual freshwater flow cycle in the Hudson River contains two periods of rapidly increasing flows: a primary period from mid-February through April and a secondary period from early October through mid-December (Figure 61). Variations in the magnitude of freshwater flows during

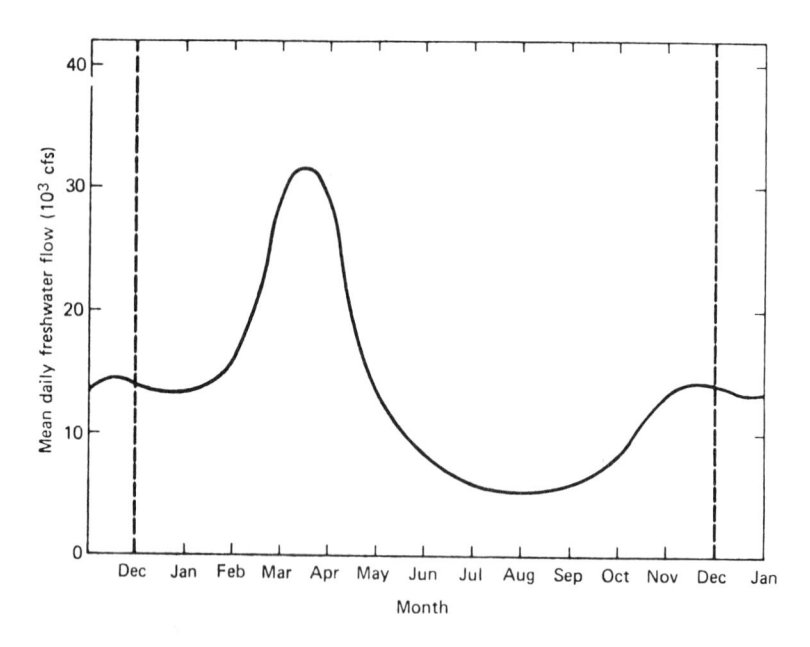

Figure 61. *Typical annual cycle of freshwater flow entering Hudson River estuary at Troy Dam (after Figure IV-5 of Texas Instruments 1976a).*

the primary period could explain yearly differences in downriver displacement of Atlantic tomcod postlarvae (assuming passive downstream transport of postlarvae).

Atlantic tomcod are about 5–6 mm in total length at hatching and immediately capable of active swimming (Booth 1967, Watson unpublished manuscript). Booth observed that newly hatched larvae swam toward the surface in bursts and then sank head first when swimming ceased. Peterson et al. (1980) observed that newly-hatched larvae were positively phototaxic and swam to the surface where they swallowed air, presumably to fill their swim bladders. Therefore, even though most Hudson River tomcod eggs likely remain in the substrate where they are spawned (Dew and Hecht 1976), newly hatched larvae appear to be capable of swimming into the water column and reaching the surface layers where they would be vulnerable to rapid downriver displacement during years of relatively high freshwater flow.

Using 1975–1979 data, we examined the hypothesis that high freshwater flow in the Hudson River estuary during February–April directly influenced the extent of downriver displacement of Atlantic tomcod postlarvae (Table 47 and Figure 62). Downriver displacement of postlarvae in 1976, 1977, and 1978 was directly associated with weighted freshwater flow during February–April, but the 1975 and 1979 data points suggest that other variables, perhaps water temperature, were also important influences on larval transport and downstream distribution. Daily water temperatures in February and March of 1975 and 1979 were slightly higher than in 1976, 1977 or 1978 (TI 1980a, 1980e, 1981b). These higher late winter water temperatures in 1975 and 1979 may have accelerated the development of tomcod eggs and larvae, and thereby extended the period of time larvae were available for downriver transport before the start of ichthyoplankton sampling in late March. If this occurred, a relatively large percentage of the postlarval population would have been transported downriver into the Yonkers sampling region before late March in 1975 and 1979, even though freshwater flows during February–April were not particularly high in those years. These hypotheses remain to be tested.

JUVENILES Juveniles were first collected in March. Densities peaked during the first three weeks in May, when juveniles were concentrated primarily in the Tappan Zee and Yonkers regions (Figure 63). Juveniles remained relatively abundant for two or three weeks after catches of postlarvae declined. Yearly variation in both spatial and temporal distributions of juveniles, although not presented

Table 47. *Data used to evaluate the hypothesis that during years of relatively high early spring freshwater flows Atlantic tomcod postlarvae in the Hudson River are displaced farther downstream than in years of relatively low flows.*

YEAR	EXTENT OF DISPLACEMENT[a]	AVERAGE DAILY FRESHWATER FLOW[b] FEBRUARY-APRIL	AVERAGE DAILY FRESHWATER FLOW[b] HIGH-FLOW WEEK	AVERAGE DAILY FRESHWATER FLOW[b] LOW-FLOW WEEK	WEIGHTING FACTOR[c]	INDEX OF FLOW POTENTIAL FOR DISPLACING LARVAE[d]
1975	64.52	2.309	3.893	1.159	3.359	7.759
1976	18.55	3.267	6.039	1.680	3.595	11.730
1977	69.51	2.913	8.911	0.681	13.085	38.134
1978	15.80	2.329	4.243	0.994	4.268	9.946
1979	73.95	2.961	5.099	0.981	5.198	15.400

[a] Spatial distribution index for postlarvae in Yonkers region, based on standing crops from last week in March through about mid-June (see Materials and Methods for details).
[b] Cubic feet/sec ($\times 10^4$) at USGS gauging station at Green Island (km 243), February-April.
[c] Average daily freshwater flow for high-flow week divided by average flow for low-flow week.
[d] Average daily freshwater flow (February-April) times the weighting factor.

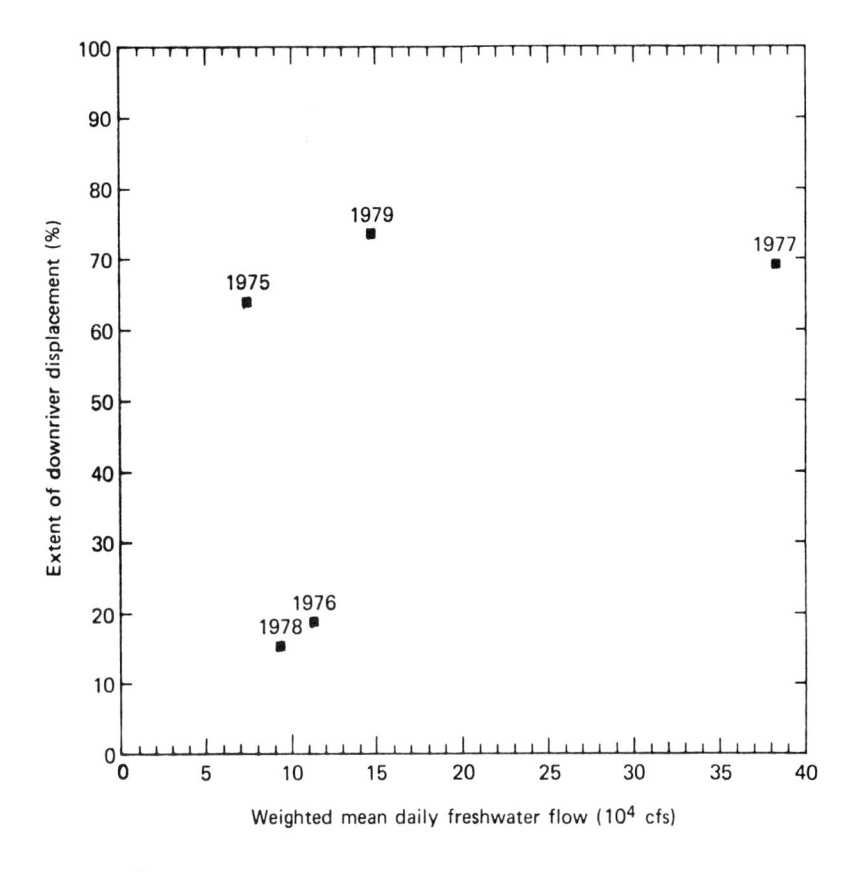

Figure 62. *Extent of downriver displacement of Atlantic tomcod postlarvae versus weighted mean daily flow entering Hudson River estuary during February–April at Troy Dam.*

here, mirrored the yearly variation in postlarval distributions (Figure 60). Catches of juveniles in the Tucker trawl and epibenthic sled declined rapidly during June, possibly because of increasing gear avoidance.

The distribution of juveniles from mid-April through November (as reflected in bottom trawl catches) varied from year to year (Figure 64). Juveniles, like postlarvae, were collected as far upriver as the Albany region, but the juveniles were most abundant between the Tappan Zee and West Point regions (km 39–88). Bottom trawls were not used in the Yonkers region (km 19–38) during all years; therefore, this region was excluded from yearly comparisons of juvenile distribution.

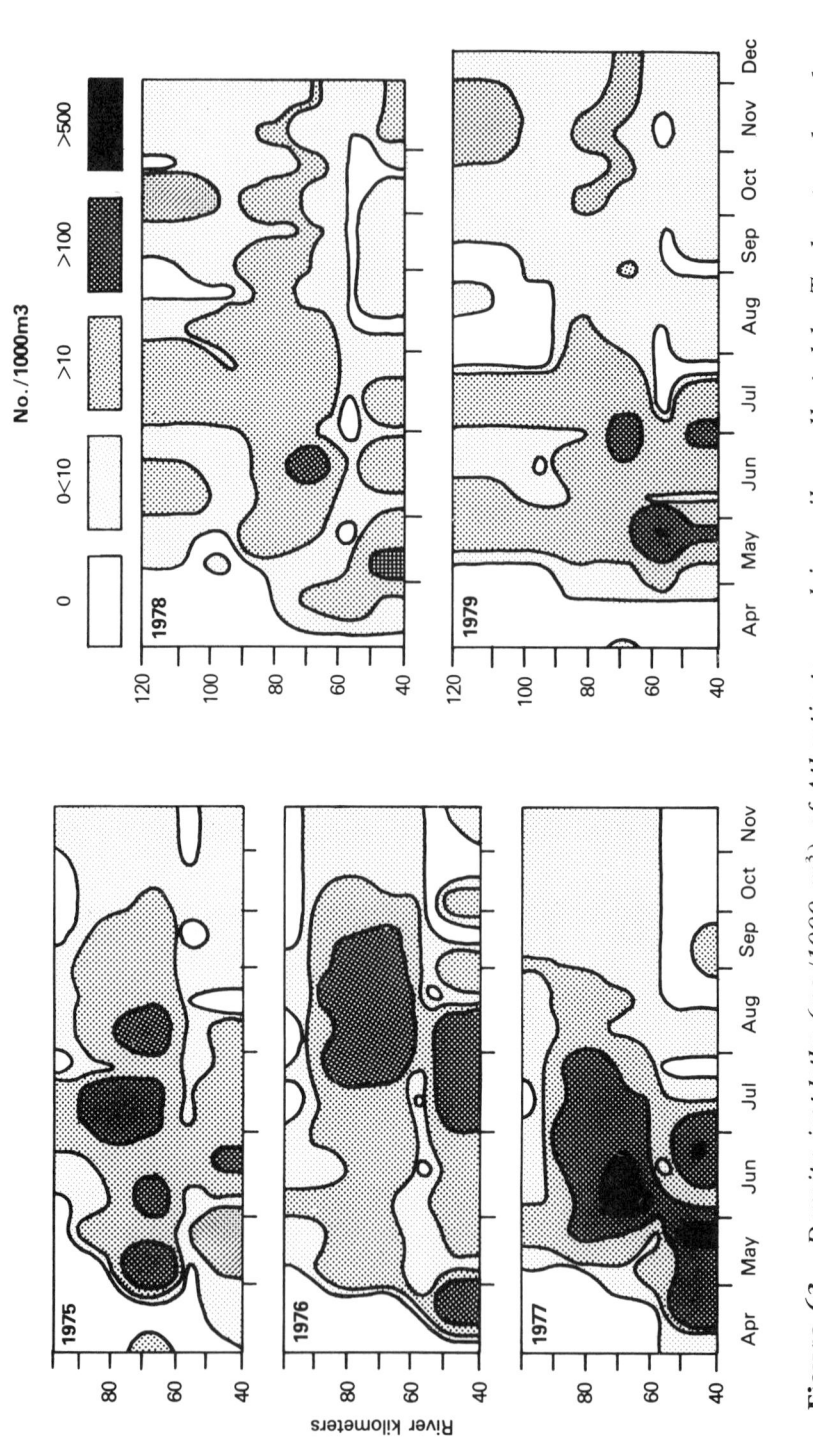

Figure 63. Density isopleths (no./1000 m³) of Atlantic tomcod juveniles collected by Tucker trawl and epibenthic sled, Hudson River estuary, 1975–79.

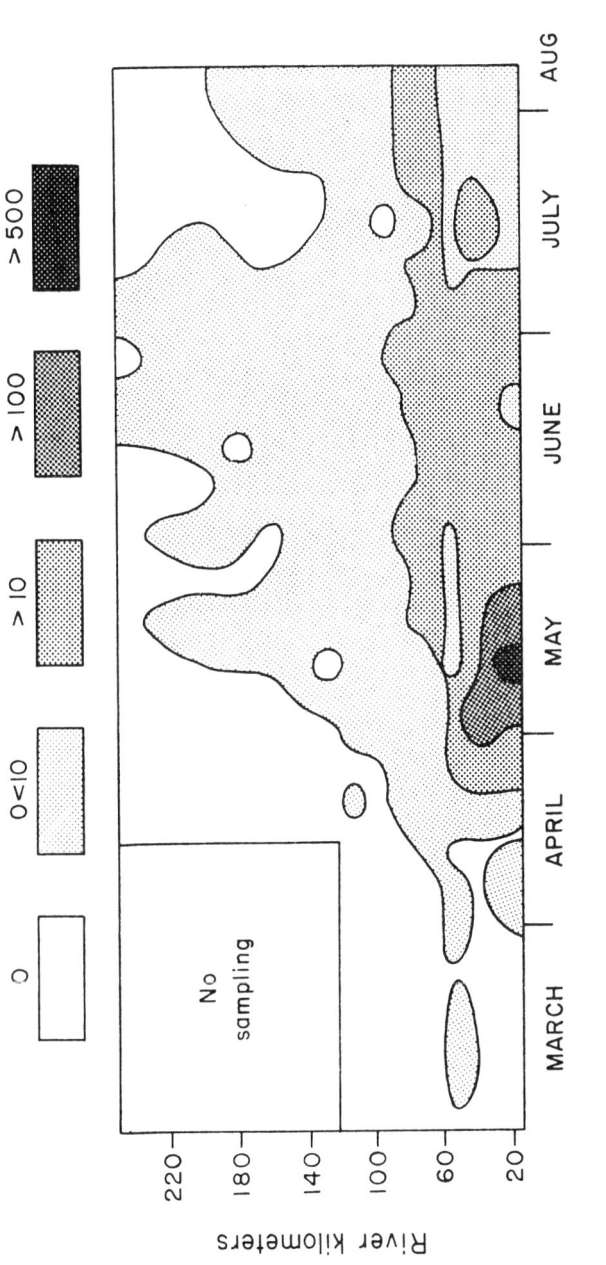

Figure 64. *Catch-per-unit-effort isopleths of Atlantic tomcod juveniles collected by bottom trawl, Hudson River estuary.*

As with postlarvae, an unknown portion of the juvenile tomcod population was located downriver of our study area. Dew and Hecht (1976) reported concentrations of postlarvae and juveniles between the Tappan Zee Bridge (km 43) and Manhattan (km 5) during April and May 1975. TI (1976h) collected larvae and juveniles during April–July 1976 in the vicinity of Liberty and Ellis Islands in New York Harbor. Juveniles were concentrated in the downstream regions of our study area (km 38–61) in April and May (Figure 60). In early summer, juveniles moved upriver, and by July the highest catches were typically in the Indian Point and West Point regions (km 62–88). This upriver movement of juveniles appears to be associated with the upriver movement of the salt front. Freshwater flows in the Hudson estuary are lowest during summer (Figure 61) and salinity increases in the downriver half of the study area as the summer progresses. Juvenile Atlantic tomcod prefer brackish water (Dew and Hecht 1976, TI 1980a) and apparently move with the cooler brackish water as it intrudes upriver.

Timing of the upriver movement of juveniles varied among years. In 1976, juveniles were abundant in the most downriver region of the study area well into August, and the summer upriver movement occurred at least a month later than during other years. The delayed upriver movement in 1976 may have been a response to higher than normal summer freshwater flows which retarded the upriver intrusion of brackish water (Table 47). In 1975, the upriver movement of juveniles occurred earlier than in other years. Juveniles were concentrated in the Yonkers and Tappan Zee regions (km 19–53) during May (Figure 63), but by late May–early June, juveniles were most abundant upriver (Figure 64) in the Indian Point and West Point regions (km 62–88). Freshwater flows in spring and early summer of 1975 were similar to 1977–79, so environmental factors other than freshwater flow and salinity may also influence juvenile distribution and movements during the summer. In 1975, water temperatures declined sharply in mid-April and were several degrees below the average water temperatures in other years through mid-May (TI 1981b). Perhaps these lower temperatures somehow contributed to the relatively early upriver movement of juveniles in 1975.

The role of summer water temperatures in the biology of Hudson River Atlantic tomcod deserves further investigation, especially since the Hudson population is near the southern boundary of the species' range. Maximum water temperatures in the Hudson River estuary usually exceed 26°C in late July or early August (TI 1981b). EA (1978a) reported an upper incipient lethal temperature of 26.8°C for juvenile Atlantic tomcod acclimated at 24.0°C. The preference of Atlantic tomcod for bottom areas (Pearcy and Richards 1962, TI 1980a) and brackish water, where cooler temperatures occur, may tend to minimize some of the potentially adverse effects of high summer temperatures in the Hudson estuary. TI (1980e) showed that the growth of juvenile tomcod was inversely related to water temperature. Summer temperatures also appear to be an important determinant of year-class success (TI 1981b).

ACKNOWLEDGEMENTS

We are indebted to the men and women who labored under occasionally hazardous and frequently uncomfortable conditions on the Hudson River estuary to collect the data presented in this paper. Richard H. Peterson provided helpful comments and a critical review of the manuscript. Special thanks are extended to Thomas H. Peck for his inspiring enthusiasm for *Microgadus tomcod*, and to Kathy Schmidt who prepared several illustrations. Beverly Knee patiently typed several versions of the manuscript. Data analyses for this paper were completed by the three authors while employed by Texas Instruments Incorporated.

Part VI
Food Chains

9. Food Habits of the Amphipod Gammarus tigrinus in the Hudson River and the Effects of Diet Upon Its Growth and Reproduction

Gerald V. Poje, Stacey A. Riordan, and Joseph M. O'Connor

INTRODUCTION

Estuaries in temperate zones are the sites of complex interactions between freshwater and marine conditions. They are subject to highly predictable changes in naturally occurring physical and chemical conditions, such as: temperatures that reflect seasonal changes in solar radiation; current directions that change with seasonal runoff patterns and tidal cycles; dissolved oxygen that fluctuates with temperature, depth, organic nutrient load, and phytoplankton productivity; and variations in freshwater flow that alter salinity profiles and the concentration of organic particulates (Vernberg and Vernberg (1976).

Ecologically and evolutionarily, most estuarine animals are considered stable and efficient users of this environment (Carriker 1967). The wealth of nutrients and high rates of recycling foster higher productivity of biomass than in either freshwater rivers or adjacent marine environments. However, in contrast to these other systems, species diversity of estuaries is characteristically low (Remaine and Schlieper 1971). True estuarine species occupy the entire length of the estuary, dominating the fauna in waters of low and variable salinity. Typically, the macrobenthic fauna is dominated by small crustaceans. In a survey of the Knysha estuary in South Africa, amphipods and decapods were the dominant animals on an individual basis, but they accounted for only 9% of the 310 species collected (Day 1967). In studies of estuaries bordering the North Atlantic Ocean, small amphipods constituted a major part of the macrofauna (Cronin et al. 1962, Spooner 1947).

Several amphipod crustaceans of the genus *Gammarus* are particularly well adapted to estuarine environments. Members of this genus are often the dominant macroinvertebrates of shallow-water and inshore coastal areas of the eastern and western North Atlantic coasts (Bousfield 1969). In the Hudson River estuary, abundances of *Gammarus* as high as 52,000 individuals per 1000 m^3 have been recorded in plankton tows; seasonally, more than 30% of all animals in the oligohaline parts of the river belong to this genus (New York University Medical Center 1975).

The population dynamics of gammarids have been documented for several European and American species (Eastern Atlantic: Dennert et al. 1968, Girisch et al. 1974, Smit 1974; Western Atlantic: Gable and Croaker 1977, Reese 1975, Van Dolah et al. 1975). Growth, moulting frequency, fecundity, and heartbeat as a function of temperature and salinity have been described for several European species (Dorgelo 1974, Kinne 1960). Ginn (1977) measured growth rates and fecundities of individual *G. daiberi* native to the Hudson River estuary. Additional aspects of reproductive potential have been investigated both in natural populations (Cheng 1942, Steele and Steele 1973, 1975) and in laboratory cultures (Clemens 1950, Poje 1977, Sexton 1928). Brooding behavior and embryonic development have also been examined (Sheader and Chia 1970).

Certain gammarids are particularly adapted to estuarine situations with fluctuating salinities. Lockwood (1962) determined that *C. daiberi* produced urine isotonic with the blood in seawater, while in more dilute media the urine was hypotonic to the blood. Werntz (1963) found that the brackish water species *G. tigrinus* could regulate the salt concentration of its urine over a wider range of salinities than either freshwater or saltwater congeners.

On an ecological level, the importance of *Gammarus* as a food resource for highly productive estuarine fisheries has been recognized (TI 1975c). The diets of gammarids vary; in the Black Sea, the brackish water species *G. locusta* has a diet including such diverse items as diatoms, rooted macrophytes, polychaetes, and copepods (Greze 1968). A field survey of several Canadian freshwater lakes has revealed the ability of *G. lacustris* to compete as a microzooplankton predator with the ubiquitous larvae of the dipteran *Chaoborus*, and to eliminate it from several systems (Anderson and Raasveldt 1974). In contrast, a laboratory investigation demonstrated the capability of *G. pulex* as a detritivore to reduce the growth rate and survival of syntopic caddisfly larvae (Trichoptera) when they were cocultured on a diet of leaf litter (Nilsson and Otto 1977).

Gammarids are well adapted to utilization of cellulose-based detritus. In a survey of twenty-six species of freshwater invertebrates, Monk (1976) demonstrated that the highest levels of cellulose were found in these amphipods. *Gammarus mucronatus* feeds both macrophagously and microphagously on living macroalgae and detritus in seagrass communities (Zimmerman et al. 1979). Both *G. minus* and *G. pseudolimnaeus* have distinct preferences for detrital particles that have been colonized by bacterial and fungal flora (Barlocher and Kendrick 1973, Kostalos and Seymour 1976); assimilation efficiency is greatest on a diet of fungal mycelia (Barlocher and Kendrick 1975).

The focus of this study is the American species *Gammarus tigrinus*. It is an estuarine species endemic to Western Atlantic tributaries from southern Labrador and the Gulf of St. Lawrence south to Albemarle Sound, the Pamlico River, and sporadically southward to Florida (Bousfield 1958, 1973). It is the most abundant amphipod in North Carolina estuaries (Williams and Bynum 1972). The species also possesses a remarkable ability to extend its range. After accidental importation, it successfully invaded northern European estuaries and displaced native species. Presently it is found in the English midlands and northern Ireland to the Netherlands and Germany (Holland 1976, Pinkster 1975).

Since *G. tigrinus* is one of the most abundant and widely distributed estuarine organisms, details of its biology are fundamental to an understanding of nutrient cycling and productivity in estuaries. As part of a baseline study to assess the ecological importance of *G. tigrinus*, the following aspects of its biology were investigated:

1. length-weight relationship;
2. age-specific growth rates;
3. feeding ecology of the species and the resultant effects of diet on growth and reproductive parameters; and
4. reproductive biology and the effect of female size on fecundity.

METHODS AND MATERIALS

Organisms and Culture Methods

Adult *Gammarus* spp. were sorted from floating mats of *Myriophyllum spicatum* in the Hudson River at Verplanck, New York, approximately forty miles above the river mouth. They were identified to species with the key by Bousfield (1973) and cultures were started with identifiable adult *G. tigrinus* (>7.0 mm). Organisms were maintained in 10-gallon aquaria in water from Sterling Lake in Tuxedo, New York (Table 48). Salinity was brought to 2 by adding commercial sea salt (Forty-fathoms Marine Mix) and held at this concentration by maintaining a constant tank level with distilled water. Laboratory temperature was 22°C, and the tanks were kept in continuous light.

The animals were fed a mixed diet of aquatic macrophytes (*Myriphyllum spicatum* and *Anarchis* sp.), filamentous algae (*Mougotia* sp.), diatoms (*Nitzchia* sp.), and palmella-type blue-green algae. Biweekly feedings of commercial fish food (Tetramin) supplemented the diet with animal protein.

Animals were measured to the nearest 0.01 mm with a calibrated ocular micrometer. Weight, to the nearest microgram, was determined after oven-drying at 60°C for twenty-four hours, with a Model 4400 Cahn Electrobalance.

Growth in Culture

The length-weight relationship was determined from measurements of 199 culture animals selected so that each 0.5-mm size class

Table 48. *Chemical analysis of Sterling Lake water.**

	mg/l
Total alkalinity	4.0
Hardness	18.0
Total nitrogen	0.12
Sodium	1.8
Potassium	1.0
Calcium	4.5
Magnesium	1.3
Sulfate	11.0
Chloride	4.0
Iron	0.03
Manganese	0.01
Zinc	0.01
Copper	0.06
Strontium	0.03
Salinity	29.6
Oxygen	7.8–11.0
Carbon dioxide	0.5–3.5
pH	7.2

*Values from Bath, 1973.

from 1.5 mm to 13 mm was represented by approximately ten specimens. Curves were calculated from individual data points and means and standard deviations were determined for lengths and weights in each 0.5-mm size class.

To determine growth rates, twenty gravid females were placed together and their young collected over a period of two days. The young ($n = 175$) were held in 15 liters of culture medium, and a sample of ten individuals was selected randomly at each of eight evenly spaced time intervals over a period of fifty-seven days. Each individual was weighed and measured, and regressions were calculated from individual data points for age versus length and age versus weight. A polynomial regression of length, weight, and age was determined from mean values.

Reproductive Capacity

The number of young released versus the length of the female was determined by individual analysis of 175 females. The mean number of young produced was obtained for each 0.5-mm size class from 4.5 mm to 11.5 mm, and these data were used to calculate a standard power curve. Observations of reproductive behavior were made on this group of animals.

Dietary Effects

Three diets were chosen to test the effects of nutrition on growth and reproduction. Fallen maple leaves were collected in October 1979. Leaves were soaked in distilled water for forty-eight hours, then air dried at room temperature. The water was replaced at frequent intervals during this procedure to minimize bacterial digestion and prevent accumulation of tannic acids.

The second food source was living parthenogenic female *Daphnia pulex*. These cladocerans (approximately 3.5 mm long) were collected from spring-bloom populations in 1976 and maintained in culture.

The third diet was composed of diatoms (*Nitzchia* sp. and macrophytes (*Myriophyllum spicatum*). This herbivorous diet approximated the culture conditions and the substratum from which the amphipods were originally collected.

Growth

Thirty-three adult *G. tigrinus* were collected from culture and placed individually in 10-cm diameter bowls with 200 ml culture medium and food. As each animal moulted, its head-to-abdomen length was measured and the individual was transferred to a bowl containing one of the three diets. The organism remained on the diet for one intermoult period and was measured again after the second moult. The increases in lengths were compared by the paired t-test (Sokal and Rohlf 1969).

Reproduction

Dietary effects on reproduction were assayed using the three sources. Thirty-three nonovigerous females were randomized into individual culture bowls containing 200 ml of culture medium and food. Males were introduced, and duration of amplexus and day of female moult were noted. The males were then removed and the measured ovigerous females were placed on one of the three diets.

The interval until release of young (second moult of female) was recorded, as was the number of young released. Males were then reintroduced and the duration of amplexus noted. After copulation, the males were removed and the females were measured. Duration of intermoult and number of young were determined at the third moult. The data were tested for significance by a paired t-test (Moult 2 versus Moult 3). The difference in numbers of young produced between moults on each diet, differences in duration of amplexus between moults, and differences in duration of brood time were compared by one-way analysis of variance (Sokal and Rohlf 1969).

RESULTS

Growth

In cultures supplied with an unlimited diverse food supply, *G. tigrinus* had a length-weight relationship defined by the power curve:

$$y = 0.0066 x^{2.732}$$

Where: y = weight in mg and x = length in mm (n = 199).

Although statistically valid confidence intervals are not available for a curvilinear regression, means and deviations for lengths and weights were computed for each 0.5-mm size class. A conservative estimate of the degree of variation in the data over the size classes selected is indicated by relative standard deviations (RSD: the ratio of the standard deviations to their respective means). For all size

FOOD HABITS

Table 49. *Mean lengths and weights, their standard deviations for* Gammarus tigrinus *at sampling dates after release from females.*

DAY AFTER RELEASE	MEAN LENGTH (mm)	STANDARD DEVIATION	MEAN WEIGHT (mg)	STANDARD DEVIATION	N
1	1.45	±0.083	0.017	±.003	10
7	2.00	±0.346	0.026	±.010	10
14	2.75	±0.443	0.007	±.032	10
21	3.69	±0.974	0.209	±.134	10
28	4.81	±0.824	0.400	±.170	10
35	5.59	±0.974	0.623	±.281	10
42	5.92	±0.766	1.226	±.658	10
49	7.78	±1.039	1.677	±.629	10
57	9.59	±0.943	3.030	±.612	10

classes ($n = 20$), the RSD for weight data was only 0.175, and for the length data, 0.033.

The lengths and weights of ten animals were measured on each of nine dates after initial release from females to define growth in terms of age (Table 49). This relationship is expressed by the following linear regression:

$$y = 0.090 + 0.143 x$$

Where: y = length in mm and x = age in days after release.

The dietary regimen of the amphipods has a variable effect on the individual's growth (Table 50). For *Gammarus* that were fed a diet of maple leaves the increase in length was 0.30 ± 0.18 mm at the second moult. For those feeding on algae and macrophytes the increase was 0.47 ± 0.20 mm. On the carnivorous diet, the mean

Table 50. *Average increases in length, standard deviations, paired t-statistics, and significance* of groups of animals before and after placed on diets.*

DIET	MEAN INCREASE	DEVIATION	t	df	P
Detritus	0.30	0.18	5.56	10	<.001
Herbivorous	0.47	0.20	7.82	10	<.001
Carnivorous	1.60	0.48	11.11	10	<<.001

*Significance = probability that there was no difference in means before and after placed on diet.

increase was 1.60 ± 0.48 mm. When compared to values at the moult prior to being placed on the respective diets, all animals showed significant increases in length ($P < .001$).

Analysis of variance of data at the initiation of the dietary regimes (first moult) demonstrated that animals were randomly distributed in the test, since there were no differences in length between the groups. At the second moult, after an average of eight days on their respective diets, the animals on the carnivorous diet had a mean increase in length greater than the others by a factor of 4 ($p < .01$; Table 51).

Reproduction

Males possess one pair of breeding papillae on the ventral surface of the seventh peraeon. In addition, their gnathopods grow at a rate faster than either females or immatures. The second gnathopod is more massive than the first. The ultimate indicator of maturity, functional testes, is not readily discernible, since the testes are visible only after staining.

Among female *G. tigrinus* reproductive maturity is signaled by several anatomical changes. The ovaries become functional and filled with dark green ova. These organs are paired and extend from the first to the seventh peraeon, lying above the intestines and below the heart. Peraeopods two to five develop brood plates which are setose sheets of cuticle projecting ventro-medially from the body.

In the estuary, late fall and winter populations of *G. tigrinus* have a reproductive hiatus. In these colder temperatures, growth is reduced to a minimum and reproduction ceases. Amplectic pairs are found again in late spring. This precopulatory condition consists of the larger male, presumably responding to a chemical stimulus, grasping the dorsal surface of the female. The most commonly observed amplection in the laboratory is one in which the left second gnathopod is reversed and hooked under the posterior portion of the fifth peraeon segment of the female. The right first gnathopod is hooked behind the head of the female. Other combinations of positions of the right and left gnathopods are also observed. The duration of amplexus varies with the temperature, but usually ranges from twenty-four to seventy-two hours in laboratory cultures at 22°C. Feeding occurs during amplexus, but the female remains passive; the male forages for food. Pairs amplex regardless of the female reproductive state. At 22°C, among most pairs the female is already ovigerous.

Table 51. *Mean lengths, standard deviations, one way analysis of variance, F-statistics, and probabilities that animals show the same response to different diets.*

DIET	MEAN LENGTH	DEVIATION	N		SS	df	MS	F	P
				COMPARISON OF LENGTHS AT FIRST MOULT					
Detritus	8.10	0.71	11	Treatments	2	0.564	0.282	0.473	N.S.*
Herbivorous	7.82	0.91	11	Errors	17.898	30	0.597		
Carnivorous	7.83	0.68	11	Total	18.462	32			
				COMPARISONS OF LENGTHS AT SECOND MOULT					
Detritus	8.40	0.69	11	Treatments	8.648	2	4.323	7.662	.003
Herbivorous	8.29	0.93	11	Errors	16.931	30	0.564		
Carnivorous	9.43	0.59	11	Total	25.579	32			

*Not significant.

Copulation occurs after the female moults. While her body is pliable, the male shifts the female to a position perpendicular to his body length. He pushes the female's peraeopods forward and her pleopods backward with his left and right peraeopods, thus exposing the brood pouch. The female responds by arching her body to bare the region of the oviducts. He receives the sperm from the genital papillae onto his pleopods, which he then thrusts into the brood pouch with several quick forward motions. The act is completed by a kicking motion of the male uropods that breaks the amplexus and releases the female.

Within an hour after copulation, the female extrudes ripened ova through the oviducts into the brood pouch where they are fertilized. The ovoid ova (0.26 mm long × 0.11 mm wide) are not rigid and can be elongated for passage through the oviducts (0.09 mm diameter). In the brood pouch the eggs (0.42 mm long × 0.32 mm wide) harden and holoblastic cleavage proceeds. The young hatch as immatures and are retained by the female until she moults.

In the estuary, breeding occurs from late spring until fall. Overwintering females and females released in the spring are capable of producing several broods during the season; at 22°C, a female can release a brood every eight to ten days. Female *G. tigrinus* were isolated from cultures and maintained individually when they were observed to be ovigerous. They were measured and allowed to release young. The relationship between female length and number of young produced was determined from mean values for sixteen size classes from 4.8 mm to 11.5 mm. The power curve regression was:

$$y = 0.026x^{4.147}$$

Where: y = number of young produced and x = female length in mm.

Diet primarily affected one of the three parameters of the reproductive process. Gammarids feeding as detritivores on maple leaves showed no significant difference from preexperimental duration of intermoult or in number of young produced. There was, however, a significant ($p < 0.05$) decrease in duration of amplexus when the animals paired with a male on the same diet (Table 52).

Amphipods feeding as herbivores showed no change in duration of amplexus, but there was an increase of 3 in the average number of young produced. ($P < 0.05$; Table 52).

Table 52. *Mean differences, t-statistics for comparisons, and probabilities that diet change affected parameters of reproduction.*

A. Amplex duration (days) before experimental diet compared to that while on diet.

	MEAN DIFFERENCE	df	t	p
Detritus	0.64	10	1.88	<0.1
Herbivorous	0.36	10	0.94	N.S.
Carnivorous	0.18	10	0.48	N.S.

B. Duration of intermoult (days) before experimental diet compared to that while on diet.

	MEAN DIFFERENCE	df	t	p
Detritus	−0.27	10	−1.40	N.S.
Herbivorous	−0.18	10	−0.61	N.S.
Carnivorous	−0.08	10	−0.43	N.S.

C. Number of young released before experimental diet compared to that while on diet.

	MEAN DIFFERENCE	df	t	p
Detritus	−1.18	10	1.17	N.S.
Herbiverous	3.00	10	2.74	<.02
Carnivorous	4.82	10	4.72	<.001

Amplexus and intermoult durations were unchanged on the carnivorous diet; however, there was a statistically significant mean increase of 4.82 young ($p < .001$; Table 52).

Analysis of variance (Table 53) revealed no change in duration of intermoult or duration of amplexus, but animals on the detritus diet produced fewer young than those on the other diets.

DISCUSSION

Epibenthic and planktonic amphipods of the family Gammaridae are some of the most successful organisms of temperate estuarine and coastal marine ecosystems, particularly in the North Atlantic. This is due to the success of the members of this genus in maximizing their niche breadth to include virtually every aspect of the widely varying environments of oligohaline estuarine areas. These amphipods are highly productive on diets as diverse as macrophytic detritus, filamentous algae, and microzooplankters. Diverse substrata do not present limiting factors to the populations, since gammarids are distributed

Table 53. *Means, standard deviations, one way analysis of variance tables, F-statistics, and probabilities that animals show the same response feeding on the three diets.*

DIET	MEAN DIFFERENCE	s		SS	df	MS	F	p
			Difference in duration of amplexus (before and during diet) (days)					
Detritus	−0.64	1.12	Treatments	1.15	2	0.58	0.39	N.S.*
Herbivorous	−0.36	1.29	Errors	44.73	30	1.49		
Carnivorous	−0.18	1.25	Total	45.88	32			
			Difference in duration of intermoult (before and after diet) (days)					
Detritus	0.27	0.65	Treatments	1.27	2	0.64	0.98	N.S.
Herbivorous	−0.18	0.98	Errors	19.45	30	0.65		
Carnivorous	0.09	0.75	Total	20.73	32			
			Difference in number of young produced (before and after diet)					
Detritus	−1.18	3.34	Treatments	208.24	2	104.12	8.74	<.001
Herbivorous	3.00	3.63	Errors	357.27	30	11.91		
Carnivorous	4.82	3.37	Total	565.52	32			

*Nnot significant.

as epibenthos on sands, gravels, and muds (Spooner 1947). Reproduction occurs over a wide range of temperatures and salinities, and maternal protection is provided for the eggs. These characteristics encourage high abundances of *Gammarus* over a long season and throughout the estuary.

Growth

The culture conditions used in these experiments approximate ambient summer conditions in many temperate estuaries. At 22°C, and with an unlimited supply of aquatic flora, animals grew at an average rate of 1 mm per week, approximately doubling their weight at the same time. Embryonic development required eight to ten days.

Ginn (1977) demonstrated that the sympatric species, *G. daiberi*, grew to 6 mm in sixty days at 25°C, considerably slower than *G. tigrinus*. The European euryhaline species, *G. zaddachi*, has a faster growth rate, achieving a mean length of 6 mm in approximately twenty days at 20°C culture temperature (Kinne 1960).

Although growth rates vary among these three species, all indicate the potential for rapid increase in biomass. The differences become less significant, however, when the size at sexual maturity is considered. *Gammarus tigrinus* reproduces at 5 mm, *G. daibei* at 4 mm, and *G. zaddachi* at 7 mm. In each case, sexual maturity is reached in twenty-eight to thirty days, regardless of the species and slight differences in culture temperatures. One of the keys to the abundance of *Gammarus* is this ability to reach reproductive maturity in one month's time.

Diet has a pronounced effect on the growth rate but *Gammarus tigrinus* can use diverse food sources and still achieve high rates of growth. Detritus and algae were assimilated to about the same extent and resulted in length increases of 0.3 mm and 0.5 mm over a period of one moult, but as carnivores they grew 1.6 mm per moult. This exceeds the growth rate for mass cultured organisms, suggesting that animal protein is necessary for optimum growth.

Reproduction

The size of *Gammarus tigrinus* eggs is comparable with congeneric species (Steele and Steele 1975) and the duration of egg development corresponds to that reported for other species of *Gammarus* (Clemmens 1950, Ginn 1977, Kinne 1971) and other amphipods (Shyamasundari 1976). Steele and Steele (1975) demonstrated that egg size in crustaceans was related to a number of factors which determine population dynamics, including size at maturity, number of broods, number of eggs per brood, timing of the release of young, and the fitness of the young. Amphipod eggs are relatively large and the development time is relatively short. In addition, the young are released as immature adult forms. This limits brood size and frequency in amphipods, but confers a greater ability for the young to cope with their environment (i.e., the young are larger, better developed, and more able to occupy the adult niche).

This egg-size and development relationship may explain, in part, why *Gammarus* and other amphipods are especially successful in higher latitudes. In these regions environmental temperatures are low and optimum conditions for the growth of the young are temporally limited; only one brood of young may develop successfully each season. Fitness of the young is an overriding reason for faunal dominance.

Conversely, at lower latitudes the environment can support several broods per year and species producing large numbers of rapidly developing small eggs are selectively favored, even though the young may be less capable of survival (Van Dolah and Bird 1980).

In addition to their large eggs, *Gammarus* fecundity is maternal-size dependent, a characteristic that varies among species (Table 54). Although each species can produce up to thirty young, estuarine species have a greater reproductive potential (Ginn 1977, Hynes 1955).

Table 54. *Power curve regression of female length ($\bar{X}$) versus number of young produced (Y), coefficients of correlation (r^2); approximate size at onset of maturity (X_m); young production at 10 mm length (X_{10}) for several species of* Gammarus.

SPECIES	SALINITY PREFERENCE	POWER CURVE	r^2	X_m	X_{10}
G. lacustris[a]	fresh water	$Y = 0.421 \times X^{1.615}$	0.99	9.5	17.35
G. fasciatus[b]	fresh water	$Y = 0.144 \times X^{2.266}$	0.99	5.5	26.57
G. zaddachi[c]	estuarine	$Y = 0.004 \times X^{3.362}$	0.99	7.0	9.21
G. tigrinus	estuarine	$Y = 0.003 \times X^{4.147}$	0.96	5.0	42.08
G. palustris[d]	estuarine-marine	$Y = 0.029 \times X^{3.129}$	0.98	4.0	39.03

From data presented in:
[a]Hynes, 1955
[b]Clemens, 1950
[c]Kinne, 1960b
[d]Van Dolah et al., 1975

The oligohaline *G. lacustris* and *G. fasciatus* had mean fecundities of 28.2 and 36.0, respectively, at their upper size limits. The marine salt marsh species *G. palustris* showed mean production of approximately thirty-two young at their longest body lengths. The two estuarine species, *G. zaddachi* (from Kiel, northern Germany) and *G. tigrinus* (from the Hudson River) had upper mean fecundities of fifty-one and fifty-seven young, respectively.

Our dietary results, combined with the behavioral observation that these amphipods will rapidly evacuate their digestive tracts of plants and detritus in order to consume the preferred animal diet, indicate that *Gammarus* population dynamics in the estuary may reflect the seasonal availability of the planktonic fauna upon which they prey.

SUMMARY

Among the factors that make *Gammarus tigrinus* one of the most abundant organisms in temperate oligohaline waters are the following:

1. It has an efficient osmoregulatory system that enables it to accommodate to extreme fluctuations in salinity (Werntz, 1963). This allows it to thrive in a habitat that is physically and chemically restrictive for most organisms.

2. Its reproductive ability is unequaled by freshwater or marine amphipods. At favorable temperatures, *G. tigrinus* matures rapidly and quickly increases its population size. In areas where it has recently expanded its range (e.g., Holland and northern Germany), *G. tigrinus* successfully competed with native species (such as *G. zaddachi*) which do not attain maturity at as small a size.

3. *Gammarus tigrinus* displays flexibility in utilizing diverse food sources. Estuaries are charactertized by seasonal fluctuations in quality and quantity of productivity. Nutrient supplies are usually not restrictive in estuaries but most available organic nutrients are distributed among allochthonous particles from terrestrial primary productivity, phytoplankton, microzooplankton, and the microdetritivores. Although growth and reproduction proceed on diets of detritus or algae, growth rates, and production increase significantly when animal protein is available. During this study there were no obvious dietary distinctions attributable to a particular size class of gammarids. Newly released young consume small cladocerans as voraciously as adults, and both can maintain themselves on detritus. Since the characters distinguishing *Gammarus* species are slight and usually not associated with

functional feeding morphology (Bousfield 1973), it is conceivable that the smaller but more abundant *G. tigrinus* could utilize these invaded estuaries more efficiently, thereby displacing larger native species. Future studies of competition among sympatric species of *Gammarus* under various dietary regimes would provide useful information about the niche breadth of "omnivorous" epibenthic organisms.

The ability of gammrids to change to an animal diet when it is available may play a major role in energy flow and community structure in estuarine ecosystems. For example, *G. tigrinus* in culture have been observed preying upon eggs and larvae of striped bass and, although this predation varied between observations, generally consumed one or two yolk-sac larvae daily (eighty observations on individual *Gammarus*). Striped bass and many other fish species utilize estuaries as nurseries for egg production and larval growth. Since abundances of *Gammarus* usually exceed those of ichthyoplankton by several orders of magnitude (New York University Medical Center 1975), these amphipods may play a significant role in cropping the early life history stages of fishes that spawn in estuaries, especially forms with demersal eggs or whose planktonic eggs and larvae occur predominantly at or near the bottom.

Part VII
Pollution

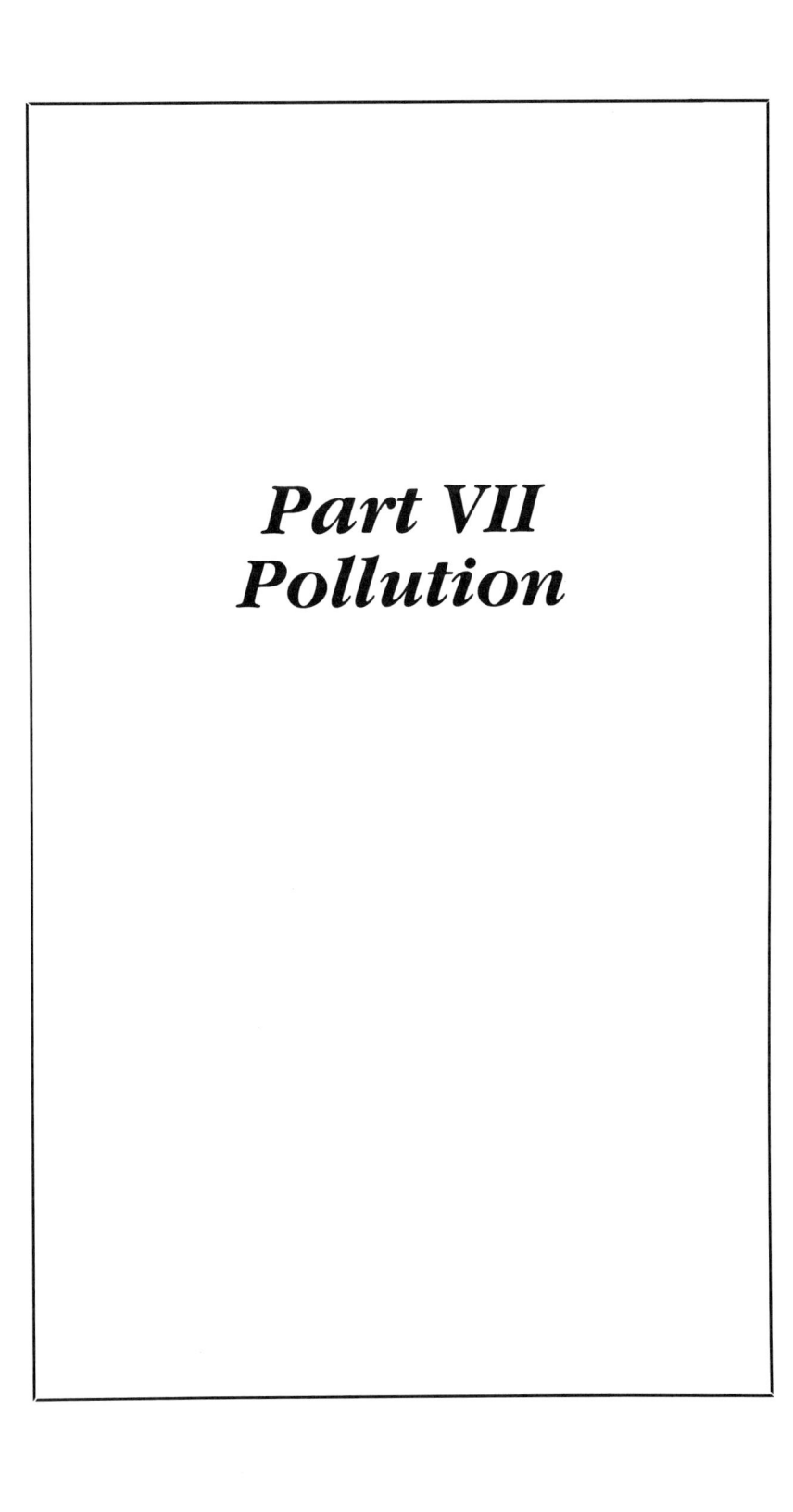

10. Heavy Metals in Finfish and Selected Macroinvertebrates of the Lower Hudson River Estuary

Stephen J. Koepp, Edward D. Santoro, and Gerard DiNardo

INTRODUCTION

Heavy metals are of particular concern in the aquatic environment because of their persistence and potential toxicity. Metal contaminants within the Hudson River can be traced to both industrial and municipal discharges (USEPA 1978). Specific sources include pulp and paper mills, petroleum storage and refining plants, manufacturers of pharmaceuticals, plastics and resins, cement and concrete products, and gypsum and asbestos materials, as well as sewage treatment plants (Mehrle et al. 1982). In the past, contaminant discharges, including those of heavy metals into the lower Hudson River, have been cited as a factor contributing to poor water quality, reduced fisheries, and decreased recreational potential of the adjacent apex of the New York Bight (Mueller et al. 1975, O'Connor and Stanford 1979, Steimle et al. 1982).

There are few published reports on metal levels in the aquatic fauna of the lower Hudson River Estuary. Evidence of cadmium accumulation has been observed in blue crabs (*Callinectes sapidus*) collected in the Foundry Cove region of the Hudson River (Kneip and O'Connor 1980). McCormick and Koepp (1979) reported mercury and zinc contamination of both resident and migratory finfish and shellfish collected seasonally in Newark Bay and Upper New York Bay. Mehrle et al. (1982) identified cadmium, lead, selenium, and arsenic as major metal contaminants in young-of-the-year striped bass (*Morone saxatilis*) from the lower Hudson River. However, Armstrong and Sloan (1980) noted that mercury levels in the flesh of New York State finfish have decreased since 1970, including a 32% decline for largemouth bass (*Micropterus salmoides*) from the lower Hudson River.

In conjunction with a statewide survey of metals in the estuarine biota of New Jersey in 1978–1979, funded by the New Jersey Department of Environmental Protection (Ellis et al. 1981), we took seasonal samples of finfish, crustaceans, and bivalves at four New Jersey sites: Alpine Park (RM 19) on the Hudson River, George Washington Bridge (RM 12), Caven Cove (RM −2), and the Military Ocean Terminal at Bayonne (RM −3) in upper New York Bay. This survey had the dual objectives of assessing long-term and short-term human health implications associated with possible metal contamination and of providing a data base of information on the occurrence of heavy metals for future comparison.

METHODS

Ten monthly collections were made between 1 June 1978 and 31 May 1979 (Figure 65). Inshore sampling gear included 50-foot and 100-foot nylon beach seines and minnow and crab traps. Offshore collections were made with hoop, fyke nets and gill nets, and hook and line. All organisms were measured to the nearest 0.5 cm, weighed to the nearest gram, and refrigerated until return to the laboratory where individual or pooled samples were dissected and portions were stored in zip-lock bags at −20°C. Small fish generally were ground whole, while only edible fillets were analyzed for fish more than 150-mm fork length. These fillets consisted of the left side of the fish from just behind the operculum to the base of the tail, including the bones of the rib cage and the left pelvic fin but excluding the skin, vertebral column, and the dorsal, anal, and caudal fins. Similar fillets from the right side were stored for archival purposes. Whenever possible, several specimens of the same species were pooled as a single composite sample. Only similarly sized specimens whose fork lengths differed by less than 0.5 cm were combined.

METAL ANALYSIS

Samples were homogenized by blending and ultrasonic disruption, and duplicate portions were placed into glass jars, one for analysis and one to be kept for archival purposes. Prior to mercury analysis, the sample was thawed and rehomogenized in an ultrasonic blender, after which an aliquot was removed and placed into a 300 ml BOD bottle. 25 ml of a sulfuric-nitric acid mixture was added, and the

HEAVY METALS IN FINFISH

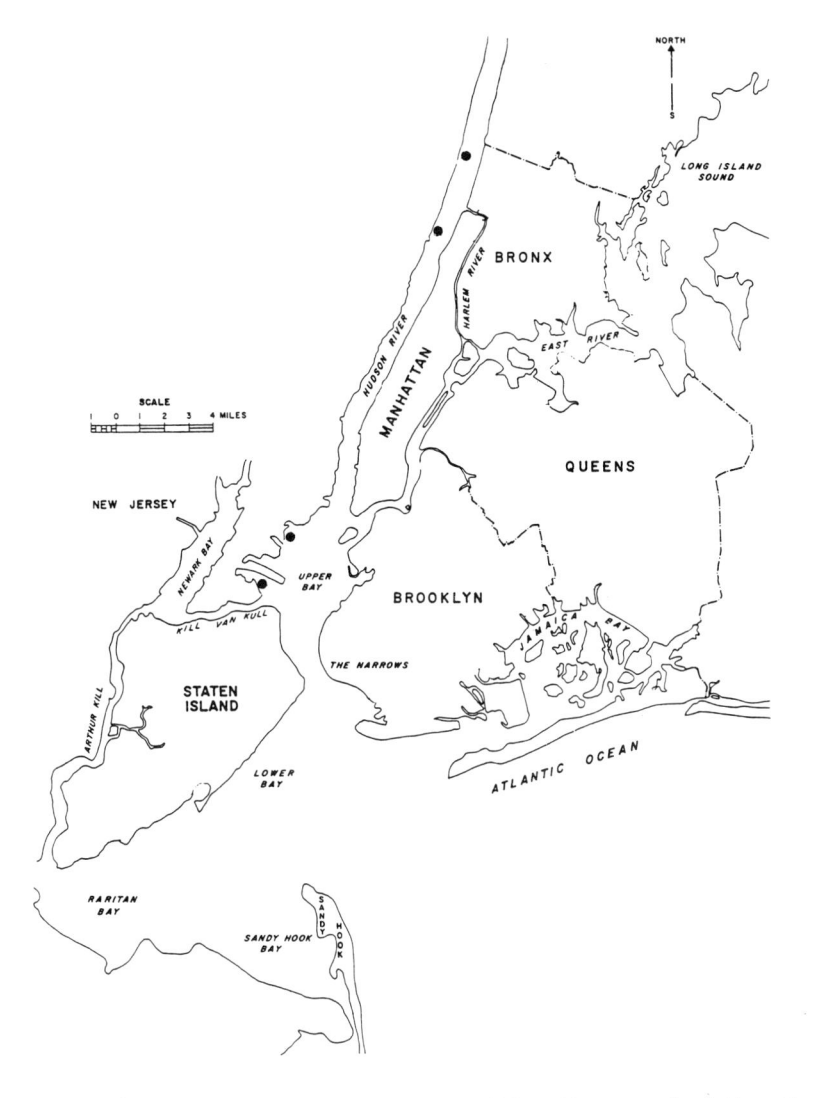

Figure 65. *Map of the lower Hudson River Estuary depicting the location of the four sampling stations (dark crescents) used in the study.*

sample incubated in a water bath at 30°C until all solids were dissolved. The sample was then removed from the water bath, cooled to 4°C, and potassium permanganate was added until a permanent blue color appeared. The digested samples were then allowed to stand overnight at room temperature, after which they were analyzed for mercury using the cold-vapor technique (Hatch and Ott, 1968).

For cadmium, lead, copper, and zinc, 10 grams of the tissue homogenate was acidified with 25 ml of 20% sulfuric acid and dried in an oven at 110°C.

The dried samples were then transferred to a cold clean muffle furnace, heated to 270°C for three hours and stored overnight at 450°C. Following ashing, the samples were washed in nitric acid, evaporated to dryness, and held in a muffle furnace for thirty minutes. Finally, the cooled dried samples were resuspended in a warm aqueous nitric acid solution, diluted, and an aliquot removed for analysis. Cadmium, lead, copper, and zinc were determined by flame atomic absorption spectroscopy and arsenic was measured by atomic absorption spectroscopy using the gaseous hydride furnace method (Grieg and McGrath 1977).

RESULTS

The mean concentrations of the six metals are presented in Tables 55 through 58. These data were derived from three hundred analyses representing a total of two thousand four hundred specimens. Due to the limited sample size of each monthly collection, results have been summarized for the entire sampling year. With the exception of mercury (and to a lesser extent, zinc and copper), metal concentrations were similar to levels reported for the same species throughout New Jersey (Ellis et al. 1981).

Mercury

The mean mercury concentration for juvenile striped bass collected near the Bayonne Military Ocean Terminal (RM −3) was 0.69 ppm (Table 58), as compared with the average of 0.36 ppm for all of our stations. White perch (*Morone americana*), 9 cm to 19 cm total length, averaged 0.40 ppm of mercury for all stations. Fillet samples of white perch from Alpine State Park (RM 19) and the George Washington Bridge (RM 12) averaged 0.67 ppm and 0.69 ppm of mercury, respectively (Tables 55 and 56). Concentrations of mercury in this species have been observed to increase with age (Ellis et al. 1981).

Table 55. *Trace metal residues (ppm wet weight) in finfish and crustacea collected in the Hudson River at Alpine, New Jersey (River Mile 19).*

SPECIES	SAMPLE	# ANIMALS (ANALYSES)	AVERAGE LENGTH (cm)	Hg	Cd	Pb	Zn	As	Cu
Striped bass	whole	29 (9)	12.0	0.28±.21	0.05±.13	0.01±0	25.7±9.6	0.18±.16	3.6±0.3
Striped bass	filet	4 (4)	26.9	0.37±.21	0.01±0	0.01±0	13.2±14.6	0.21±.26	3.6±3.2
White perch	whole	4 (2)	11.6	0.23±.02	0.01±0	0.01±0	14.0±10.2	—	0.8±1.0
White perch	filet	11 (7)	18.3	0.67±.44	0.15±.20	0.22±.3	14.8±10.3	0.28±.29	7.6±5.6
Bluefish	whole	5 (1)	11.0	0.38	0.19	1.10	88.8	0.41	20.5
Bluefish	filet	4 (2)	18.7	0.50±.10	0.53±.04	0.10±.1	13.4±1.8	0.25±.10	21.2±2.5
Northern weakfish	whole	2 (2)	12.0	0.30±.02	0.01±0	0.01±0	14.9±3.6	0.31	4.2
Winter flounder	whole	11 (2)	7.2	0.10±.08	0.01±0	0.01±0	33.1±9.2	0.12±.15	0.6±0.8
Hogchoker	whole	8 (1)	11.6	0.14	0.01	0.01	9.7	—	1.3
Hogchoker	filet	17 (2)	11.2	0.51	0.01	0.01	47.0±46	0.53	24.6±3.2
Channel catfish	filet	1 (1)	24.0	0.65	0.01	0.01	32.1	0.48	6.2
Bay anchovy	whole	63 (5)	7.4	0.16±.10	0.01±0	0.01±0	24.5±15.7	0.14±.14	13.2±20.7
Atlantic tomcod	whole	32 (2)	10.3	0.27±.05	0.01±0	0.01±0	17.4±9.4	0.01±0	7.0±8.2
Atlantic tomcod	filet	1 (1)	16.8	0.22	0.01	0.01	38.8	0.15	3.2
Silversides	whole	9 (3)	10.6	0.24±.09	0.01±0	0.01±0	34.4±10.4	0.09	4.6±5.0
Atlantic herring	whole	126 (2)	5.7	0.17±.05	0.01±0	0.01±0	35.9±18.8	0.26±.36	11.9±1.8
Blueback herring	whole	40 (5)	9.7	0.24±.20	0.01±0	0.01±0	26.3±12.8	0.39	5.5±3.9
Menhaden	filet	12 (9)	27.2	0.11±.10	0.02±.03	0.01±0	17.3±19.3	0.16±.11	6.6±11.0
Alewife	whole	5 (1)	7.5	0.01	0.01	0.01	12.1	0.12	0.3
American eel	filet	11 (9)	47.6	0.25±.25	0.01±0	0.01±0	33.1±40.3	0.38±.17	5.4±8.5
Blue claw crab	whole	35 (5)	4.8*	0.25±.20	0.01±0	0.01±0	50.2±32.7	0.19±.06	30.9±16.2

*carapace width

Table 56. Trace metal residues (ppm wet weight) in finfish and crustacea collected in the Hudson River at the George Washington Bridge (River Mile 12).

SPECIES	SAMPLE	# ANIMALS (ANALYSES)	AVERAGE LENGTH (cm)	Hg	Cd	X̄ TOTAL (PPM) ± SD Pb	Zn	As	Cu
Striped bass	whole	84 (6)	9.0	0.22±.08	0.01±0	0.01±0	31.6±14.8	0.10±.05	8.5±12.5
Striped bass	filet	6 (3)	20.4	0.45±.40	0.01±0	0.01±0	30.4±18.9	0.32±.40	5.2± 4.2
White perch	whole	3 (1)	9.0	0.14	0.01	0.01	24.1	—	2.3
White perch	filet	5 (5)	19.0	0.63±.30	0.26±.60	0.07±.10	34.9±30.6	0.20±.10	6.2± 4.3
Bluefish	whole	4 (2)	11.3	0.23±.01	0.01±0	0.01±0	43.7±41.5	0.15	4.4± 6.2
Bluefish	filet	1 (1)	19.5	0.21	0.01	0.01	4.1	0.24	14.5
Winter flounder	whole	5 (3)	8.7	0.12±.10	0.01±0	0.01±0	34.0±9.1	—	0.6± 0.8
Winter flounder	filet	1 (1)	14.9	0.50	0.01	0.01	21.7	0.21	0.01
Hogchoker	whole	15 (5)	12.5	0.22±.13	0.08±1.8	0.01±0	21.9±31.7	0.22±.03	8.9± 5.2
Catfish	filet	1 (1)	27.5	0.10	0.01	0.01	8.9	—	0.4
Brown bullhead	filet	1 (1)	28.2	0.47	0.01	0.01	2.7	0.09	1.5
Atlantic tomcod	whole	2 (2)	13.4	0.17±.03	0.01±0	0.01±0	18.4±12.9	0.12	1.1± 0.3
Atlantic tomcod	filet	10 (3)	20.8	0.19±.20	0.01±0	0.01±0	5.3±0.6	0.13±.10	0.5± 0.5
Silversides	whole	45 (2)	9.0	0.52±.30	0.01±0	0.01±0	36.5±8.2	0.41	16.6±10.7
Mummichog	whole	9 (3)	7.0	0.14±.09	0.01±0	0.01±0	23.4±16.5	0.93	4.7± 6.7
Atlantic herring	whole	30 (1)	5.4	1.06	0.01	0.01	34.2	0.36	10.8
Blueback herring	whole	2 (1)	7.5	0.18	0.01	0.01	33.4	—	0.8
Menhaden	filet	7 (5)	23.6	0.14±.10	0.01±0	0.01±0	7.7±3.4	0.22±.20	2.0± 1.5
Alewife	whole	3 (3)	17.6	0.32+.12	0.01±0	0.01±0	8.6±11.0	0.23±.14	2.0± 1.2
American eel	filet	13 (7)	128.9	0.36±.25	0.01±0	0.01±0	27.6±17.3	0.87±.43	5.5± 4.1
Blue claw crab	whole	5 (1)	6.3*	0.15	0.01	0.01	14.0	0.34	23.8
Grass shrimp	whole	68 (2)	4.0	0.14±.08	0.01±0	0.01±0	26.2±15.8	—	20.1± 7.8

*carapace width

Table 57. Trace metal residues (ppm wet weight) in finfish, crustacea and bivalves collected in Upper New York Bay at Caven Cove (River Mile −2).

SPECIES	SAMPLE	# ANIMALS (ANALYSES)	AVERAGE LENGTH (cm)	Hg	Cd	Pb	Zn	As	Cu
Striped bass	whole	6 (3)	10.9	0.52±.30	0.21±.30	0.01±0	39.4±13.1	0.95±1.0	16.4± 3.9
Striped bass	filet	2 (2)	15.8	0.43±.10	0.01±0	0.01±0	23.4± 5.3	0.11	2.1± 2.5
White perch	filet	1 (1)	17.1	0.52	0.01	0.01	5.3	0.01	0.01
Bluefish	whole	66 (4)	9.4	0.38±.30	0.38±.60	0.50±1.0	80.4±45.1	0.10±.10	20.1±18.5
Bluefish	filet	9 (6)	40.2	0.41±.06	0.01±0	0.01±0	13.0± 5.5	1.20	1.9± 1.1
Winter flounder	whole	3 (2)	10.9	0.05±.05	0.48±.70	0.01±0	23.1± 8.9	0.26±.03	0.8± 1.1
Winter flounder	filet	6 (1)	6.6	0.12	0.01	0.01	0.5	0.42	18.7
Atlantic tomcod	whole	1 (1)	11.4	0.48	0.01	0.01	16.4	0.01	9.1
Atlantic tomcod	filet	3 (3)	17.8	0.15±.10	0.01±0	0.01±0	16.4±15.9	0.20±.01	0.5± 0.9
Silversides	whole	150 (10)	9.6	0.28±.20	0.29±.70	0.01±0	25.1±11.5	0.10±.08	4.7± 7.9
Atlantic mackerel	whole	1 (1)	11.1	0.12	0.01	0.01	55.5	0.45	13.7
Mummichog	whole	98 (4)	6.3	0.13±.10	0.01±0	0.01±0	38.0±15.2	0.60±.70	9.6±15.2
Striped killifish	whole	50 (3)	7.4	0.13±.05	0.01±0	0.01±0	32.1±17.0	0.16±.03	13.5± 2.2
Atlantic herring	whole	21 (1)	6.4	0.36	0.01	0.01	34.3	—	128.4
Blueback herring	whole	106 (3)	8.4	0.21±.20	0.14±.20	0.01±0	19.0±10.5	0.75±.80	2.9± 4.3
Menhaden	whole	10 (1)	7.8	0.30	0.01	0.01	44.1	—	2.2
Menhaden	filet	11 (4)	23.4	0.26±.08	0.13±.20	0.01±0	13.3± 3.9	0.28	7.2±10.1
Hickory shad	filet	3 (1)	18.6	0.19	0.01	0.01	8.8	0.01	1.9
Blue claw crab	whole	9 (3)	11.7*	0.17±.09	0.01±0	0.01±0	29.7±16.2	0.57±.50	19.3±32.9
Grass shrimp	whole	23 (1)	3.3	0.01	0.01	0.01	49.4	0.01	22.2
Ribbed mussel	viscera	8 (1)	4.0	0.14	0.01	0.01	20.3	0.46	0.01
Soft clam	viscera	12 (2)	5.6	0.32±.20	0.01±0	0.01±0	15.9±15.3	0.30±.13	7.8± 3.2

*carapace width

Table 58. Trace metal residues (ppm wet weight) in finfish, crustacea and bivalves collected in Upper New York Bay adjacent to the Military Operations Terminal (River Mile −3).

SPECIES	SAMPLE	# ANIMALS (ANALYSES)	AVERAGE LENGTH (cm)	Hg	Cd	X̄ TOTAL (PPM) ± SD Pb	Zn	As	Cu
Striped bass	filet	5 (2)	23.9	0.69 ± .18	0.35 ± .50	3.4 ± 4.8	61.9 ± 72.8	0.42	11.7 ± 1.6
Bluefish	whole	23 (4)	11.9	0.33 ± .20	0.01 ± 0	0.01 ± 0	27.9 ± 43.0	0.22 ± .30	7.8 ± 12.7
Winter flounder	whole	3 (2)	9.4	0.33 ± .30	—	—	47.8	0.04	—
Atlantic tomcod	whole	2 (1)	10.7	0.82	0.01	0.01	50.3	—	14.0
Summer flounder	filet	1 (1)	18.8	0.09	0.01	0.01	46.0	0.41	2.2
Northern weakfish	whole	2 (1)	7.8	0.33	0.01	0.01	0.6	—	0.01
Silversides	whole	79 (6)	8.8	0.47 ± .40	0.01 ± 0	0.01 ± 0	38.8 ± 27.3	0.30 ± .20	5.9 ± 8.2
Mummichog	whole	41 (7)	8.9	0.36 ± .30	1.90 ± 4.2	0.14 ± .30	40.6 ± 18.7	0.16 ± .10	3.6 ± 4.1
Stickleback	whole	9 (1)	6.0	0.19	0.01	0.01	29.1	0.14	0.9
Blueback herring	whole	39 (3)	4.9	0.21 ± .30	0.01 ± 0	0.01 ± 0	94.0 ± 47.0	0.25 ± .08	2.8 ± 2.3
Menhaden	filet	5 (2)	21.3	0.99 ± 1.1	0.01 ± 0	0.01 ± 0	47.2 ± 5.7	0.60	30.1 ± 28.2
Alewife	whole	6 (2)	10.2	0.35 ± .11	0.01 ± 0	0.01 ± 0	57.6 ± 6.2	0.01	3.1 ± 2.0
American eel	filet	2 (2)	41.8	0.21 ± .16	0.01 ± 0	0.01 ± 0	44.7 ± 52.7	0.46	3.7 ± 3.8
Blue claw crab	muscle	6 (5)	15.1*	0.55 ± .30	0.11 ± .20	0.01 ± 0	45.1 ± 34.0	0.19 ± .20	16.3 ± 12.4
Blue claw crab	liver	4 (2)	14.1*	0.21 ± 0	0.05 ± .06	0.01 ± 0	72.5 ± 28.1	—	44.4 ± 4.4
Grass shrimp	whole	1200 (6)	3.7	0.15 ± .04	0.02 ± .03	0.01 ± 0	35.9 ± 48.5	0.27 ± .29	16.9 ± 13.5
Ribbed mussel	viscera	32 (2)	4.8	—	0.01 ± 0	0.01 ± 0	8.9 ± 12.6	0.07 ± .11	3.0 ± 0
Soft clam	viscera	77 (3)	3.4	0.35 ± 0	0.01 ± 0	0.01 ± 0	52.4 ± 34.4	0.24 ± .03	13.5 ± 12.5
Hard clam	viscera	15 (1)	2.3	0.55	0.01	0.01	102.3	0.03	22.0

*carapace width

Young-of-the-year bluefish (*Pomatomus saltatrix*) averaged 0.35 ppm of mercury overall, whereas larger specimens collected near Caven Cove (RM −2) averaged 0.41 ppm (Table 57). The only other species with concentrations of mercury greater than 0.30 ppm were: blue crab (*Callinectes sapidus*), 0.31 ppm; Atlantic silversides (*Menidia menidia*), 0.34 ppm; and Atlantic menhaden (*Brevoortia tyrannus*), 0.36 ppm. The highest individual mercury concentration for any sample, 0.99 ppm, was found in fillet samples of Atlantic menhaden collected near the Military Ocean Terminal (Table 58).

Cadmium

Two species, mummichog (*Fundulus heteroclitus*) and bluefish, had cadmium residue averages (all stations) above 0.20 ppm. However, the overall average for the mummichog, 0.95 ppm, may be misleading, since only samples collected near the Military Ocean Terminal had detectable amounts of this metal (Table 58). In contrast, cadmium was found in bluefish from all stations. Cadmium concentrations in excess of 0.2 ppm were occasionally observed among striped bass (0.35 ppm at RM −3) and Atlantic silversides (0.29 ppm at RM −2).

Lead

Lead concentrations were at or near the detection limit (0.01 ppm) for most species, and the highest levels of lead were found in predators such as the bluefish (0.92 ppm) and striped bass (0.26 ppm).

Zinc

Mean zinc concentrations ranged from 14.0 ppm to 44.0 ppm. Blue crabs (44.0 ppm), blueback herring (*Alosa aestivalis*) (42.0 ppm), and mummichog (39.0 ppm) had the highest averages. In general, specimens of any species from the upper New York Bay had greater average concentrations of zinc than the same species from stations in the lower Hudson River.

Copper

Average copper concentrations ranged from 0.20 ppm for the alewife to 42.0 ppm for blue crab. A single pooled whole-body sample of Atlantic herring from Caven Cove (RM −2) had 128.4 ppm copper. The overall average for this species is 41.0 ppm.

Arsenic

Only a few species had arsenic concentrations in excess of the New Jersey statewide average. American eels (*Anguilla rostrata*) at

0.70 ppm and mummichogs at 0.52 ppm had the highest average concentrations of arsenic in the lower Hudson River estuary. Several samples of striped bass and bluefish from Caven Cove had arsenic levels in excess of 1.0 ppm (Table 57).

Trends and Trophic Levels

The species most frequently encountered at a majority of the stations were assigned to trophic categories on the basis of feeding preferences of different size classes as determined from the literature and supplemented by our own food studies.

The following categories were recognized:
 1. Piscivores: bluefish, striped bass larger than 9 cm, and white perch larger than 12 cm.
 2. Invertivores: winter flounder (*Pseudpleuronectes americanus*), hogchoker (*Trinectes maculatus*), and white perch smaller than 12 cm.
 3. Planktivores: Atlantic menhaden, blueback herring, Atlantic herring (*Clupea harengus*) and alewife (*Alosa pseudoharengus*).
 4. Omnivores: Atlantic tomcode, silversides, and mummichogs of all sizes.
 5. Scavengers: blue crab and American eel.
 6. Detritivores: grass shrimp (*Hippolyte* sp.).
 7. Filter feeders: ribbed mussel (*Modiolus demissus*) and blue mussel (*Mytilus edulis*).

Histograms of average metal concentrations at four stations are presented in Figures 66 and 67. Accumulation of mercury is apparent among piscivores, invertivores, and omnivores from all stations. With the exception of detritivores, all trophic groups show mercury residues well above 0.20 ppm for species collected near the Military Ocean Terminal. Detectable levels of cadmium and lead were generally confined to piscivores and, to a lesser extent, to omnivores and invertivores collected in Upper New York Bay. Although the values essentially fall within normal background levels, piscivores collected near the Military Ocean Terminal showed greater accumulation of lead than the same species at the other three stations.

Trophic histograms for zinc, copper, and arsenic are presented in Figure 67. Both zinc and copper concentrations are significantly greater for all trophic groups in Upper New York Bay than in the two lower Hudson River stations. Arsenic concentrations are variable and show no clear pattern either for trophic groups or between locations.

HEAVY METALS IN FINFISH

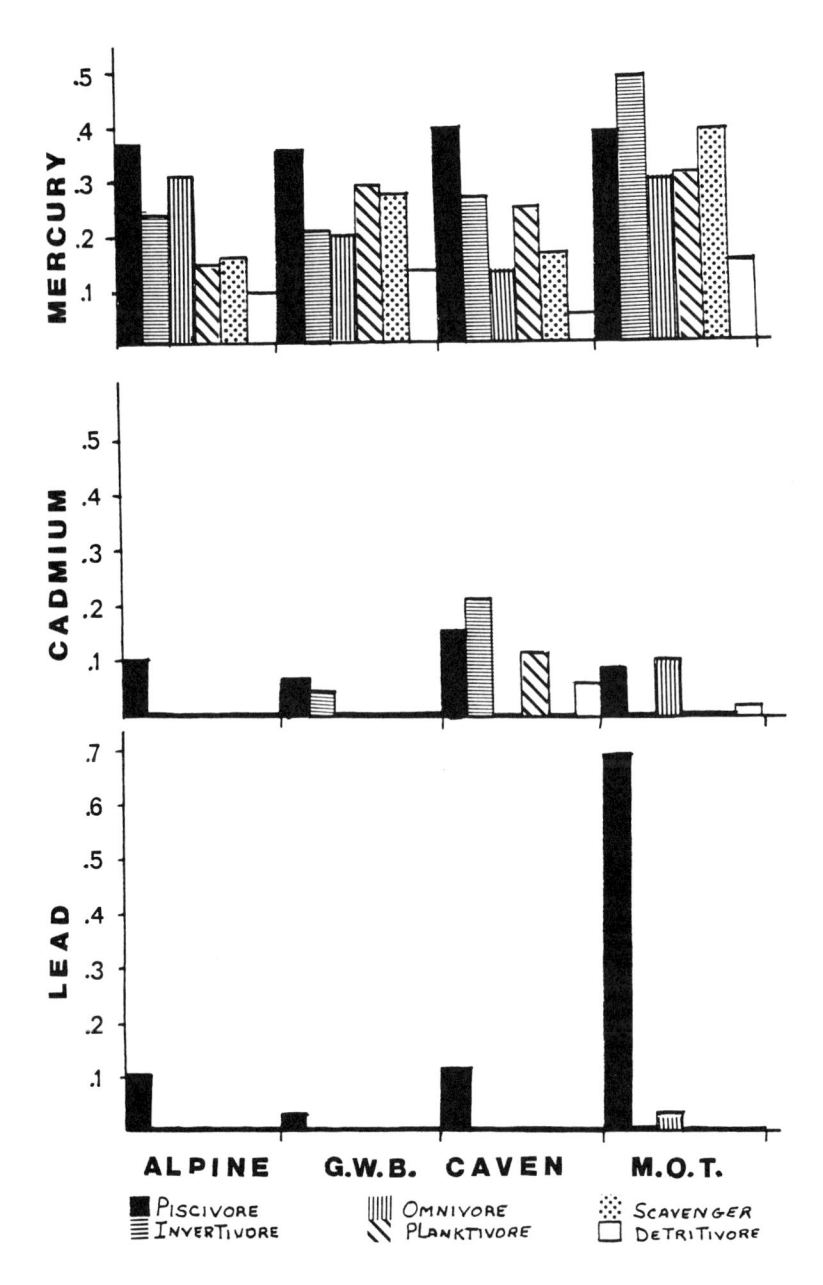

Figure 66. *Trophic level contributions (in ppm wet weight) of mercury, cadmium and lead in aquatic fauna collected at four stations in the lower Hudson Estuary.*

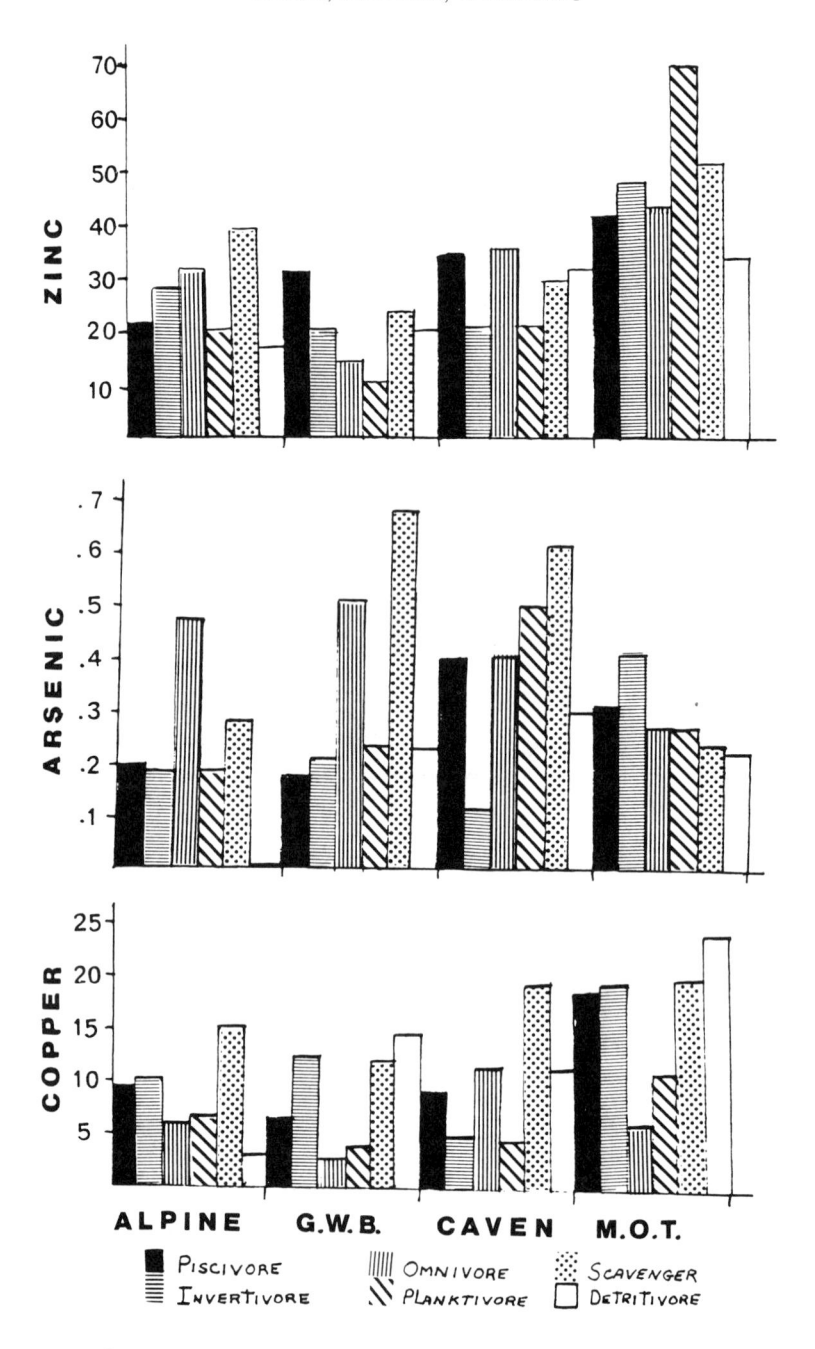

Figure 67. Trophic level contributions (in ppm wet weight) of zinc, arsenic, and copper in aquatic fauna collected at four stations in the lower Hudson Estuary.

DISCUSSION

Several instances of cadmium contamination have been reported among organisms in the lower Hudson estuary. Kneip and O'Connor (1980) noted high cadmium levels in blue crabs collected from several sites adjacent to Foundry Cove. Mehrle et al. (1982) reported age-dependent uptake of cadmium in young-of-the-year striped bass collected along the lower Hudson River. Interestingly, no blue crabs analyzed for the present study contained cadmium in excess of 0.10 ppm. The dramatic differences between our findings and those of Kneip and O'Connor probably reflect local availability of this metal, emigration of crabs from cleaner waters, or selective depuration. Conversely, the accumulation of cadmium in young-of-the-year bluefish and striped bass noted in our samples may indicate a lack of ability of these predatory species to eliminate this metal by internal physiological processes.

The bioaccumulation of mercury is well documented for a wide range of species and localities (Leland et al. 1976, Whittle et al. 1977). Adult bluefish and striped bass from New York Bight waters have been reported to accumulate mercury in proportion to body weight, with average concentrations ranging from 0.50 ppm to 1.0 ppm (Reish et al. 1980). Similar concentrations have been reported for juvenile striped bass, bluefish, and adult white perch collected in Newark Bay and upper New York Bay (McCormick and Koepp 1980).

Although the FDA guidelines of 1.0 ppm was not exceeded in samples collected during this study, the detection of average mercury concentrations near 0.40 ppm in young-of-the-year striped bass and bluefish must be viewed with concern. The propensity of white perch to accumulate mercury with age should be investigated further. Such research must be coupled with dietary data if biomagnification of mercury in this species is to be understood properly.

The evaluation of zinc and copper residues in biota collected from contaminated waters must be approached with caution. Both metals are essential elements in biological systems and as such are subject to internal regulation (O'Connor and Rachlin 1982). Nevertheless, there are clear indications of bioaccumulation over background levels for both zinc and copper in finfish and invertebrates from Upper New York Bay as compared with those captured at the two Hudson River stations. In particular, average concentration of zinc in striped bass, bluefish, blue crabs, and soft clams (*Mya arenaria*) collected near the Military Ocean Terminal are consistent with previously reported levels for the Upper New York Bay (McCormick and Koepp, 1979).

The apparent relegation of accumulated lead to a few predatory finfishes (bluefish and striped bass) is consistent with published reports concerning this metal. Using partition coefficient analysis, O'Connor and Rachlin (1982) suggested an inability on the part of such large predators to regulate tissue lead residues, thereby resulting in the accumulation of this metal. A similar response was suggested for cadmium in these species.

With the exception of arsenic levels in menhaden and of mercury in white perch, metal concentrations did not correlate statistically with the organisms' growth parameters. Although this in part may reflect a limited sample size for some species, there is precedent for this lack of correlation in the literature. Eustace (1974) reported small differences in muscle concentrations of zinc and copper in thirty-nine species of fishes collected in the Derwent estuary, Tasmania. These differences were attributed to feeding habits and could not be correlated with fish size. Montgomery (1974) reported no detectable differences in lead concentration between size classes of herring collected in a single haul-seine tow. Generally, poor correlation was noted between metals other than cadmium and growth of young-of-the-year striped bass from the lower Hudson River (Mehrle et al. 1982).

Trends in metal accumulation are better correlated with trophic levels than with individual growth patterns, although such factors as size, movements, and internal regulation and distribution modify these trends. Piscivores showed evidence of enhanced accumulation of mercury, zinc, copper, lead, and, to a lesser extent, cadmium, in Upper New York Bay, as compared with the lower Hudson River.

ACKNOWLEDGEMENTS

The authors thank the following individuals at Montclair State College for their technical assistance: Nancy Dorato, Frank Barbone, Gary Buchanan, Justine Paffman, David Calligaro, and Darvene Adams. Special thanks to Dr. Su Ling Cheng of New Jersey Institute of Technology for performing the chemical analyses. This research was supported by a grant from the New Jersey Department of Environmental Protection.

11. Recent Dissolved Oxygen Trends in the Hudson River

Jeffrey A. Leslie, Karim A. Abood, Edward A. Maikish, and Pamela J. Keeser

INTRODUCTION

Estuaries have long been used by industry and municipalities as a water source and waste receiver; the Hudson River is no exception. This use of the river often degrades the quality of the water, thereby affecting aquatic life and other uses of the water. To determine the extent of the impact, governmental agencies, utilities, industries, colleges, and others have monitored water quality. These data are collected for specific reasons but are all too often buried in an appendix to a technical report or permit application. Few efforts have been made to tap these data to present a comprehensive overview of water quality in the Hudson.

The purpose of this paper is to examine recent trends in water quality data in order to (1) identify environmental problem spots, (2) evaluate the effects of environmental management programs, (3) determine whether previous conclusions are supported or rejected by the new data, and (4) define the need for future analyses.

For this paper, several comprehensive and easily accessible data bases were selected to provide a general idea of how riverwide dissolved oxygen (DO) profiles changed in the Hudson River during the 1970s. The riverwide profiles are used for an overall comparison; in-depth comparisons are made at two stressed areas of the river, New York City and Albany. Dissolved oxygen was chosen because it can be used as an indicator of pollution and is critical to biological processes. Historically, it was one of the first water quality parameters to be measured; thus, there is a relatively large historical data base for the Hudson River.

RIVERWIDE PROFILES

Most of the DO data considered in this paper were collected in conjunction with biological studies of the fish populations of the Hudson or waste allocation studies. A large portion of the sampling during the 1970s focused on striped bass and white perch. These species spawn in the spring and become juvenile recruits by the end of the summer. Therefore, spring and summer water quality data are relatively abundant, but fall and winter data are sparse. Summer through early fall is also the period when waste loads have the maximum impact on the river DO; therefore, studies were made frequently during this period. In the following sections, actual DO data collected during the summer and spring are compared with projected DO profiles.

Summer

In the summer, the Hudson River is characterized by low freshwater flows, warm water temperature, and low DO levels. This is the critical season for aquatic and marine life; not only are the river's resources severely stressed, but the newly spawned larvae are sensitive to environmental perturbation. It is no surprise that a large portion of annual sampling is done at this time.

Figure 68 presents DO levels plotted against milepoint (MP), that is, location in miles above the Battery (MP 0). The 1967 results are from a four-day helicopter survey conducted in August by QLM over the entire length of the river (LMS 1976f). Summer data from various governmental agencies, utilities, and private industry were averaged by milepoint to provide the 1970–72 values. The dash-dot line represents the results of a predictive computer model that assumed all point discharges along the river complied with the "best practicable technology" (BPT) by 1977, as required by the Federal Water Pollution Control Act (FWPCA) of 1972 (LMS 1976f). The 1978 results are averages by milepoint of data collected during Hudson River Field Week II (HRFW II), a one-week riverwide sampling effort by industries, utilities, consultants, governmental agencies, colleges, and others, and coordinated by the Hudson River Research Council (HRRC 1980). The stepped solid line indicates the New York State minimum acceptable DO level for given reaches of the river (6 NYCRR 701,702,858.6,864.60).

As one measure of comparability among years, freshwater flows (Q) at the head of the estuary (Green Island, New York) are also presented in Figure 68. The 1967 and 1978 values are the monthly averages for August of the respective years. The 1970–1972 flow is the mean of the monthly average flows for August from the three years considered. For the 1977 BPT model run, a flow was chosen to simulate low-flow conditions (LMS 1976f). Freshwater flows affect the

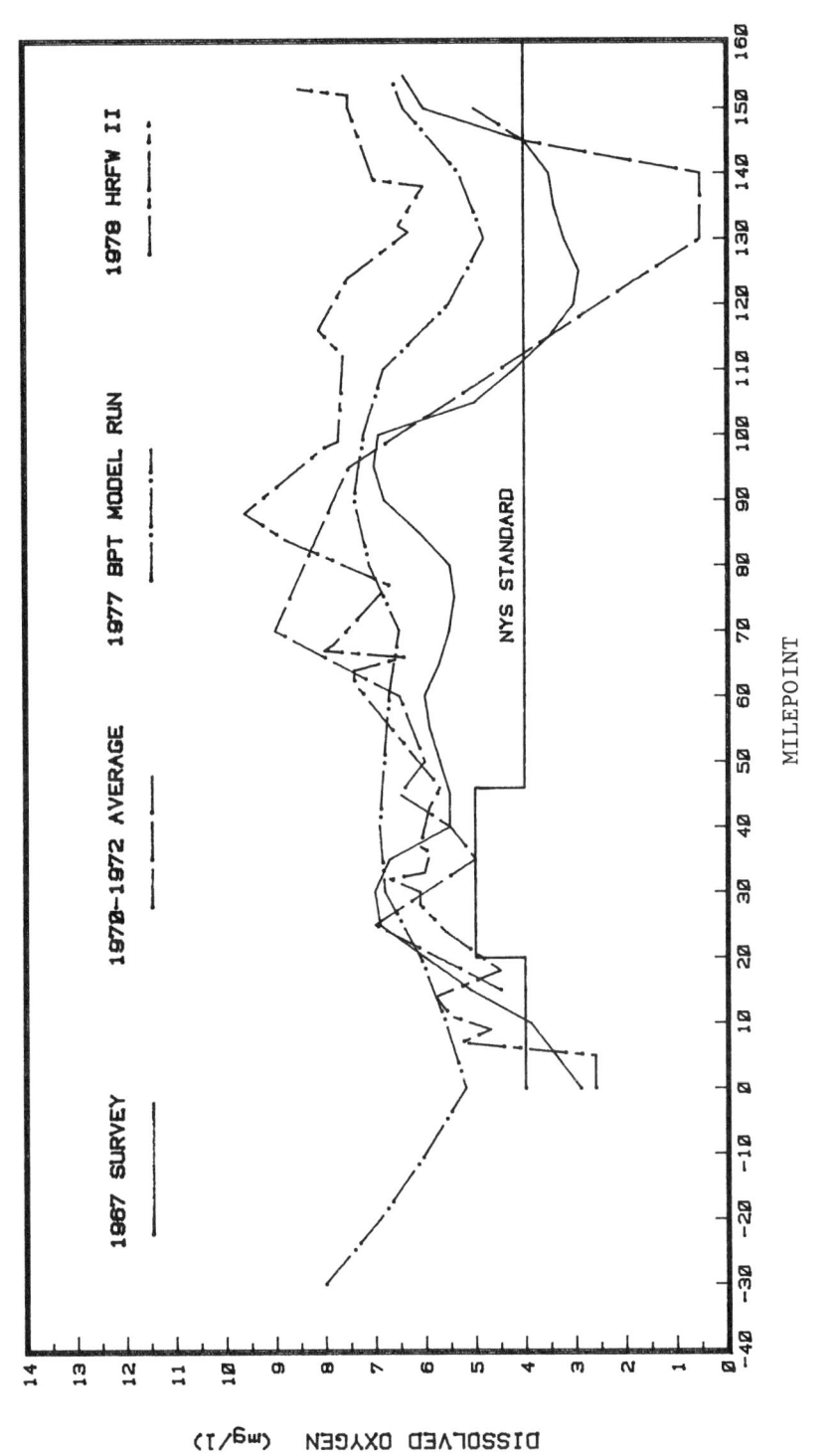

Figure 68. *Summer Hudson River dissolved oxygen data. (1967, Q = 5,749 cfs; 1970–1972, Q = 6,823 cfs; 1977, Q = 5,400 cfs; 1978, Q = 5,976 cfs)*

available dilution and the movement downstream of oxygen-consuming discharges, as well as temperature and salinity, which in turn affect the DO saturation level, that is, the maximum amount of DO that the water can retain. Other parameters affecting DO levels include waste loads, temperature, and overall climatic conditions.

Although detailed cause/effect comparison among these four profiles have not been made, a number of overall reach and riverwide observations based on these profiles are made below. The fact that the flows corresponding to the 1967, 1977, and 1978 profiles are quite similar reduces the influence of this parameter on these observations.

The most obvious feature in Figure 68 is the sag in the region just downstream of Albany (MP 120–150). DO levels were below the state standard in 1967 and approached 0.0 mg/l in 1970–72. Improvement to above the state standard was projected with the advent of BPT by 1977 under river-flow conditions approximating the 1967 conditions. In 1974, two waste treatment plants came on line in the Albany area, treating about 80% of the city's domestic wastes. As the HRFW II data in Figure 68 indicate, the DO levels observed in 1978 were above the standard and higher than the projected BPT level for similar river-flow conditions.

In the New York City area (MP 0–10), the 1967, 1970–72, and 1978 lines in Figure 68 are almost identical and indicate the presence of another sag below the state standard. Construction delays of two major sewage treatment plants (STPs)—the North River STP in Manhattan and the Passaic Valley STP in New Jersey—are partially responsible for the failure to achieve the projected 1977 BPT level.

In the mid-river region, the DO levels during the 1970s are irregular, but for all practical purposes they are the same. There appears to have been a slight increase over the 1967 level, but profiles for this region are well above the state standard.

Figure 69 compares actual 1978 HRFW II (HRRC 1980) field data with predicted DO levels at various waste load and levels of treatment presented by LMS (1976f). The 1973 and 1977 lines indicate projected DO levels under actual waste-load conditions for each year. The second target year set by the FWPCA was 1983, when all discharges were supposed to have been treated using the "best available technology" (BAT). The basic difference between BPT and BAT is improved industrial waste treatment. Since most of the discharges in the Hudson River estuary are from domestic sewers, BPT and BAT are essentially the same. Elimination of Discharge (EOD) was targeted

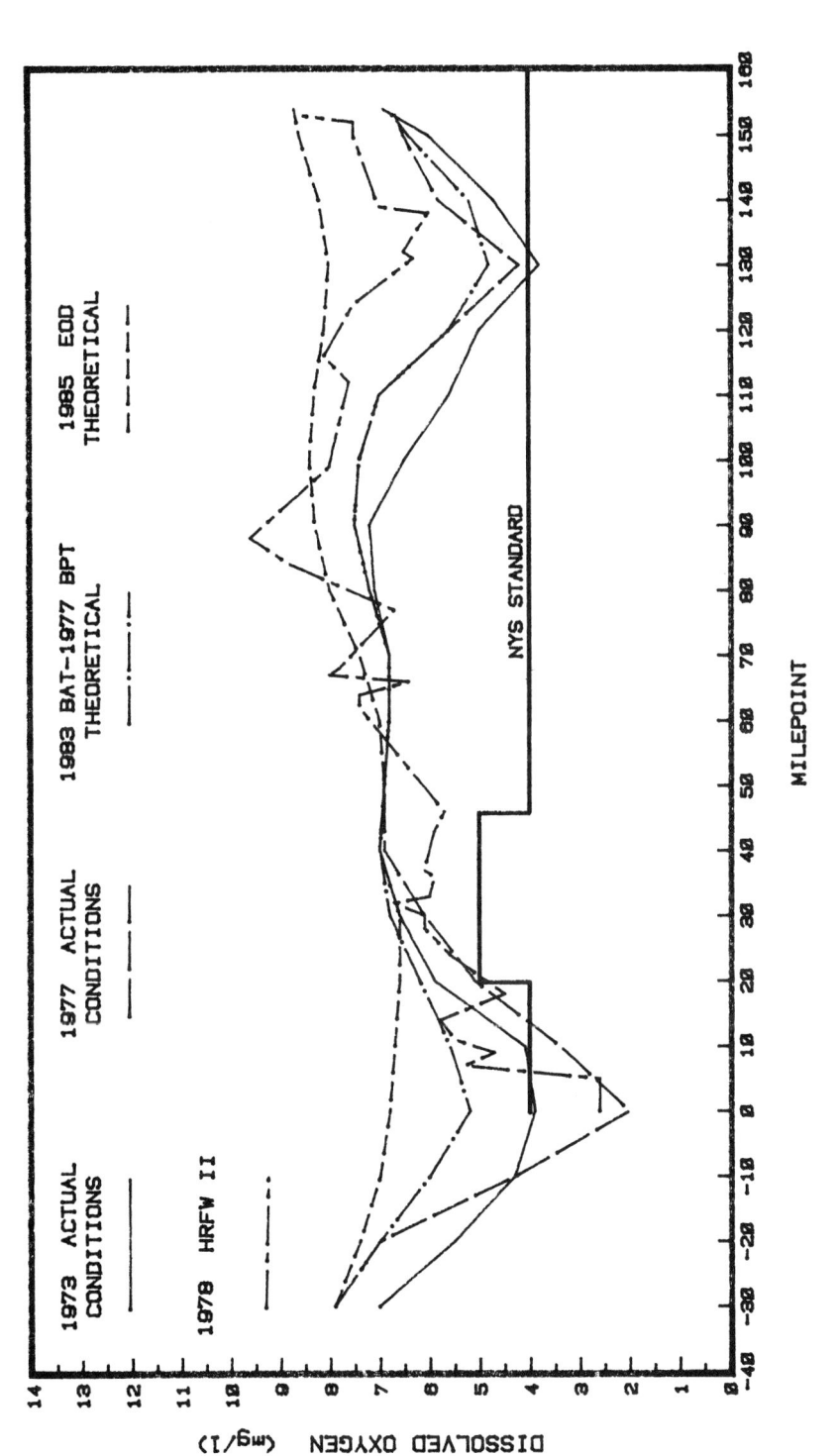

Figure 69. Predicted summer dissolved oxygen profiles in the Hudson River. (1978 HRFW, Q = 5,976 cfs; model run, Q = 5,400 cfs)

for 1985 and would have permitted only the discharge of "clean" water to the river. Figure 72 indicates that this level of treatment was projected to have a more dramatic effect on DO levels in the northern reaches of the estuary, probably because of the elimination of discharges north of the Troy Dam, which defines the upper boundary of the estuary.

In the Albany area, the predicted DO levels using actual 1977 waste loads are almost identical to the predicted levels using theoretical BAT-BPT waste loads (Figure 69). The DO levels measured during HRFW II (i.e., after the STPs came on line) are considerably higher than the projected levels and well above the state standard.

No change is expected in the mid-river region until EOD is achieved, and then only in the region north of MP 60. The HRFW II data show general agreement with the 1977 actual and BAT-BPT theoretical predicted levels, but they are highly variable and must be cautiously interpreted. Since the HRFW II data were collected by many independent participants, differences in sampling and analytical techniques may have introduced artifacts in the data. Nevertheless, the mid-river DO level is well above the state minimum.

In the New York City area, the predicted DO levels under 1977 actual conditions (without the North River or Passaic Valley STPs) and the DO levels measured during HRFW II show general agreement. Both sag below the state minimum and are below the 1983 BAT-1977 BPT predicted level.

Spring

In contrast to the critical summer period, spring is characterized by high freshwater flows and lower temperatures, resulting in high DO levels. Unfortunately, data are somewhat sparse for this period.

Figure 70 shows spring data analogous to the summer data presented in Figure 68. Note that the first Hudson River Field Week (HRFW I) was conducted in April, one year before HRFW II. The 1970–72 and 1978 flows were calculated the same as they were for the summer, except that April data were used. The flow for the 1977 BPT model run was chosen to simulate low springtime flows (LMS 1976f).

The most striking feature is that both the projected and measured DO levels are everywhere above 7.0 mg/l. Although the Hudson estuary is not classified as such, these DOs exceed the minimum level required for trout-spawning fresh water. The data for the Albany and New York City areas are limited, but sags in the 1970–72 DO profiles

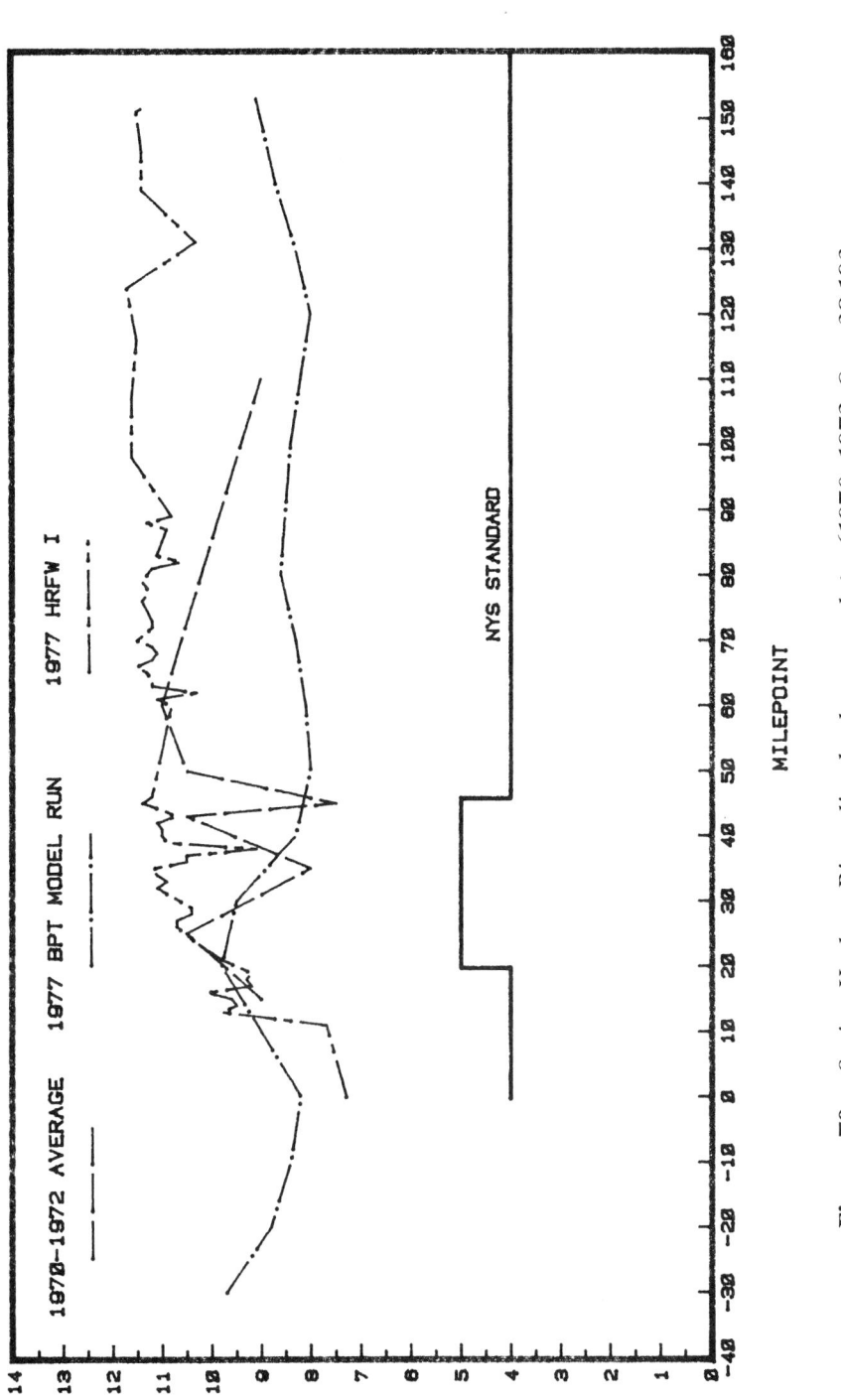

Figure 70. Spring Hudson River dissolved oxygen data (1970–1972, Q = 38,193 cfs; 1977 BPT, Q = 19,000 cfs; 1977 HRFW, Q = 40,560 cfs)

are indicated in both areas; however, they are less pronounced than those observed in the summer, and it appears that the minimum values would be well above the standard. The HRFW I data collected after the Albany STPs came on line indicate a sag in the New York City area similar to that of 1970–72, but no appreciable sag in the Albany area.

The 1970–72 and HRFW I flows (Figure 70) increased approximately eightfold over their counterparts in Figure 68, while the 1977 BPT flow increased only fourfold. The model runs were made to simulate critical low-flow conditions to give "worst-case" results. On the other hand, the 1970s were characterized by spring flows above the long-term average flow. Low spring freshwater flows like those observed in the early 1980s probably resulted in a DO profile more like the 1977 BPT results shown in Figure 70. It is doubtful that DO levels would approach the state standard even under drought conditions. Again, waste loads, temperature, and other climatic conditions, not covered in the overview may have some impact on DO levels.

THE ALBANY AREA

The previous section indicated that a closer examination of the seasonal variation in DO levels would be worthwhile. Figure 71 presents dissolved oxygen values recorded by the New York State Department of Environmental Conservation (NYSDEC) at Glenmont, New York (just south of Albany, MP 142), from 1970 through 1979. The figure indicates that during the 1970s, the DO level was well above the state standard for the better part of the year, sagging close to this value only during August, September, and October. Prior to 1974, the recorded minimum values were at or below the standard, with 1970 and 1971 being exceptionally bad years. After the STPs came on line in 1974, the recorded minimum values were all above the state standard, and they were greater than 5.5 mg/l after 1976.

The overall water quality of the area also seems to have improved. A survey conducted before the STPs came on line found a single healthy golden shiner. Large numbers of young blueback herring and an American eel were found "swimming slowly at the surface, gulping air...." (QLM 1971a). In 1975, fifty-two species representing twenty different families were found in the area (LMS 1975f). This diverse community is indicative of a more stable environment.

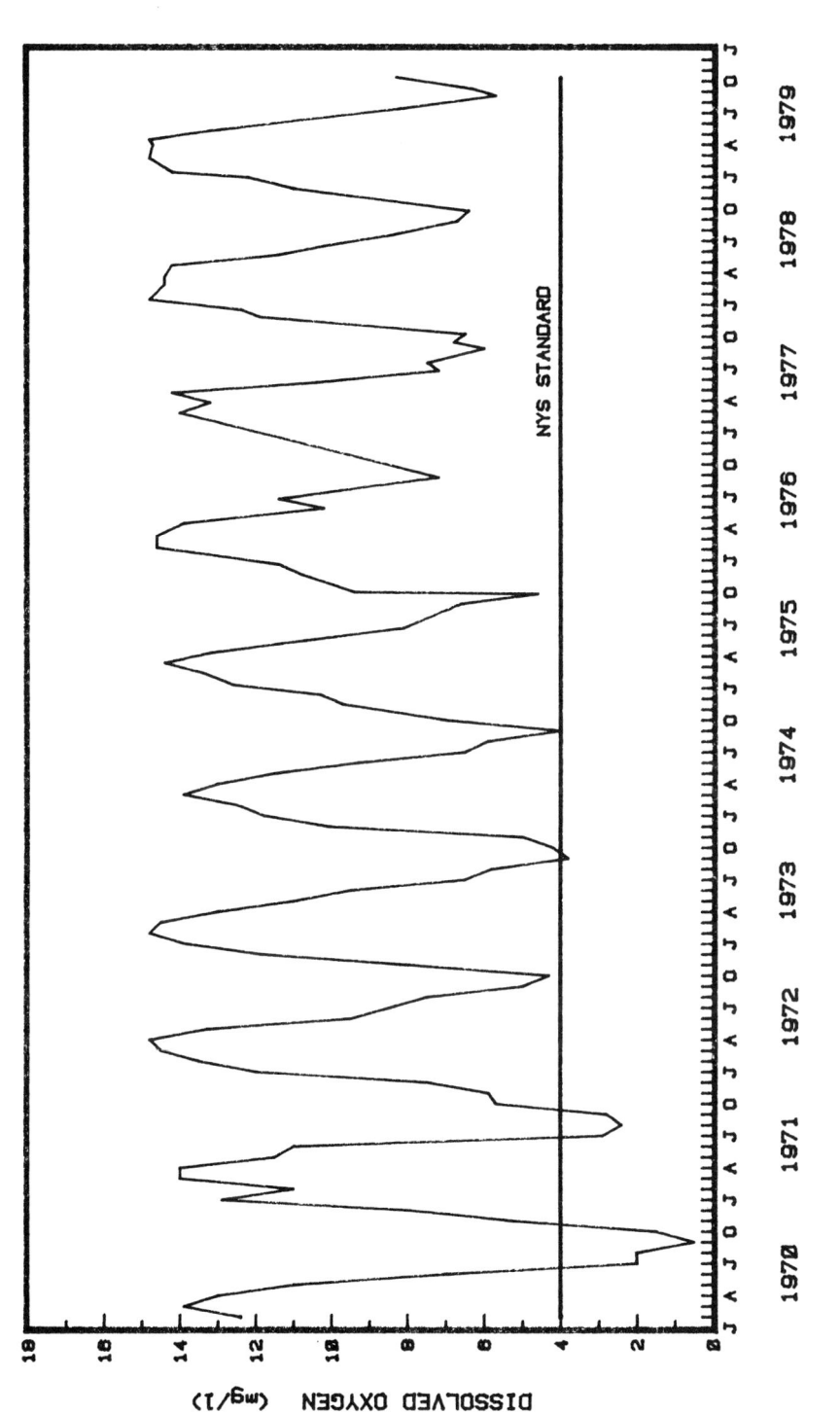

Figure 71. *NYSDEC dissolved oxygen values at Glenmont, N.Y., 1970–1979.*

Although the freshwater flows of the 1970s were unusually high (Table 59), they were high both before and after the STPs came on line. The DO maximums occurred in late winter/early spring and are consistently in the 13.0 mg/l to 15.0 mg/l range, both before and after the STP came on line. Therefore, there is a strong indication that the STPs are responsible for the improvement in the area's water quality and that their greatest effect seems to occur during the critical summer months.

The DO profiles for 1970, 1973, 1978, and 1979 appear in Figure 72 on an expanded horizontal scale. It becomes more evident that even in bad years like 1970, the DO level is well above the state standard for most of the year. It also appears that the 1978 and 1979 minimum DO values are about 2.0 mg/l higher and occur earlier in the summer than the 1973 minimum. The freshwater flows shown are the average of the monthly mean flow for August, September, and October in Table 59.

The lower graph in Figure 71 shows the percent DO saturation for 1970, 1978, and 1979. The DO saturation profiles tend to standardize the data with respect to the effects of temperature and salinity, which are also dependent on freshwater flows. The vertical line at 30% represents New York State's minimum requirement for water not primarily for recreational contact, shellfish culture, or fishlife propagation (Class II). Although not applicable to the Albany area, this standard is presented as a frame of reference for interpretation.

In all three years, DO saturation was 80%–90% in the winter and began to sag in late spring. In 1970, the DO saturation sagged below the Class II requirement between June and October. In contrast, the 1978 and 1979 levels never sagged below 70%.

THE NEW YORK CITY AREA

Figure 73 presents data collected by the Interstate Sanitation Commission (ISC 1978, 1980) at the Narrows (MP −6) from 1974 to 1978. The upper graph indicates that although New York City DO levels consistently sag below the state standard in summer, they are considerably higher than the standard for the better part of the year. In fact, readings of 10 mg/l–12 mg/l in the winter and early spring are not uncommon.

The bottom graph presents temperature data for the same period to give some indication of the relationship between DO and temperature. Not only does temperature affect the solubility of dissolved

Table 59. *Freshwater flows at Green Island*

MONTH	1918–1976 LONG-TERM AVERAGE FLOWS (cfs)	1971	1972	1973	1974	1975	1976	1977	1978[a]	1979[a]	RANK OF FLOWS AS TO WETNESS (1918–1979) 1979
Jan	12879	9002	13410	26210	22010	19070	14740	7956	26310	20162	10
Feb	12474	12110	10930	20460	18640	19370	31260[b]	8032	14120	11848	28
Mar	22344	20220	26860	29410	20730	23680	31690	43540	21870	44274[c]	2
Apr	31155	37270	37960	30960	30170	25580	36760	40560	33560	37100	21
May	19480	35240	40520[c]	27600	22960	20000	31800	16020	18730	19571	25
Jun	9791	7334	29630[b]	13050	8791	12970	15220	7325	9954	8318	33
Jul	7040	6233	18380[c]	10390	11780	7464	15280	5735	4643	4643	44
Aug	5608	8929	7616	5591	6359	8966	14630[b]	5439	5976	5227	33
Sep	6289	9315	6309	4791	10390	17030[c]	9573	14410	6195	7800	15
Oct	8125	7811	7291	5650	9049	23400	23230	30140[b]	8450	11340	10
Nov	12485	7291	26150[c]	8280	17180	22500	17930	23440	8020	16594	16
Dec	13979	17000	27010[c]	26420	19380	18780	14080	26450	10720	15328	23
Mean	13462	14830	21017[c]	17410	16433	18211	21311[b]	19086	14041	16885	9

[a] Preliminary data.
[b] Highest recorded monthly average flow since 1918.
[c] Second highest recorded monthly average flow since 1918.
From LMS 1980.

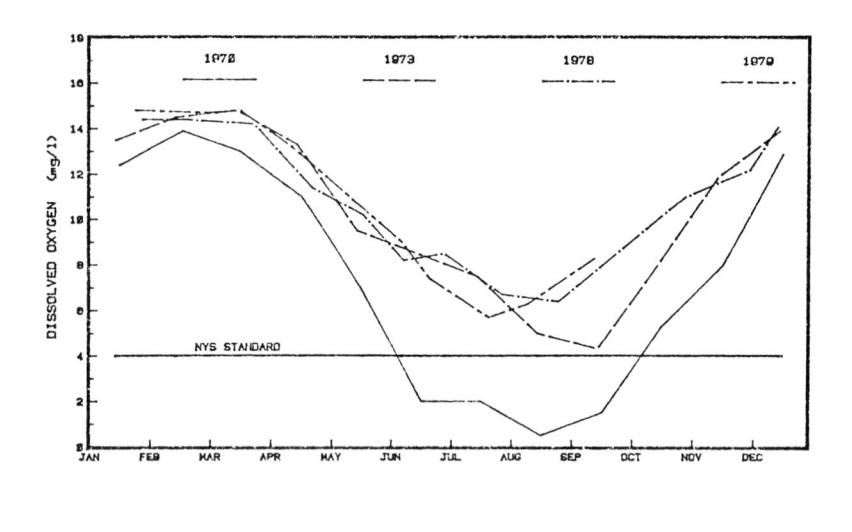

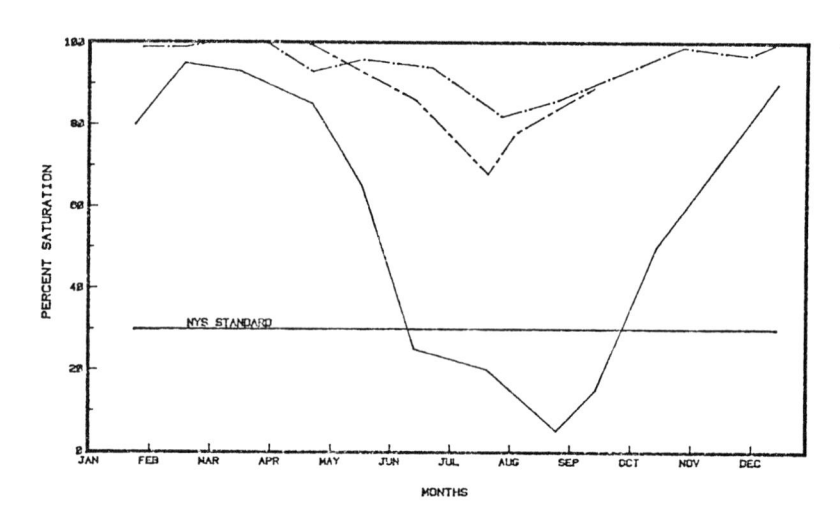

Figure 72. *Monthly dissolved oxygen values and percent saturation measured at Glenmont, N.Y. (1970, Q = 6,091 cfs; 1973, Q = 5,344 cfs; 1978, Q = 6,874 cfs; 1979, Q = 8,122 cfs)*

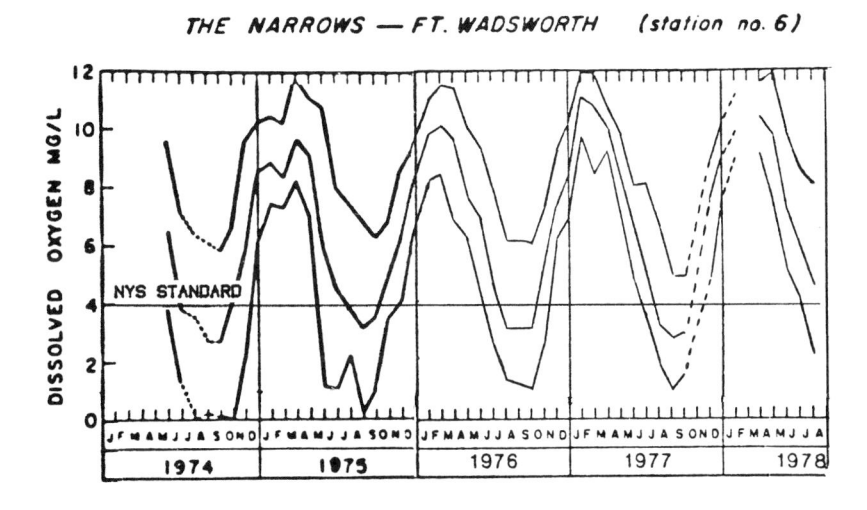

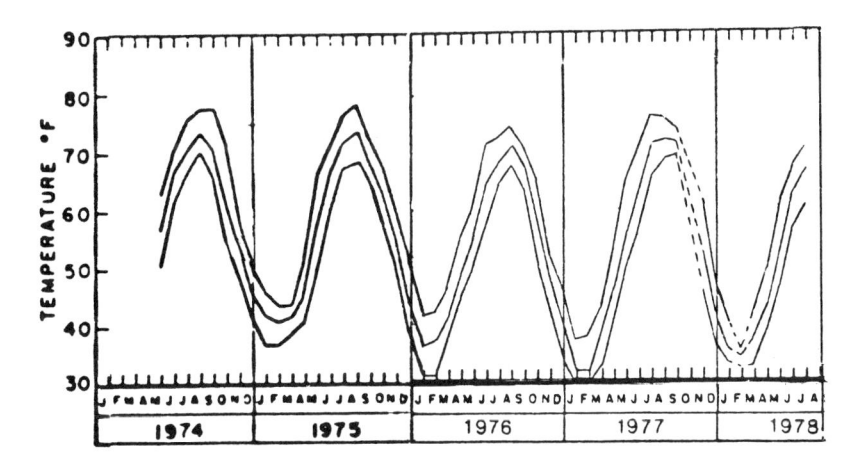

Figure 73. *Historical DO and temperature data. Top line—maximum monthly value; center line—average of the daily average values; bottom line—minimum monthly value.*

oxygen, but rising temperature tends to increase biological activity, which in turn increases oxygen demand on the system.

The depth-averaged DO concentrations measured from April 1979 through April 1980 at four stations between milepoints 0 and 9 (LMS 1980f) are presented in Figure 74. All four stations show a similar pattern: high values (9.0 mg/l–11.0 mg/l) in winter and spring, sagging below the standard in the summer, and climbing again in the fall. Figure 75 shows the same pattern in DO saturation for surface, mid-depth, and bottom measurements at a single station in the same region. Note that the surface level did not quite sag below the Class II standard shown as a frame of reference. The higher DO saturation at the surface is most probably attributable to the exchange of oxygen at the air-water interface and the stratified estuarine flow regime that typifies the New York City region during the summer.

The section of the Hudson that flows by New York City has often been characterized as having extremely low DO levels and being unable to support fish life. This condition has historically given the impression that this reach of the Hudson is a grossly polluted and unusable water body. The data presented have shown that dissolved oxygen concentrations in the area are well above the state standard for most of the year and fall below that level only during the summer months. Transient fish do indeed frequent the area in the noncritical months (Aleveras, 1973).

SUMMARY AND CONCLUSIONS

The data presented in this paper indicate that DO levels in the Hudson River show a distinct seasonal variation. In the 1970s, the DO levels were in the 9.0 mg/l–12.0 mg/l range over most of the river during the winter and spring, when flows were high and temperatures low. A riverwide decrease in DO was observed during the summer, with sags close to or below the state standard observed at Albany and New York City. The mid-river levels were comparatively higher than and showed seasonal variations similar to those at the head and mouth of the estuary. Studies such as the New York City 208 study (Hazen and Sawyer 1979) often focus on the critical summer period. Conclusions and environmental management programs are often formulated based on this worst-case condition. Care must be taken not to extrapolate these conclusions through time or space without sufficient data. The effects of environmental variables, particularly freshwater flow, on DO levels also must be considered when making management decisions.

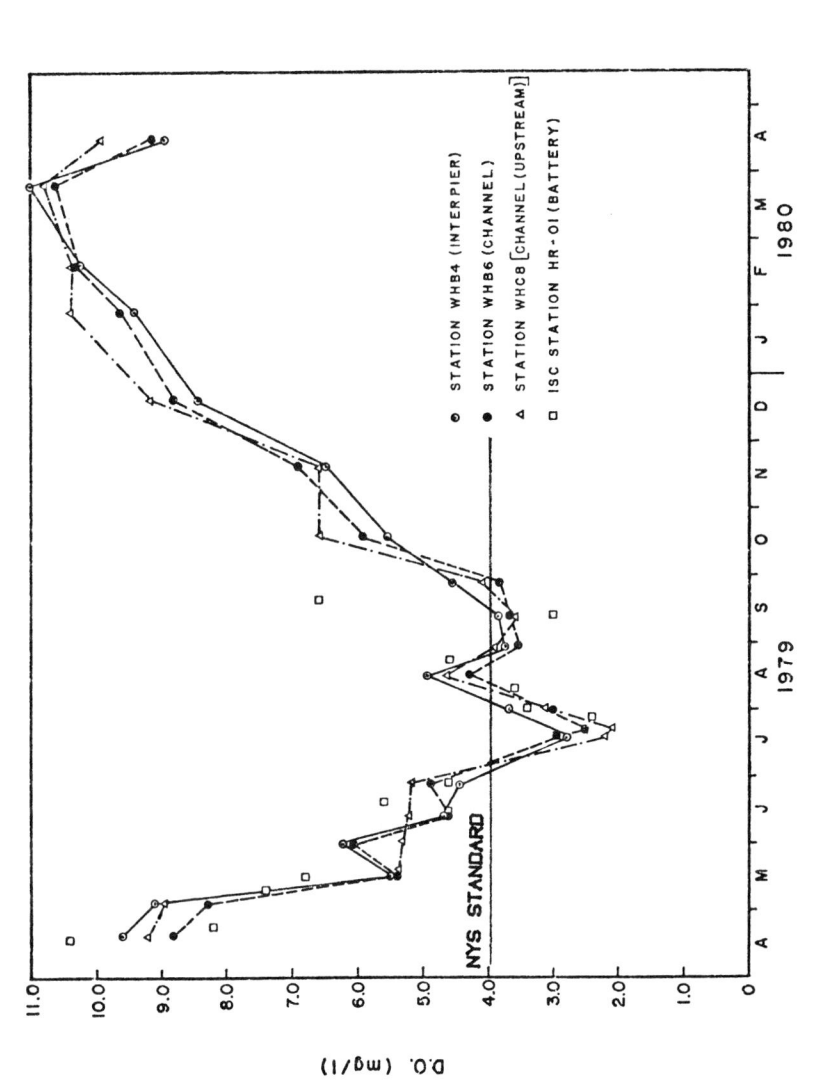

Figure 74. Mean dissolved oxygen concentration vs time, West Side Highway Project, 1979–1980. (mean of depths)

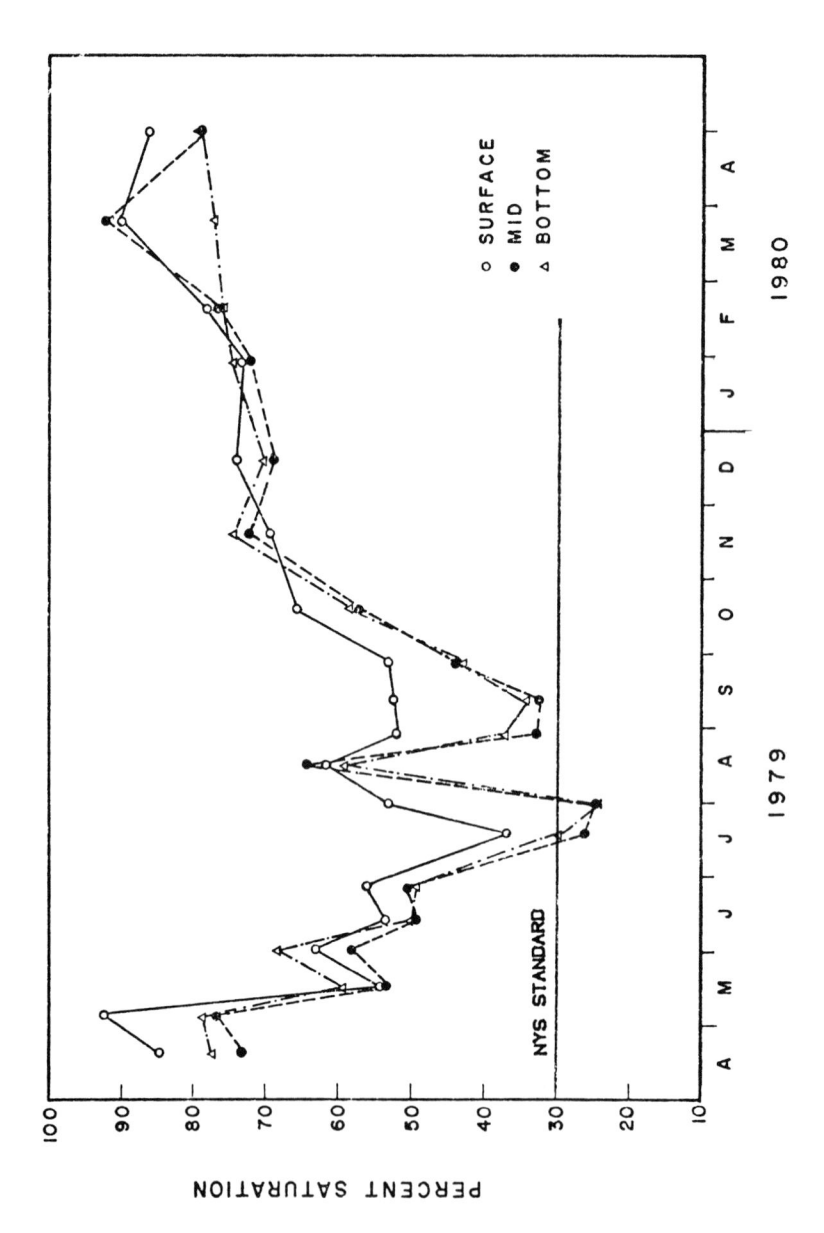

Figure 75. *Percent DO saturation at Station WHA3 (channel), West Side Highway Project, 1979–1980.*

The overall DO outlook for the Hudson River is encouraging. In the area north of New York City extending as far north as the Federal dam at Troy, municipal treatment facilities and elimination of other discharges have resulted in DO levels that are more than capable of supporting aquatic life throughout the year. Even in the New York City area, stressful DO conditions exist only during the summer months; during the rest of the year, the DO is above stressful levels. The Hudson River is very much alive and does not fit the septic, polluted label given to it a few decades ago.

RECOMMENDATIONS

Encouraging as the DO picture of the Hudson River may be, other chemical parameters also govern the health of the river. These include the more traditional parameters such as biochemical oxygen demand, nitrates, sulfates, and phosphates, as well as the toxic substances that have recently generated so much concern. Some of these parameters may not be as encouraging. Municipal STPs are designed to remove oxygen demand and some nutrients, not toxic substances. All of these parameters must be examined on a regular basis. Even then, accumulation of chemicals in sediments and tissues of organisms will not be detected by traditional water-quality monitoring programs; techniques such as bioassays and elutriate testing may be required to examine these potential problems.

An extensive data base exists, scattered throughout various reports by governmental agencies, the utilities, and private industry. There is a need to consolidate these data into a single riverwide data base for ready access. Before inclusion in the data base, the sampling and analytical procedures for each datum would have to be evaluated for acceptability, and questionable data would have to be eliminated. Some attempt would have to be made to normalize the data for environmental factors such as freshwater flows and changing waste loads. Only then could seasonal, annual, riverwide, and local trends be analyzed statistically.

Admittedly, preparation of such a consolidated data base would be a herculean task, but considerable benefits could be realized. Problem areas could be defined for parameters that have been traditionally monitored, and a baseline could be established to facilitate future comparisons for those parameters that have only recently received attention. A solid data base is one of the most useful tools for formulating and evaluating environmental management programs and legislation.

12. PCB Patterns in Hudson River Fish: I. Resident Freshwater Species

R. W. Armstrong and R. J. Sloan

INTRODUCTION

Massive contamination of Hudson River sediments and biota from prolonged industrial discharges at Ft. Edward and Hudson Falls, New York, has stimulated the most comprehensive monitoring of fish for PCB contamination ever undertaken (Hetling et al. 1978, Horn et al. 1979). A summary of the chronology of key events related to this major environmental perturbation is given in Table 60.

As a condition of the 1975 Settlement Agreement signed by General Electric Corporation (GE) and the New York State Department of Environmental Conservation (NYDEC), the active discharge of PCB from the GE capacitor manufacturing facilities was sharply curtailed in 1976 and terminated in June 1977 (Hetling et al. 1978). Under the same agreement, funds were made available for monitoring concentrations of PCB in fish collected annually at various locations below the discharge site (NYDEC 1976b).

The Hudson River fish-monitoring project, fully implemented in the spring of 1977, includes studies on PCB patterns in resident freshwater fishes and in migrant and marine species. This report deals with trends and patterns of PCB contamination in resident freshwater species and complements a companion paper in this volume.

METHODS

Collection and Preparation of Fish

All specimens were collected by the Bureau of Fisheries, Division of Fish and Wildlife, NYDEC, by electroshocking and gill nets. Whenever possible, thirty individuals of each species were collected. Annual collections were taken at the same time each year (± two weeks). The fish were tagged, measured, and weighed, then frozen whole in individual polyethylene bags. Later they were thawed and standard fillets were prepared, each consisting of an entire side of each fish

Table 60. *Timing of events related to the Hudson River PCB problem*

YEAR(S)	EVENT
1947(?)–1976	GE discharge of waste PCB at Fort Edward
1973	Dam at Ft. Edward retaining major PCB deposits removed
1975	PCB-contaminated biota linked to discharge
1976	Restrictions placed on commercial and recreational fishing in Hudson River
1976	GE/NYDEC Settlement Agreement
1976–1977	Last major spring flood stages in Hudson River
1977	PCB fish-monitoring project initiated
1979	PCDF detected in some Hudson River fish
1981	PCDD detected in some Hudson River fish
1982	"Hot-spot" sediment dredging scheduled

from the operculum to the base of the tail, with the scales removed. Brown bullhead fillets were skinned. The fillets were retagged, refrozen, and shipped via air priority freight to Raltech Scientific Services, Madison, Wisconsin, for analysis.

Analytical Procedures

At the laboratory, the fillets were ground and homogenized; then, weighed samples of the tissue were dried with anhydrous sodium sulfate and extracted with three portions of petroleum ether. An aliquot of the partially evaporated extract was run on a Florisil column, standardized to elute PCB Aroclors. A sample of the eluted extract was then injected into a gas chromatograph fitted with an electron-capture detector and standardized with Aroclors 1221, 1242, and 1254 obtained from the Monsanto Corporation. The column packing was 3% OV-1. Repeated tests confirmed that the Aroclor 1242 pattern obtained from fish extract is indistinguishable from that produced by Aroclor 1016. The limit of detection was 0.1 ppm, wet basis, for each Aroclor. Quantification was done by comparing several peak heights and areas to those produced by the respective Aroclors. The percent by weight of tissue soluble in petroleum ether was calculated from the mass of residue remaining after evaporation of a 5 ml portion of the original petroleum ether extracted to dryness at 40°C. This latter figure reflects the overall percentage of lipid material in the tissue.

A quality control protocol consisted of two components. Periodic performance samples were analyzed on a rotating basis by Raltech Scientific Services, other private contractors, and several laboratories of the NYDEC and the New York State Department of Health. In addition, three of each twenty samples were controls: a reagent blank, a duplicate analysis, and a spiked recovery sample.

Data Analysis

For reasons discussed below, all PCB concentrations (total PCB and specific Aroclors) have been converted to a lipid basis by dividing each analytical concentration of PCB in micrograms per gram wet tissue (ppm) by the fractional lipid content of the tissue sample. Resulting concentrations are referred to as lipid-based and are in units of micrograms per gram of lipid.

Temporal changes in total PCB and its Aroclors are calculated for given species and locations on a percentage basis from lipid-based PCB concentrations. PCB half-life calculations are based on the assumption that declines follow a simple first-order rate law:

$$\ln\frac{C}{C_o} = -kt = -\frac{0.693}{t_{1/2}} t$$

in which C = lipid-based PCB concentration at time t
C_o = lipid-based PCB concentration in initial year data
k = rate constant (year^{-1})
t = time (year)
$t_{1/2}$ = time (years) for a 50% decline in concentration

For those data sets including more than two years of measurements, a regression fit of the rate law was used to obtain a half-life and an average annual decline.

Half-life values obtained from these calculations do not describe the time required for a 50% PCB decline to occur in a given fish. Rather, they reflect the time required for a 50% lipid-based decline to be observed in a comparable population of the same species and residence collected in different years. Such half-lives depend on both the rate of PCB accumulation and its rate of depuration.

RESULTS AND DISCUSSION

Lipid-correlated PCB Concentrations

The earliest published PCB analyses on fish taken from the Hudson River below Fort Edward (Spagnoli and Skinner 1977) showed that species such as American eel with high lipid content contained substantially higher PCB concentrations than relatively lean species such as largemouth bass, yellow perch, and brown bullhead. Enhanced accumulation of PCB by lipid-rich fish is consistent with the distinctly hydrophobic character of PCB molecules, which exhibit octanol-to-water partition coefficients in the range 10^4 to 10^6 (National Research Council 1979).

Data collected since 1975 shows that lipid-dependence of PCB contamination of Hudson River fish even causes variation between individual fish of the same species. For example, PCB variation in 1979 collections of yearling pumpkinseeds is highly correlated ($P <$ 0.01) with the lipid content of individuals from the same location (Figure 76).

Highly significant PCB-lipid correlations appear to be the norm for Hudson River resident fish collected below Fort Edward, even when the collections contain a range of size classes. Correlations between total PCB and lipid in largemouth bass collected at Stillwater in 1977 and 1980 are shown in Figure 77. PCB-lipid correction coefficients are over 0.90 for both collections. By contrast, there are no significant correlations between PCB level and either length or weight. The highest PCB concentration (235 ppm) in the 1977 sample was in an 11-inch fish and the lowest value (6.2 ppm) was in a 12-inch specimen.

Average PCB correlations covering the entire array of data sets for resident species collected in two or more years at specific locations are summarized in Table 61. Specific values of the individual correlation coefficients for these collections are listed in Table 62. Of the thirty-two collections, twenty-eight had highly significant ($P < .01$) PCB-lipid correlations. Only two small collections, each with fewer than ten fish, did not show significant correlations ($P < .05$) between PCB and lipid. Conversely, only eight of the collections had significant correlations between PCB level and fish length.

Lipid Dependence and River Flow

Good correlations between PCB levels and lipid content have also been found in fish from the PCB-contaminated Furans River in France (Keck and Raffenot 1979), and similar lipid-dependent PCB concentration patterns have been observed in fish from several other New York stream systems, including the Hoosic River and the Valatie Kill in Rensselaer County, and the Grass River at Massena (NYDEC unpublished). A major dominant source of PCB has been identified, or at least is highly suspected, in each of these systems.

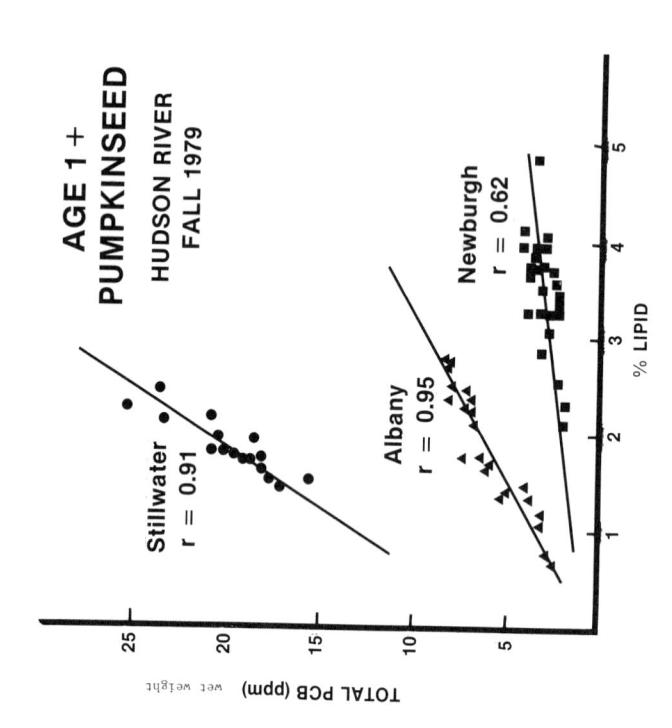

Figure 76. Variation in total PCB concentration (wet basis) with lipid content of yearling pumpkinseed collected in 1979 at three Hudson River locations: (●), Stillwater; (○), Albany; (■), Newburgh.

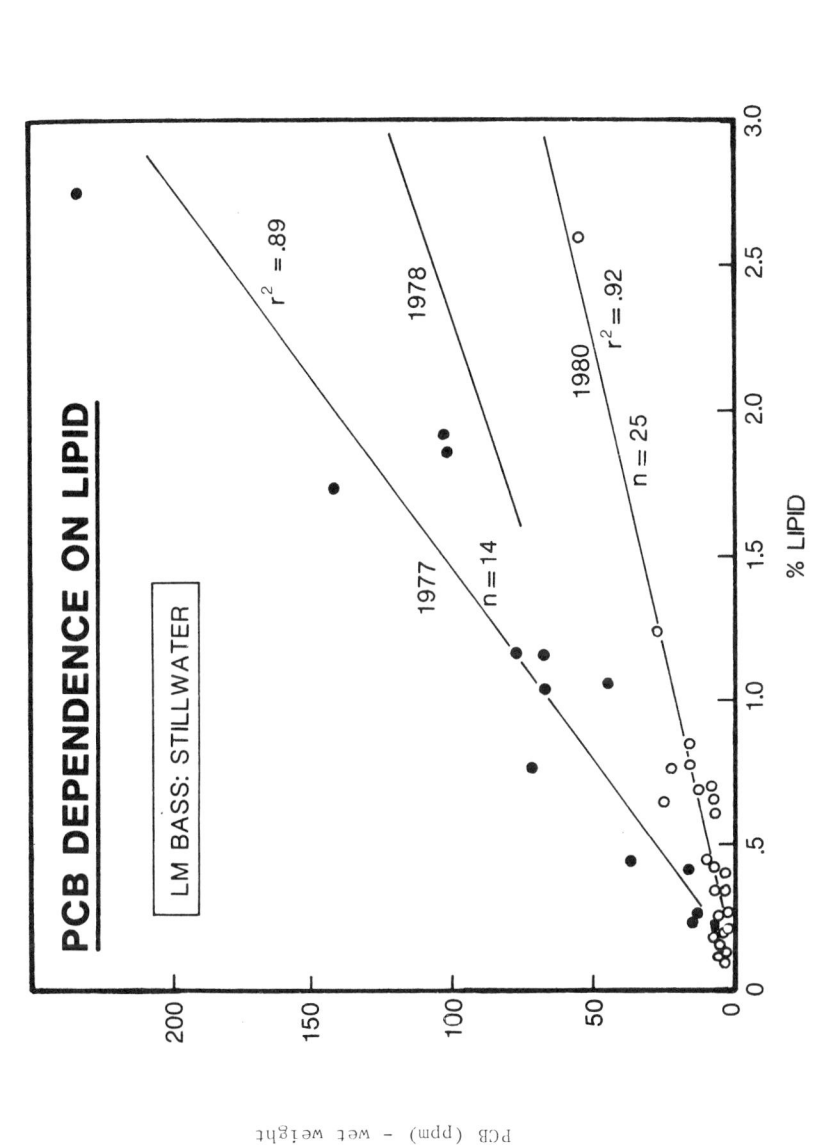

Figure 77. Relationships between total PCB concentration (wet basis) and lipid content for largemouth bass collected at Stillwater in 1977 (●), and in 1980 (○). A portion of the regression line for the 1978 sample is also included.

Table 61. *Average correlation coefficients relating total PCB concentrations to length and lipid content of resident fish collected at several Hudson River locations.*

LOCATION	SPECIES[a] ANALYZED 1977	1978	1979	1980	AVE. r̄, TOTAL PCB VS. LIPID	VS. LENGTH[b]
Above Glens Falls	—	—	Pksd	Pksd	0.09	—
Stillwater	BB Gold LMB YP	Gold LMB	BB Pksd	BB Gold LMB Pksd YP	0.83	0.27
Albany/Troy	BB WP	BB WP	BB Pksd	BB Pksd WP	0.79	0.05
Catskill	LMB YP	LMB Rbs	—	LMB RbS YP	0.61	0.46
Newburgh	—	—	Pksd	Pksd	0.53	—

[a]Species abbreviations are identified in index.
[b]Lengths of yearling pumpkinseed not determined.

Significant PCB-lipid correlations are not generally found in fish from large lakes such as Cayuga Lake (Bache et al. 1972, NYDEC unpublished) and Lake Ontario (NYDEC unpublished). Thus a lipid-correlated PCB contamination of fish appears to be a characteristic of river systems with a dominant source of contamination. Apparently accumulation of PCB in individual fish living downstream from the source is driven by the relatively homogenous and unidirectional flux of PCB which tends to produce concentrations that are closely correlated with the lipid content of the individual fish. If so, it is reasonable to expect that these correlations will decrease with increasing downstream distance from the source as the gradient declines and the unidirectional movement of PCB is perturbed by erratic secondary sources, tributary mixing, and tidal fluctuations. Table 61 shows this to be the case although the correlations are significant for resident species collected as far south as Newburgh.

Similar lipid-based accumulation patterns might also result from concerted homogeneous fluxes of virtually any hydrophobic contaminant. Support for this "impacted riverine" fish contamination model is developed further in this report and in the following paper focused on migrant and marine species in the Hudson River estuary.

Temporal Trends in PCB Contamination

Temporal trends in levels of PCB contamination can be inferred from annual collections of fish with equivalent lipid levels. Since lipid content cannot be determined in the field, one must adjust analytical results based on wet tissue by calculating the concentration of PCB per unit weight of lipid.

PCB (μg PCB/g lipid)

= PCB (μg PCB/g flesh)/Lipid fraction (g lipid/g flesh)

Averages of lipid-based PCB concentrations will be independent of individual variations in lipid contents and provide a better measure of trends in PCB levels in the populations. The general lack of correlation between PCB concentration and fish size implies that fish do not need to be of the same size or age to be used in temporal trend analysis. An ultimate test of age-independence, however, would require monitoring of several age classes, including immature stages, over several years.

PCB Trends in Mature Resident Fishes

Comparative samples of mature freshwater resident species have been obtained since 1977, primarily at three locations: Stillwater, approximately 25 miles south of Fort Edward; the Albany-Troy area below the Federal dam; and Catskill, about 85 miles south of Fort Edward. A detailed tabulation of PCB concentrations and related data is presented in Table 62, and trends in lipid-based PCBs are shown in Figures 78–80. An overall summary of annual changes in PCB concentrations and half-life values is presented in Table 63.

At all locations, there were prominent declines in total PCB between 1977 and 1980 for all species. The average annual decline in total PCB for fish from these locations is 34.0 ± 12.6%. It is evident that these decreases are mostly due to declines in Aroclor 1016, the mixture used almost exclusively by GE in the last several years of active discharge (NYDEC 1976a). Decreases in Aroclor 1016 concentrations are uniform for all species and locations, corresponding to an average annual decline of 47.3 ± 10.0%, and an average half-life of 1.15 ± 0.38 years. Analysis of year-by-year declines in Aroclor 1016 gives no indication that the annual decreases of recent years are any less than those observed between 1977 and 1978; that is, declines in Aroclor 1016 continue to mimic a true first-order process.

Table 62. *Selected parameters related to PCB concentrations in standard fillets of mature Hudson River fish collected 1977–1980.*

LOCATION	SPECIES	YEAR	NUMBER	% LIPID	TOTAL PCB[a] (ppm, wet)	LIPID-BASED PCB (ppm) ARO1016	ARO1254	TOTAL PCB[a]	CORR. COEFF. (r) TOTAL PCB-LIPID	TOTAL PCB-LENGTH
Stillwater	BB	1977	30	4.74	106.5 ± 49.2	1908 ± 799	388 ± 253	2508 ± 1056	0.52	0.28
	BB	1979	30	0.77	8.97 ± 12.26	734 ± 359	589 ± 567	1336 ± 854	0.93	0.00
	BB	1980	30	1.23	12.34 ± 6.56	694 ± 190	750 ± 290	1479 ± 466	0.88	0.33
	Gold	1977	16	10.25	559.4 ± 506.8	3961 ± 3065	589 ± 467	5255 ± 3700	0.89	0.56
	Gold	1978	29[b]	7.93	273.6 ± 237.4	2684 ± 1278	565 ± 330	3571 ± 1645	0.83	0.70
	Gold	1980	30	6.70	72.62 ± 55.42	537 ± 326	660 ± 424	1206 ± 654	0.79	0.14
	LMB	1977	14	1.08	70.72 ± 62.04	4470 ± 1589	1114 ± 333	6010 ± 2020	0.94	−0.22
	LMB	1978	30	3.84	153.08 ± 81.57	3135 ± 1175	915 ± 413	4318 ± 1588	0.72	0.41
	LMB	1980	25	0.54	10.44 ± 13.83	840 ± 347	868 ± 379	1735 ± 722	0.96	0.33
	YP	1977	29[b]	0.38	12.60 ± 8.85	2555 ± 1295	851 ± 353	3725 ± 1690	0.83	−0.26
	YP	1980	7	0.07	0.84 ± 0.60	450 ± 171	507 ± 272	957 ± 420	0.79	0.72

Albany/Troy	BB	1977	30	4.34	37.90 ± 27.90	676 ± 422	185 ± 115	904 ± 511	0.57	0.10
	BB	1978	11	5.06	25.16 ± 10.46	359 ± 117	101 ± 38	515 ± 146	0.68	−0.40
	BB	1979	22	2.31	7.15 ± 9.20	169 ± 88	136 ± 75	306 ± 139	0.89	−0.10
	BB	1980	21	1.42	2.09 ± 1.66	96 ± 63	88 ± 64	206 ± 135	0.87	0.52
	WP	1977	30	9.14	118.4 ± 73.2	1066 ± 840	182 ± 146	1365 ± 976	0.66	−0.10
	WP	1978	30	9.03	85.4 ± 41.1	715 ± 187	171 ± 87	948 ± 229	0.87	0.10
	WP	1980	30	5.16	16.04 ± 9.87	122 ± 72	182 ± 91	316 ± 129	0.71	0.24
Catskill	LMB	1977	27	1.26	29.56 ± 19.33	1732 ± 959	671 ± 500	2436 ± 1170	0.79	0.70
	LMB	1978	18	1.83	28.96 ± 21.17	1034 ± 649	539 ± 450	1600 ± 1056	0.62	0.59
	LMB	1980	20	0.42	1.08 ± 0.69	119 ± 76	183 ± 133	350 ± 223	0.83	0.66
	RbS	1978	20	0.93	4.08 ± 2.42	247 ± 132	195 ± 117	458 ± 231	0.56	0.31
	RbS	1980	20	1.66	2.63 ± 5.51	98 ± 70	223 ± 170	380 ± 287	0.90	0.28
	YP	1977	20	0.39	4.58 ± 3.19	1080 ± 741	367 ± 334	1497 ± 1081	0.71	0.67
	YP	1980	9[b]	0.23	0.54 ± 0.31	67 ± 75	164 ± 141	277 ± 168	−0.14	0.00

[a] Total PCB includes Aroclor 1221; all PCB concentrations are arithmetic means with standard deviations specified.
[b] One fish omitted as statistical outlier.

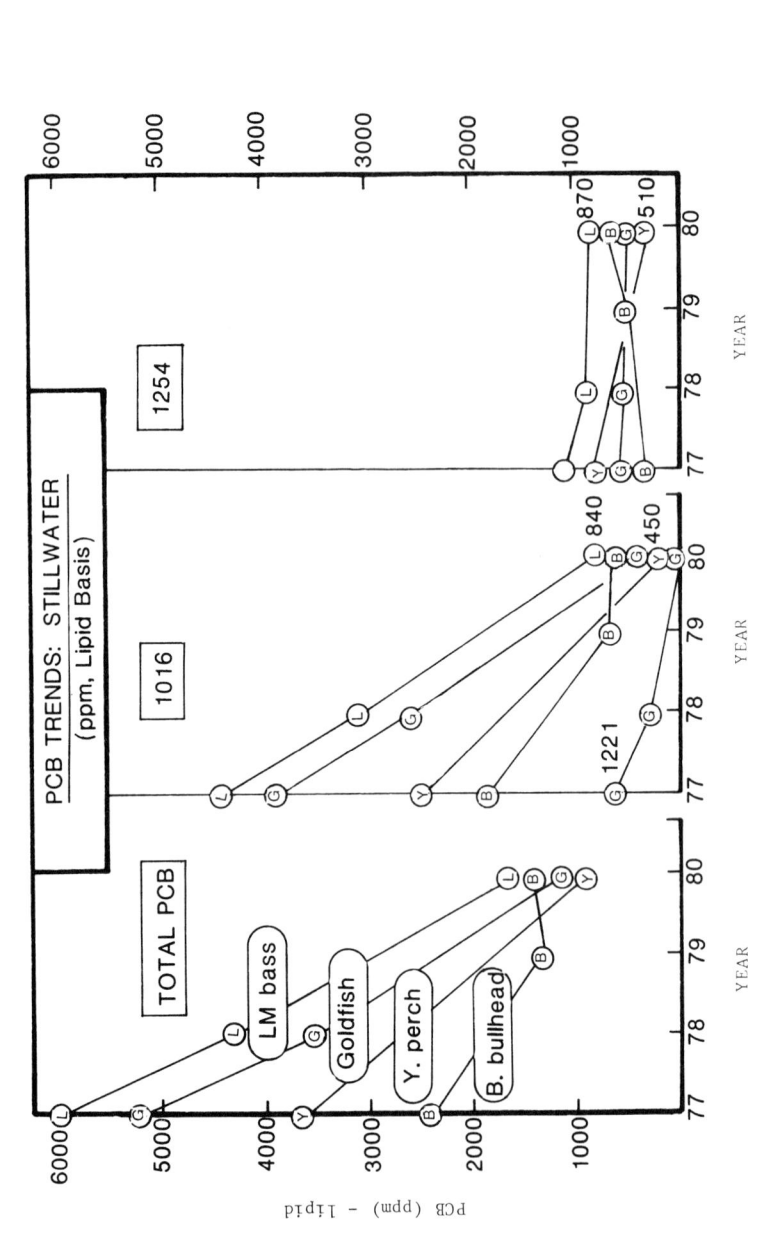

Figure 78. *Temporal changes for lipid-based total PCB and its component Aroclors for four species of resident fish taken at Stillwater from 1977 to 1980. An example of the rapid decline of Aroclor 1221 is also included.*

Figure 79. Temporal changes for lipid-based total PCB and its component Aroclors for two species of resident fish taken at Albany/Troy from 1977 to 1980. An example of the rapid decline of Aroclor 1221 is also included.

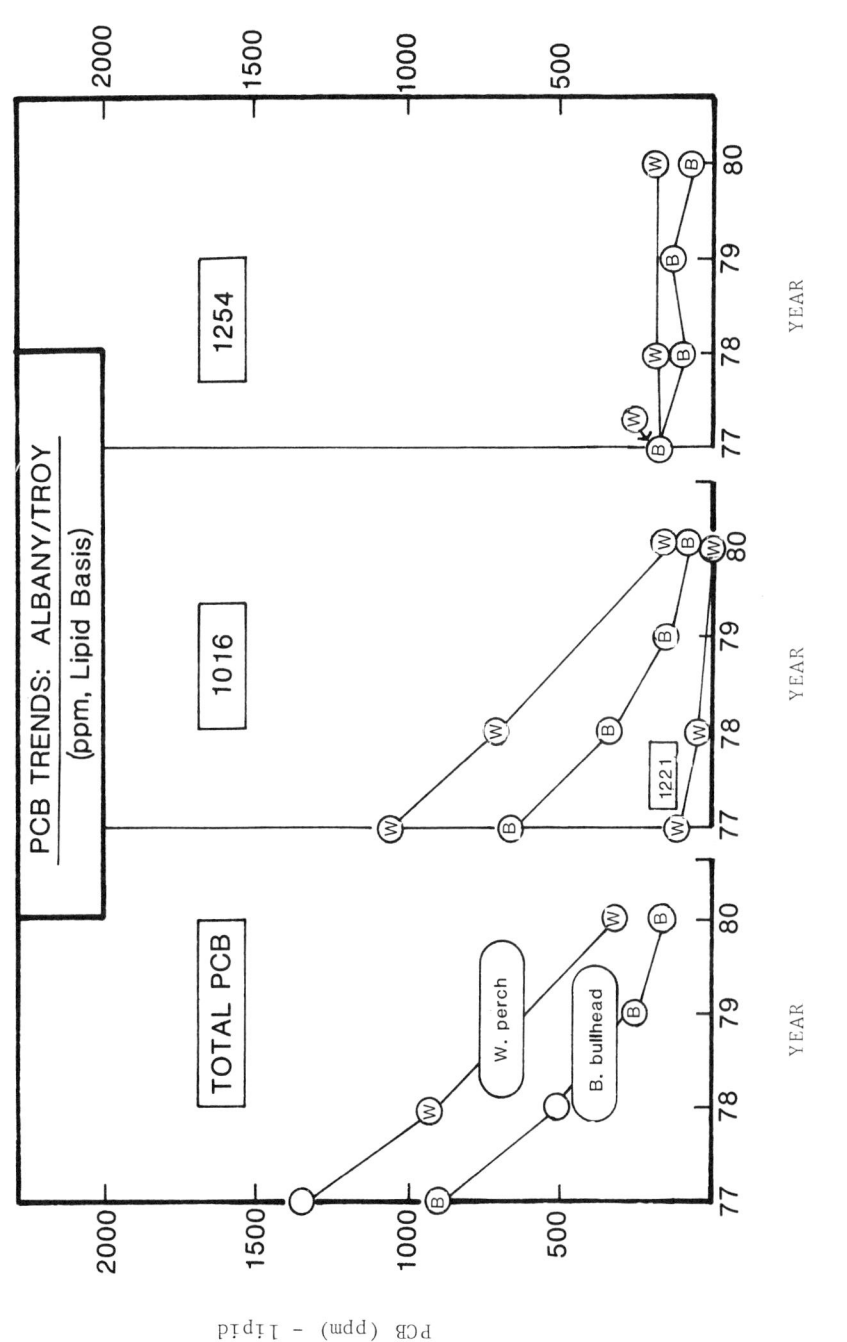

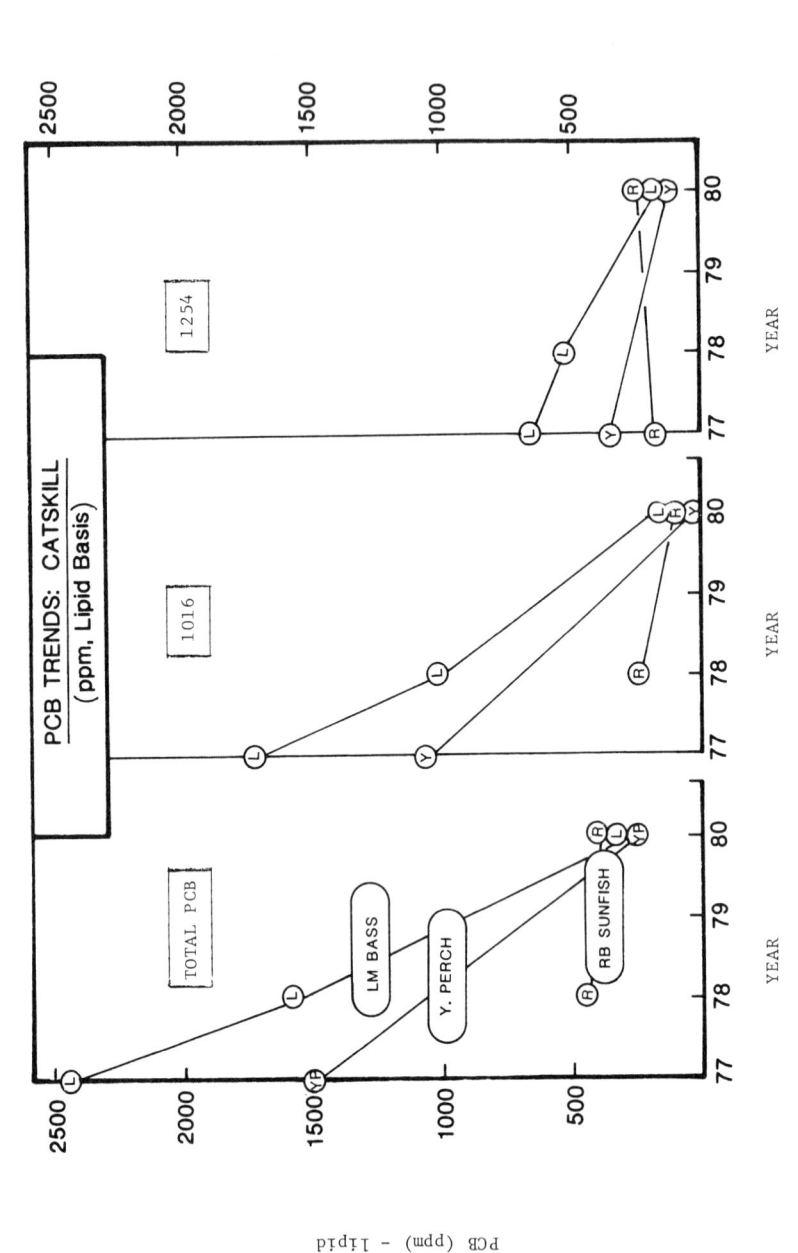

Figure 80. Temporal changes for lipid-based total PCB and its component Aroclors for three species of resident fish taken at Catskill from 1977 to 1980.

Table 63. *Calculated half-lives and annual rates of change of PCB for collections of resident Hudson River fish.*

LOCATION	SPECIES	% LIPID	DATA YEARS	HALF-LIFE (yr) ARO 1016	HALF-LIFE (yr) ARO 1254	HALF-LIFE (yr) TOTAL PCB	ANNUAL RATE OF DECLINE (%) ARO 1016	ANNUAL RATE OF DECLINE (%) ARO 1254	ANNUAL RATE OF DECLINE (%) TOTAL PCB
Stillwater	BB	2.25	77,79,80	1.94	(Increase)	3.54	−30.0	(+25)	−17.8
	Gold	7.93	77,78,80	1.00	(Increase)	1.39	−50.0	(+4)	−39.3
	LMB	2.08	77,78,80	1.21	9.23	1.65	−43.6	−7.2	−34.3
	YP	0.32	77,80	1.20	4.01	1.53	−43.9	−15.9	−36.4
	Stillwater means			1.34	—	1.47	−41.9	(+1.5)	−32.0
Albany/Troy	BB	3.17	77,78,79,80	1.05	3.59	1.40	−48.3	−17.6	−39.0
	WP	7.78	77,78,80	0.93	No change	1.39	−52.5	0	−39.3
	Albany/Troy means			0.99	—	1.40	−50.4	−8.8	−39.2
Catskill	LMB	1.15	77,78,80	0.75	1.80	1.05	−60.3	−32.0	−48.3
	RbS	1.30	78,80	1.50	(Increase)	7.45	−37.0	(+6)	−8.9
	YP	0.34	77,80	0.75	2.59	1.23	−60.3	−23.5	−43.1
	Catskill means			1.00	—	3.24	−52.5	−16.5	−33.4
Overall means, 3 locations				1.15 (±0.38)	—	2.29 (±2.07)	−47.3 (±10.0)	−6.8 (±17.5)	−34.0 (±12.6)

Patterns of change in concentration of Aroclor 1254, the more highly chlorinated PCB component, are considerably more complex for these resident trend samples. There have been few substantial declines in this Aroclor, and three trend samples (brown bullhead and goldfish at Stillwater; redbreast sunfish at Catskill) show small but significant increases in Aroclor 1254 content between 1977 and 1980. The overall average change in levels of 1254 for the entire data set, including nine mature resident species is only $-6.8 \pm 17.5\%$. This is only about one-tenth the average Aroclor 1016 decrease.

Although the much slower decrease in Aroclor 1254 is consistent with the greater stability and lower volatility of its more highly chlorinated molecules (Horn et al. 1979, Hansen 1979, National Research Council 1979), other factors may also act to retard its apparent rate of decline in fish flesh. Since Aroclor 1254 was the mixture most widely used by industrial companies, it is likely that the river continues to receive Aroclor 1254 from many low-level secondary sources including airborne contamination and as leachate and discharge from various sources on the main river and its tributaries. Another reason for the lack of decline in Aroclor 1254 levels may be related to the problem of achieving unambiguous quantification of the Aroclor components of a PCB mixture. A number of lower-chlorinated PCB homologs of a PCB mixture apparently are eliminated or depurated more rapidly by fish than are the PCBs themselves (Bache et al. 1972, National Research Council 1979). It is possible that chromatograms of "old" PCB may be interpreted in ways that underestimate Aroclor 1016 and overestimate Aroclor 1254.

There is strong evidence that higher chlorinated PCB homologs continue to contaminate Hudson River fish at levels comparable to those present several years ago. A similar persistence of Aroclor 1254 also has been reported by Wilson and Forester (1978), who observed that levels of this contaminant continued to remain in the range of 5 ppm (wet basis) in the oysters of Escambia Bay, Florida, for several years after the source of PCB had been completely abated. To the extent that a relatively nondeclining pattern of Aroclor 1254 accumulation is correlated with residual sediment contamination by this mixture, it follows that the proposed "hot spot" dredging project is necessary to aid full recovery of the Hudson River from PCB accumulation, in spite of the relatively rapid declines in Aroclor 1016.

These complexities in temporal changes in Hudson River PCB fish contamination highlight the caution that must be exercised in interpreting the magnitude of a PCB problem in terms of summed

Aroclor concentrations. Each of the 209 homologs of PCB possesses its own physical, chemical, and toxicological properties. Until PCB quantification is more precisely documented in terms of specific homolog content, the implications of temporal variations in total PCB levels on fish, other biota, and even human fish-consumers cannot be evaluated.

Aroclor 1221, composed primarily of monochlorinated PCB, has dropped below the detection limit in virtually all resident fish tested in 1980 (Figures 79 and 80). The rapid decline of this PCB component is consistent with the high relative reactivity and high volatility of monochlorinated PCB (National Research Council 1979).

PCB Trends in Yearling Pumpkinseed

A special monitoring project was initiated to discern patterns of PCB accumulation in a Hudson River fish species during a single year of exposure. One-year-old pumpkinseeds, because of their availability at a wide range of locations, were selected for this annual study. The fish were analyzed whole and composited in sets of three or four individuals to reduce variance. PCB concentrations and other data from the 1979 and 1980 collections are given in Table 64.

The data for pumpkinseed for both years shows a distinct pattern of diminishing accumulation from Stillwater to south of Newburgh. This is similar to the decreasing downriver trend shown by other resident species. Average PCB levels in these one-year-old pumpkinseeds are relatively high and highly correlated with individual lipid contents. Pumpkinseeds collected above Glens Falls constitute control samples, as they were taken above the original discharge site at Fort Edward. These samples contain low PCB in the range of 0.5 ppm (wet basis), a value only 0.2 ppm higher than the detection limit of the summed Aroclors. PCB concentrations for these control samples are not correlated with lipid content, a result typical of waterways without dominant sources of PCB.

Both at Stillwater and at Albany, the one-year-old pumpkinseeds showed significant decreases in Aroclor 1016 levels comparable to those documented for mature fish from the same locations. The small but significant 1979–1980 increase in Aroclor 1016 in Newburgh pumpkinseeds is the *only* increase in lipid-based concentration of Aroclor 1016 in any of the twelve sets of temporal trends. Correlations between PCB and lipids for the Newburgh collections were less significant than those from upriver, an indication of a more random

Table 64. Selected parameters related to PCB concentrations in yearling pumpkinseed collected at various Hudson River locations in 1979 and 1980

	ABOVE GLENS FALLS 1979	ABOVE GLENS FALLS 1980	STILLWATER 1979	STILLWATER 1980	ALBANY/TROY 1979	ALBANY/TROY 1980	NEWBURGH 1979	NEWBURGH 1980
Number	68[a]	72[a]	64[a]	75[b]	88[a]	75[b]	100[a]	75[b]
% lipid	3.19	3.90	1.86	3.20	1.85	3.94	3.49	4.91
Total PCB[c], wet basis (ppm)	<0.44 ±0.09	<0.60 ±0.10	19.91 ±2.56	21.73 ±4.22	5.89 ±1.93	16.74 ±2.38	3.03 ±0.69	4.63 ±1.46
Correlation (r), PCB-% lipid	−0.09	0.26	0.90	0.73	0.95	0.90	0.62	0.43
Total PCB[c], lipid basis (ppm)	14.1 ±3.6	15.5 ±2.8	1080 ±71	686 ±100	323 ±41	426 ±26	88 ±17	94 ±27
Aroclor 1016, lipid basis (ppm)	<3.4 ±0.8	<3.8 ±1.2	686 ±43	486 ±77	206 ±28	138 ±9	33 ±5	41 ±10
Aroclor 1254, lipid basis (ppm)	7.5 ±2.7	9.1 ±1.6	386 ±36	196 ±39	111 ±15	285 ±24	53 ±10	51 ±18

[a] Analyzed as composites of 4 whole fish.
[b] Analyzed as composites of 3 whole fish.
[c] Total PCB includes Aroclor 1221; all PCB concentrations are arithmetic means with standard deviations specified.

accumulation pattern for these fish in both data years. As the pumpkinseed monitoring program continues, this Newburgh exception to the pattern of Aroclor 1016 declines should be studied closely.

Changes in levels of Aroclor 1254 in yearling pumpkinseed between 1979 and 1980 present a perplexing pattern: a substantial decline at Stillwater, a major increase at Albany, and no significant change at Newburgh.

Although the Newburgh result is consistent with the patterns generally observed for more mature fish, the large decline in Aroclor 1254 for the Stillwater collections is surprising. At a given location and time, the ratio of the concentrations of Aroclor 1016 to Aroclor 1254 is nearly species-independent. At Stillwater in 1980, for example, four species averaged 1016/1254 ratios of 1.00, 0.94, 0.99, and 0.94. In contrast, the 1980 1016/1254 ratio for Stillwater pumpkinseed is 2.54, with clear indications that the high value stems from an exceptionally low concentration of Aroclor 1254 relative to that in other species from Stillwater. Patterns of accumulation in yearling pumpkinseed, however, may be sensitive to the precise 1016/1254 ratio characterizing a very localized domain of river sediment, and the "patchiness" of sediment contamination has been documented for this stretch of the river (Tofflemire et al. 1979).

Pumpkinseed collected in the Albany Turning Basin show an increase of 157% in lipid-based Aroclor 1254 between 1979 and 1980. The 1980 value (285 ppm) is even significantly higher than that found in upriver Stillwater pumpkinseed (196 ppm). It is known, however, that a maintenance dredging operation was carried out in late 1979 in the same general area of the Turning Basin where the 1980 pumpkinseed sample was obtained (J. Tofflemire, personal communication). The elevated levels of PCB in these fish in 1980 may be related to the resuspension of PCB-laden sediments, although it is necessary to assume that the resuspended material was characterized by a very high 1254/1016 ratio.

The pumpkinseed monitoring project provides a great deal of insight into annual changes in the PCB-accumulation potential of resident species, but decipherable patterns may only emerge after several years of additional data have been gathered. In future years, an effort will be made to supplement yearling collections with other yearclasses of pumpkinseed so that annual changes in the PCB concentrations of older fish can be compared with the changes observed in the yearlings. These data should also provide direct information on the possibilities of either age-related uptake or PCB depuration as a result of changes in contaminant availability.

PCB Declines and River Flow Conditions

An overall pattern of major PCB declines in Hudson River resident fish since 1977 is unquestionably present. It is tempting to conclude that these declines are entirely related to the total cessation of active PCB discharge at Fort Edward in 1977. Apart from the likelihood that other less prominent sources of PCB may also have been abated in recent years, the foregoing conclusion must be held in abeyance because of recent patterns of flow conditions in the river.

In October 1973 a dam located about one mile below the PCB discharge site was removed. Several years later, in 1976 and 1977, spring flows in the upper Hudson River exceeded the 50-year flood flows (Malcolm Pirnie, Inc. 1976, U.S. Geological Survey 1977). Erosion of sediments which had collected behind the dam necessitated extensive dredging to maintain the navigation channel. Data collected in 1977 and 1978 indicated that much of the displaced sediment was highly contaminated with PCB (Hetling et al. 1978). It is possible that a significant fraction of the very high PCB levels found in 1977 and 1978 fish samples may have resulted from the resuspension of PCB-laden sediments by the exceptional 1976 and 1977 spring floods. USGS hydrologists have demonstrated that total PCB concentrations in the river water rose dramatically under the high flow conditions associated with those floods (Hetling et al. 1978). It is not at all clear, however, how much of this measured concentration, which includes a large particulate fraction, is readily available for accumulation by fish.

In any event, high flows did not occur in the upper Hudson River between 1977 and 1981. Thus, the systematic PCB declines reported in this paper have unfolded entirely during a period of stable average flow conditions. If the sharp increase in PCB accumulation by Albany pumpkinseeds between 1979 and 1980 was indeed related to sediment resuspension caused by local dredging, it is surely possible that resuspension caused by future flood conditions could lead to a dramatic general reversal of the declining PCB levels of recent years. Prudent planning appears to dictate that upriver dredging—despite its potential for short-term localized impacts from resuspension—will in the long-run both lower the risk of massive floodwater-induced resuspension of PCB and perhaps also serve to hasten the decline in the current levels of Aroclor 1254.

Current Levels of PCB in Hudson River Resident Fish

The 1982 limit allowed by the U.S. Food and Drug Administration (FDA) for total PCB in the edible flesh of fish offered for sale in interstate commerce was 5 ppm, wet basis. In 1984 a reduction to 2 ppm was implemented. Although FDA limits have no direct connection to sportfish consumption, most state agencies are strongly guided by these limits in formulating consumption advisories for sportfishermen.

A listing of recent average concentrations of total PCB found in standard fillets of various Hudson River fish is given in Table 65. Since contamination levels vary with specific location, the overall averages are intended to show only general patterns of contamination.

Limited sampling of fish from the river north of Glens Falls reveals low levels of PCB contamination of fish in the range of 0.5 ppm, well below the FDA limit.

The region between Fort Edward and the Troy Dam has been closed to both commercial and recreational fishing since 1976. Although major declines in levels of PCB have occurred in fish from this general location (Table 62), virtually all species tested continue to average well over the FDA limit. Only a 1980 sample of yellow

Table 65. *Average levels of PCB in Hudson River resident fish (wet basis)*

LOCATION	SPECIES	YEAR ANALYZED	% LIPID	PCB (ppm)
Above Glens Falls	BB	1979	0.4	0.4
	Pksd	1980	3.9	0.6
Fort Edward	BB	1980	0.9	12
to Federal Dam	Gold	"	6.7	73
at Troy	LMB	"	0.5	10
(CLOSED TO	Pksd	"	3.2	20
FISHING)	YP	"	0.1	0.8
Troy to Battery	BB	1980	1.4	2.1
	Carp	1979	10.2	41
	CP	1980	0.2	1.0
	LMB	"	0.4	1.1
	NP	"	0.3	2.3
	Pksd	"	4.4	11
	Pp	"	1.2	6.2
	WC	1979	3.5	12
	WP	1980	5.2	17
	YP	"	0.2	0.5

perch with a remarkably low average lipid content of 0.1% fell below the FDA limit.

The large variety of resident species analyzed from various locations south of the Federal dam at Troy shows a wide range of PCB contamination, generally well-correlated with the average lipid content. As a rule, resident species averaging more than about 3% lipid (carp, pumpkinseed, white catfish, white perch) substantially exceed 5 ppm, whereas the leanest species (chain pickerel, largemouth bass, northern pike, yellow perch) averaged only 1–2 ppm in 1980.

It may be significant that the values obtained for the average lipid contents of many Hudson River fishes in recent years are surprisingly low in relation to the values observed in earlier years. Possibly these low fat (lipid) contents reflect the intense competition for food among increasing populations of fish in an essentially unharvested resource.

CONCLUSIONS

Standard fillets of resident Hudson River fish contain PCB concentrations that are highly correlated with their lipid contents, and are typically unrelated to the size of the fish. Lipid-based accumulation arises because of a relatively rapid PCB uptake by fish exposed to spatially uniform fluxes of lipophilic PCB.

Between 1977 and 1978, total lipid-based PCB declined an average of 34% per year in comparable populations of resident species from Stillwater south to Catskill. These PCB declines have resulted from sharply declining rates of accumulation of Aroclor 1016. Fish levels of Aroclor 1254, the more highly chlorinated mixture, did not decline appreciably in resident fish between 1977 and 1980.

Declining patterns of PCB contamination have appeared during a period of stable flow conditions, and do not reflect the potential impact of floodwater resuspension of sedimentary PCB residues.

In spite of major declines, PCB levels in the flesh of nearly all fish sampled between Fort Edward and Troy continue to exceed FDA limits as do a number of lipid-rich resident species from the lower river.

13. PCB Patterns in Hudson River Fish: II. Migrant and Marine Species

R. J. Sloan and R. W. Armstrong

INTRODUCTION

The Hudson River has a well-developed saline estuary in addition to a strictly freshwater tidal section with tributary streams. A number of marine and migrant fishes use the Hudson during various parts of their life cycles, and it is of interest to ask whether the biology of the migrant species differs sufficiently from that of freshwater species to give rise to different patterns of accumulation of chemical contaminants such as polychlorinated biphenyls. The purpose of this paper, a companion to the Armstrong and Sloan article in this volume is fourfold: (1) to determine if the PCB accumulation pattern in migrant species is similar to that of resident species; (2) to examine the reasons for the similarities or dissimilarities; (3) to evaluate particular species as indicators that can be used to describe further trends in PCB contamination of the Hudson River biota; and (4) to document current PCB concentrations in a variety of species, including those where insufficient temporal data are available to evaluate trends in PCB levels.

METHODS

Procedures related to the collection, preparation, and analysis of data are described in the preceding paper by Armstrong and Sloan. All whole organisms and organ samples were prepared separately by the fisheries staff at the New Paltz Office, Region 3, New York State Department of Environmental Conservation, prior to shipment to the analytical laboratory (Raltech Scientific Services, Madison, Wisconsin). Samples of American eel fillets were analyzed with the skin removed, the accepted U.S. Food and Drug Administration procedure for "scaleless" species. Samples of blue crab tissue (muscle, hepatopancreas) were removed from the exoskeleton before analysis.

Statistical analysis follows the procedures in Sokal and Rohlf (1969) and Steel and Torrie (1960). Percent or first-order PCB declines were developed with the procedures outlined by Armstrong and Sloan.

During the course of the Hudson River PCB Monitoring Project, quality assurance samples provided by the New York State Department of Health indicated that Raltech Scientific Services analyses were reliable. Of every twenty analyses, one was a blank to check background contamination, one was a duplicate analysis to check precision, and one was spiked with all three Aroclor mixtures (Aroclors 1016, 1221, and 1254) to check efficiency of extraction. Appendix A summarizes these internal quality control data.

RESULTS AND DISCUSSION

Species Relationships

Lipophilic partitioning of PCB is widely recognized (National Research Council 1979) and must be considered when making interspecies comparisons (Spagnoli and Skinner 1977, Skea et al. 1979). Although the association between total PCB concentration and fat (lipid) content is variable (Table 66), the correlations are generally consistent with the pattern exhibited by the resident freshwater species (Armstrong and Sloan, this volume). Of the fifty sample sets suitable for calculation of PCB-lipid correlation coefficients, twenty-five (52%) were significant ($P < 0.05$). However, only one of the twenty collections of American shad showed a significant association between total PCB concentration and lipid content, and if shad are excluded, the percentage of samples with significant PCB-lipid correlations is 83% (twenty-five out of thirty sets).

The major differences between the patterns of PCB concentrations in resident and freshwater species and that of the American shad (and, to a lesser extent, striped bass, Atlantic tomcod, alewife, and blueback herring) will be examined in greater detail in the following sections. In general, the degree of interspecific differences is not well established, probably because of the complexity and incomplete knowledge of PCB exposure sequences in most waterways that have been studied. Large lacustrine and marine systems have many PCB sources and complex mixing patterns and, since fish species differ in life history and migratory patterns, variation in exposure can result in dissimilar accumulation patterns in different species. Dustman et al. (1971) note that there is generally some increase in PCB contamination at higher trophic levels. Our data indicate that within an

Table 66. *Hudson River PCB data summary for migrant/marine fish.*

SPECIES	LOCATION	YEAR	NUMBER[a]	% LIPID	PCB (−ppm) − WET BASIS[c] TOTAL	ARO 1016	ARO 1254	TOTAL PCB[c] (ppm) − LIPID BASIS	CORRELATION COEFFICIENT (r) OF TOTAL PCB TO LIPID	LENGTH
BC-M[b]	Below Newburgh	1976	26(2)	0.35	<0.75	—	—	204	—	—
BC-M		1979	17	0.30	<0.50±0.45	<0.10±0.002	0.34±0.44	179±115	0.29	—
-HP[b]			17	4.09	6.70±5.49	0.71±0.56	5.91±5.06	152±70	0.75	—
AS										
Immature										
-Adult	Indian Pt.	1980	6	2.40	2.80±2.02	0.65±0.66	2.06±1.55	280 ±391	0.83	0.65
	Catskill	1981	1	15.1	4.96	<0.20	4.76	31.5	—	—
SNS										
-Fillet	Indian Pt.	1980	1	1.11	1.83	0.19	1.54	165	—	—
-Liver			1	5.81	7.10	0.67	6.33	122	—	—
-Gonad			1	20.0	29.6	2.62	25.9	148	—	—
BbH	Mohawk R. (Lock 7)	1979	30	4.1	2.50±0.95	1.06±0.47	1.34±0.60	75.1±39.4	0.67	0.19
	Albany/Troy	1978	3(1)	7.8	3.91	1.78	1.67	49.9	—	—
		1980	40(1)	5.6	1.81	0.72	0.95	32.3	—	—
Alw	Albany/Troy	1978	5(1)	5.2	5.64	3.73	1.40	109	—	—
		1979	18	8.3	3.98±1.28	1.77±0.65	1.67±0.51	50.1± 11.8	0.86	0.47
	Catskill	1979	13	7.0	2.16±0.99	0.66±0.57	1.35±0.56	45.0± 39.9	−0.05	−0.33
	Saugerties	1979	18	5.8	2.41±1.47	0.76±0.58	1.40±0.71	44.0± 18.2	0.50	0.33
	Kingston	1979	24	8.2	2.50±1.04	0.69±0.45	1.71±0.84	31.7± 12.8	0.46	0.54
		1980	40(2)	9.3	3.02	0.70	2.22	32.8	—	—
	Newburgh	1979	20	8.1	2.60±1.12	0.61±0.44	1.84±0.73	33.8± 13.6	0.56	0.34
AmS	Albany/Troy	1980	4[e]	6.2	1.72±1.52	0.96±1.04	0.63±0.44	26.6± 7.7	0.93	0.04

Table 66. *Hudson River PCB data summary for migrant/marine fish. (Continued)*

SPECIES	LOCATION	YEAR	NUMBER[a]	% LIPID	PCB (–ppm) – WET BASIS[c] TOTAL	ARO 1016	ARO 1254	TOTAL PCB[c] (ppm) – LIPID BASIS	CORRELATION COEFFICIENT (r) OF TOTAL PCB TO LIPID	LENGTH
	Catskill	1980	18♂	12.2	2.38±1.02	0.95±0.50	1.05±0.47	20.3± 10.3	0.02	0.46
			12♀	10.0	0.92±0.35	0.21±0.12	0.52±0.21	10.4± 5.8	–0.06	–0.46
	Poughkeepsie	1977[d]	12♂	—	7.04±2.88	—	—	—	—	0.31
			21♀	—	5.51±2.23	—	—	—	—	0.44
		4/20/78	15♂	16.5	3.98±1.90	2.89±1.54	0.85±0.46	24.2± 10.6	0.27	0.60
			14♀	13.3	1.66±0.86	0.90±0.59	0.53±0.30	12.4± 5.7	0.43	0.15
		5/5/78[d]	13♂	21.4	4.21±1.79	2.63±1.50	0.90±0.50	20.7± 9.2	0.34	0.07
			16♀	19.0	1.63±0.77	1.06±0.55	0.36±0.23	8.6± 4.0	0.35	0.25
		5/16/78	8♀	18.5	3.25±2.46	2.15±1.92	0.79±0.48	17.3± 7.1	0.70	0.91
		5/9/80[a]	5♂	14.6	2.46±1.21	1.02±0.79	1.16±0.61	19.4± 15.1	0.31	0.59
			24♀	11.6	1.20±0.41	0.36±0.28	0.64±0.23	12.0± 7.2	–0.08	–0.06
	Peekskill	1978[d]	10♀	15.8	2.23±1.16	1.19±0.74	0.61±0.47	15.4± 9.6	–0.43	0.38
		1980[d]	1♂	10.1	2.78	1.36	1.32	27.5	—	—
			30♀	12.3	1.22±0.79	0.22±0.18	0.54±0.17	10.5± 6.7	0.14	0.05
	Tappan Zee Bridge	1977[d]	4♂	—	3.35±1.11	—	—	—	—	0.54
			15♀	—	2.98±0.74	—	—	—	—	0.07
		4/13/78	20♂	18.5	3.28±2.13	2.14±1.73	0.67±0.37	18.1± 12.1	–0.004	0.11
			20♀	13.8	2.73±5.44	1.23±2.71	1.14±2.62	19.2± 28.8	0.25	0.20
		5/12/78[d]	15♂	17.1	3.18±1.83	1.89±1.51	0.88±0.43	17.9± 8.6	0.66	0.55
			15♀	14.2	1.46±0.48	0.49±0.37	0.60±0.12	10.4± 3.4	0.24	0.40
		1979[d]	8♂	18.4	1.54±0.68	0.71±0.33	0.84±0.40	8.7± 4.1	–0.01	0.84
			7♀	16.8	1.17±0.44	0.37±0.11	0.80±0.35	7.0± 2.1	0.53	0.30
		1980[d]	14♂	13.3	1.93±1.09	0.75±0.51	0.83±0.47	16.3± 11.1	–0.37	0.06
			16♀	12.1	1.22±0.67	0.33±0.29	0.63±0.33	10.1± 4.9	0.35	0.28

AmE	Poughkeepsie	1981	30	13.3	13.24 ± 11.81	0.93 ± 0.56	12.24 ± 11.46	129 ± 134	0.50	0.29
	Peekskill	1981	30	9.99	10.70 ± 9.68	0.73 ± 0.66	9.85 ± 9.08	184 ± 333	0.61	0.56
	Indian Point-Nyack	1978	42	12.5	73.9 ± 66.7	39.9 ± 41.6	33.2 ± 28.6	612 ± 418	0.68	0.07
		1980	6	8.5	9.07 ± 8.61	0.46 ± 0.29	8.51 ± 8.35	190 ± 64	0.86	0.86
		1981	30	12.6	10.83 ± 6.22	0.49 ± 0.33	10.23 ± 5.97	109 ± 79	0.85	0.05
	George Washington Bridge	1980	5	11.9	8.15 ± 4.30	0.53 ± 0.23	7.52 ± 4.10	71.0 ± 13.8	0.96	0.85
	Pier 40 (NYC)	1980	19	9.2	5.89 ± 2.50	0.38 ± 0.19	5.41 ± 2.33	66.0 ± 23.9	0.92	0.66
	Verrazano Bridge	1980	16	5.5	6.76 ± 12.89	0.22 ± 0.14	6.44 ± 12.79	98.4 ± 87.8	0.63	0.38
	Queensboro Bridge	1980	29	10.5	7.13 ± 8.73	0.44 ± 0.32	6.57 ± 8.52	78.2 ± 55.2	0.52	0.33
RS	Kingston	1979	25	2.3	4.07 ± 2.34	1.31 ± 0.75	2.64 ± 1.59	184 ± 70.6	0.77	−0.30
	Newburgh	1979	16	2.3	4.51 ± 2.78	1.32 ± 0.79	3.10 ± 1.93	213 ± 89.3	0.74	0.46
		1980	20(1)	2.3	4.33	1.22	3.01	185	—	—
	Indian Point	1980	25(5)	2.0	2.36 ± 0.31	0.65 ± 0.27	1.61 ± 0.12	121 ± 17.7	0.99	—
AT	Poughkeepsie	1979	8	0.18	0.46 ± 0.35	0.22 ± 0.24	0.14 ± 0.11	246 ± 87.7	0.86	−0.81
		1980	13	0.57	0.66 ± 0.21	0.25 ± 0.09	0.31 ± 0.13	119 ± 34.3	0.30	−0.35
	Indian Pt.-	1977	30	0.61	0.96 ± 0.74	0.65 ± 0.55	0.21 ± 0.20	166 ± 81.4	0.54	NA
	Haverstraw Bay	1980	14	0.45	<0.37 ± 0.08	<0.14 ± 0.05	<0.13 ± 0.05	86.8 ± 40.1	0.69	−0.34
StB	Riverwide	1978	375	7.9	18.10 ± 28.22	9.64 ± 18.32	7.70 ± 10.34	270. ± 418.	0.10	−0.03
		1979	29	5.1	7.04 ± 6.09	2.54 ± 2.93	4.40 ± 3.35	142 ± 123	0.16	−0.14
		1980	197	4.7	6.13 ± 7.43	1.68 ± 2.95	4.28 ± 4.83	168. ± 144.	0.22	0.13
		1981	205	3.5	4.81 ± 5.98	1.02 ± 2.20	3.50 ± 3.94	152. ± 186.	0.49	0.39
Blf	Peekskill	1979	16	1.5	3.15 ± 1.74	0.62 ± 0.31	2.43 ± 1.48	227 ± 84.0	0.85	0.64

[a] Individual analyses performed unless number of composite samples are noted in parentheses.
[b] M = muscle; HP = hepatopancreas
[c] Means ± standard deviations
[d] Data sets used in trend analyses.
[e] Individual analyses of 3 males and 1 female

Abbreviations: BC—Blue crab; SNS—Shortnose sturgeon; AS—Atlantic sturgeon; BbH—Blueback herring; Alw—Alewife; AmS—American Shad; AmE—American Eel; RS—Rainbow smelt; AT—Atlantic Tomcod; StB—Striped Bass; Blf—Bluefish

aquatic environment of relatively uniform contaminant flux, PCB content depends largely upon the quantity of lipid material in individual fish, regardless of species.

Direct correlations between organochlorine concentrations and length, age, or weight have been reported for many species (Bache et al. 1972, Youngs et al. 1972, Armstrong and Sloan, 1980) and form the basis for fish consumption advisories (NYDEC 1981a, Ontario Ministry of the Environment 1978). The length/weight/age pattern for organochlorine accumulation usually is pervasive, but compelling exceptions have been observed for resident species of Hudson River fish (Armstrong and Sloan, this volume). For the migrant and marine species listed in Table 66, only twelve (24%) of the fifty species or location sample sets showed significant correlations between PCB concentration and length ($P < 0.05$). One of these was a sample of Atlantic tomcod which was negatively correlated. Overall, lipid content was a better indicator of PCB concentrations in a migrant or marine fish than was length, which presumably also reflects age and weight variables. Important exceptions to the former will be discussed in other parts of this paper.

Organ Comparisons

Peterson and Guiney (1979) determined that the highest PCB residue concentrations in fish were in tissues of high lipid content. However, they also performed homolog specific analyses and found some evidence for distributional differences reflecting metabolic pathways for specific PCB compounds. Our data on PCB in various organs are less specific, since our analytical design reveals only overall PCB mixtures (e.g., Aroclors 1016 and 1254). Thus, we cannot attempt to understand selective organ partitioning of PCB types.

The limited data available on PCB in various organs is given in Table 66 for blue crab and shortnose sturgeon. When these results are expressed on a lipid basis (Table 67), organ differences in PCB concentrations are minimal because the variation is within the limits of analytical error. For example, the coefficient of variation for PCB concentrations in different tissues of shortnose sturgeon is reduced from 94% for the wet-weight values to 12% for the lipid-adjusted concentrations. These results are comparable to data obtained by the NYDEC (unpublished data) for fish and other organisms from the Hudson River, Lake Ontario, and the Valatie Kill in Rensselaer County. We believe the uniformity of PCB-lipid concentrations between organs in the same individual is owing to the fact that over longer periods

TABLE 67. *Comparisons of PCB concentrations between organs in a male shortnose sturgeon and 17 blue crabs (collected in 1980 and 1979 respectively).*

SPECIES	TISSUE	LIPID (%)	PCB (ppm) WET TISSUE	LIPID
SNS	Fillet	1.1	1.8	165
	Liver	5.8	7.1	122
	Gonad	20.0	29.6	148
	Average	9.0	12.8	145
	Standard deviation	8.0	12.0	81
	Coeff. variation	90%	94%	12%
BC	Muscle	0.3	<0.5	170
	Hepatopancreas	3.8	6.7	151

of time the internal organs are exposed uniformly by normal physiological mechanisms. However, short-term exposures could result initially in selective accumulation by specific organs such as the liver.

Much of the work on the toxicology of chlorinated hydrocarbon is directed toward specific organs. For example, Smith et al. (1979) found a high incidence of hepatomas in Atlantic tomcod from the Hudson River. They also noted that PCB concentrations in the livers of that species averaged almost 40 ppm, whereas our muscle tissue analyses showed concentrations of less than 1 ppm. However, Smith et al. did not report lipid contents, and we did not analyze the livers from our samples.

It is quite possible that consideration of the effects of long-term exposure and monitoring of environmental conditions can be made simpler when we understand fundamental patterns. The close relationship between lipid content and PCB accumulation indicates that it is not always necessary to analyze specific organs separately. The lipid-based PCB values in Table 67 indicate that the differences between organs are minimal and also that the shortnose sturgeon and blue crab have quite similar values. However, we caution that these data are based on different years for the two species and that the crabs are migrants while the sturgeon are probably resident.

Owing to the propensity of PCB to accumulate in lipid, removal of tissues such as fat deposits, will reduce the overall levels of lipophilic contaminants. Studies have shown that removal of the skin and other parts of the fillet, such as the belly flap, can substantially reduce contamination. The percent reduction of PCB concentrations for striped bass and American shad by proper trimming is 44% and 63%, respectively (NYDEC 1979, 1981a, 1981b, Skea 1979).

American Shad Roe

The roe of American shad is a culinary delicacy and is, in fact, the most valuable part of the fish. Fortunately, PCB levels in roe samples consistently have been near the detection limit. (0.10 ppm wet-weight basis) for each type of Aroclor (Table 68). In shad roe samples collected in 1978 near the Tappan Zee Bridge, the only Aroclor quantifiable was Aroclor 1016 (0.16 ppm) while Aroclors 1254 and 1221 were below the detection limit. In 1980, Aroclor 1016 also was below detection.

Spagnoli (personal communication 1978) suggested that shad eggs are formed before the females enter the estuary. If so, no appreciable accumulation would be expected since metabolic activity of the ovarian tissues is presumably reduced during the spawning run. However, the lipid content of roe is approximately one-fifteenth that of the fillet. When Aroclor 1016 is expressed on a lipid basis from the 1978 data, the roe PCB concentration is about 12 ppm, compared to a lipid-based level of 3 ppm in the fillet. Active accumulation in ovarian tissue by resident species or migrants which have longer migration periods needs further study, particularly for those populations having trouble reproducing.

Body Size and Accumulation of PCB

In a 14-day live-car experiment in the Hudson River below Fort Edward, Skea et al. (1979) found that accumulation of Aroclor 1016

Table 68. *Average PCB concentrations in American shad roe from Hudson River collections.*

LOCATION	NUMBER[a]	AVE. % LIPID	AVERAGE PCB (ppm)—wet weight based TOTAL	ARO 1016	ARO 1254	ARO '1221
Catskill						
1980	7(3)	1.38	<0.42	<0.12	<0.15	<0.15
Poughkeepsie						
1978	13	1.11	<0.31	<0.11	<0.10	<0.10
1980	7(3)	0.90	<0.32	<0.10	<0.10	0.12
Peekskill						
1980	7(3)	2.09	<0.30	<0.10	<0.10	<0.10
Tappan Zee Bridge						
1978	10	1.31	<0.36	0.16	<0.10	<0.10
1980	7(3)	0.89	<0.30	<0.10	<0.10	<0.10

[a]Number of composite samples in parentheses

was surprisingly rapid in four species of fishes. At the end of the experiment, concentrations in whole fish composite samples ranged from 1.8 ppm to 3.8 ppm on a wet-weight basis. Although it was not addressed in their paper, the data show that lipid-adjusted Aroclor 1016 concentrations, irrespective of species, are inversely related to body size. McLeese et al. (1980) also noted a similar inverse relationship with Aroclor 1254 uptake and body weight for the polychaete worm *Nereis virens* and the shrimp *Crangon septemspinosa*.

This same phenomenon is depicted in the data in Table 69 for four migrant species (American shad, alewife, blueback herring, and rainbow smelt) which enter the Hudson River to spawn. The species apparently do not feed in the river (Bigelow and Schroeder 1953) and the population is composed mostly of first-time spawners. As body size increases, correlations between lipid and PCB values decrease as do the lipid-based PCB values. The smallest species, rainbow smelt, has lipid concentrations comparable to those found in resident species such as immature bluefish and mature white perch. White perch levels of PCB may be somewhat higher owing to the upriver (Albany) location of collection. Alewife and blueback herring are about the same size, and consequently have comparable PCB concentrations. Female

Table 69. *PCB-lipid "saturation" as a function of migrant/resident characteristics related to body size.*

SPECIES/ LOCATION	YEAR	NO.	MEAN WEIGHT (g)	LIPID PERCENT	CORR. WITH PCB	PCB (ppm) WET WT.	LIPID BASED
AmS -Tappan Zee Brdg	1980						
Females		16	1948	12.1	0.35	1.2	10
Males		14	1644	13.3	-0.37	1.9	16
Alw -Newburgh	1979	20	262	8.1	0.56	2.6	34
BbH -Albany	1980	40[a]	194	5.6	—	3.2	32
RS -Newburgh	1979	16	9	2.3	0.74	4.5	213
Blf (juvenile) -Peekskill	1979	16	74	1.5	0.85	3.2	227
WP -Albany	1980	30	110	5.2	0.71	16.0	316

[a]One composite analyzed.

American shad weigh about 300 g more than the males, and we found that these smaller male shad had higher PCB levels than the females. Overall, the migrant species contained PCB levels in the following order: Female American shad, < male shad, ≃ alewife, ≤ blueback herring, < rainbow smelt.

Samples were selected for Table 69 from 1979 and 1980 because the majority of the PCB decline has occurred prior to 1979. This decline will be discussed in more detail in the next section. However, we want to emphasize that the samples listed in Table 69 were selected to be representative.

Closer inspection of other samples in Table 66 reveals the general pattern for a body size—"saturation" relationship. The term "saturation" is not intended to imply that the lipid found in fish is saturated in a true equilibrium sense. Rather, it probably represents a quasi-equilibrium state with respect to PCB concentrations in the water at the time of collection. If the migrant species depicted in Table 69 were resident, the PCB concentrations would have been higher. The potential full saturation concentration in an organism is unknown. In 1977, there were individual goldfish in the Stillwater area (below Fort Edward) with 40,000 ppm to 99,000 ppm PCB per gram lipid (NYDEC unpublished data). For the four migrant species in Table 69, accumulation of PCB may be relatively passive and may be primarily a function of variation in body size (i.e., the ratio of surface area to volume). This uptake may be due mostly to diffusion from the water through the skin, gills and other membranes along with ingestion of particulate matter. Furthermore, the body-size PCB relationship noted here could reflect higher metabolic rates for smaller fish, and concomitant higher water filtration rates resulting in relatively more PCB accumulation.

Temporal Trends

In a preliminary analysis of PCB trend data for Hudson River fish, Armstrong and Sloan (1980) indicated substantial declines. However, the data bases were limited to 1977, 1978, and 1979 analyses. In the present report, 1980 data are available, as is 1981 information for American eel and striped bass. These two additional years of data support and extend the observed decreases in PCB contamination discussed in the earlier report.

Even though PCB concentrations on a wet-weight basis have always been low in the flesh of Atlantic tomcod and blue crab, these species have also shown a lipid-based decrease in PCB (Table 70). The exact

Table 70. *Aroclor 1016/1254 ratios, half-lives, and average annual rates of decline for migrant/marine Hudson River fish.*

SPECIES	LOCATION	DATA YEARS	RATIO ARO 1016/1254 EARLY YEAR	RATIO ARO 1016/1254 RECENT YEAR	TOTAL PCB-LIPID BASED HALF-LIFE (yr)	TOTAL PCB-LIPID BASED ANNUAL RATE OF DECLINE (%)
BC	Below Newburgh	76,79	—	<0.29	5.13	−12.7
BbH	Albany/Troy	78,80	1.07	0.76	3.19	−20.0
Alw	Albany/Troy	78,79	2.66	1.06	0.88	−54.1
	Kingston	79,80	0.40	0.32	—	
AmS[b]	Poughkeepsie	77[a],78,80				
-male			3.26	0.88	1.98	−29.5
-female			2.94	0.56	1.36	−39.9
-female[c]	Peekskill	78,80	1.95	0.41	2.30	−26.0
	Tappan Zee Bridge	77[a],78,79,80				
-male			2.15	0.90	3.77	−16.8
-female			4.08	0.52	2.33	−25.7
AmE	Indian Point	78,80,81	1.20	0.05	1.21	−43.7
RS	Newburgh	79,80	0.43	0.41	4.90	−13.2
AT	Poughkeepsie	79,80	1.57	0.81	0.96	−51.6
	Indian Bay Pt.-Haverstraw	77,80	3.10	1.08	3.20	−19.5
StB[b]	Riverwide	78,79,80,81	1.28	0.31	1.57	−35.7
	Averages: ± S.D.		2.01 ±1.14	0.60 ±0.32		−27.5 ±16.6

[a] Data available for total PCB only; early year Aroclor ratio uses 1978 data.
[b] Half-life and rate of decline based on wet weight concentrations.
[c] Males were not available in 1978.

magnitude of the decrease is not known since the initial concentrations were less than 1 ppm total PCB (wet-weight basis) and the detection limits for each Aroclor mixture are relatively high (0.1 ppm). There are only two years of data available to compare at any given location for these two species, and some small samples have been composited; but even with these limitations, there are indications of significant overall correlations between PCB and lipid (although fat content is low at less than 1%). There is also an apparent decline through time in lipid-based PCB values. It is important to note that this change matches the pattern for other migrant and marine, and resident freshwater, species.

Two closely related species, alewife and blueback herring, are considered short-term inhabitants of the Hudson River and its major tributaries during their spawning migrations. As migrants, they probably do not feed much while in the river (Bigelow and Schroeder 1953). Consequently, the accumulation of PCB is mostly through the gills and skin or the result of incidental ingestion of particulate matter that may have PCB adsorbed to it. In 1978, the limited samples (two composites of eight individuals) of these two species from the Albany-Troy region had relatively low concentrations in the flesh—3.91 ppm for blueback herring and 5.64 ppm for alewife (Table 66). Since 1978, PCB concentrations have declined to less than 3 ppm for these species riverwide.

There is a fairly consistent and significant correlation between lipid content and PCB concentration in alewife and blueback herring, but the relationship between length and PCB is not as significant or consistent. Of six collections for which it is possible to calculate correlations, five were significant for percent lipid and total PCB, whereas only one was significant for length and PCB content. Although the data indicate a decrease in lipid dependent PCB concentration from 1978 to 1980 in the two species at Albany (Table 70), there was no evident decline in PCB levels in alewife from the Kingston area between 1978 and 1980. There are no 1978 alewife data to determine whether there was a decrease in PCB concentration between 1978 and 1979 for this species.

American shad present two striking departures from the pattern seen in other species. First, there is a notable lack of correlation between PCB concentrations and lipid content. Second, on both a wet-weight and a lipid basis, males have higher PCB concentrations than females (Table 66). The first difference may be related to the fact that shad are larger and do not spend much time in the Hudson

River. Consequently, a definite PCB-lipid association may not have enough time to develop. Males are smaller than females and therefore tend to accumulate PCB faster. These findings indicate a more complex pattern of PCB trends which is nevertheless similar to that of other species.

Because of the consistency in the data between locations and collection period (if more than one collection was made in a year), we present trend data on shad by location and use selected collection dates from the mid-portion of the spawning run. Figures 81 and 82 illustrate the temporal trends for American shad at Poughkeepsie and the Tappan Zee Bridge, respectively, from 1977 to 1980. As with other species, there has been a shift from a preponderance of Aroclor 1016 to Aroclor 1254. The 1977 analyses, conducted at the New York State Department of Health, did not differentiate between Aroclors, but, based on the analyses of other migrant and resident species, we presume that Aroclor 1016 would have been the predominant type of PCB. Continued declines are not apparent since 1979. The Tappan Zee Bridge collection in 1980 (Figure 82) showed an increase of PCB which was due to a reappearance of Aroclor 1221. The average levels of Aroclor 1221 in 1980 were about 0.4 ppm and 0.3 ppm for males and females, respectively, but in 1979 they were below the detection limit. This PCB mixture is primarily a mono-chlorinated biphenyl which was not widely used and has never figured prominently in the Hudson River PCB problem. In all previous discussions, it has largely been ignored since it was a minor addition. It is not known why this increase occurred in the 1980 Tappan Zee Bridge collection. The detailed percent changes for the shad are presented in Table 71 along with similar information for American eel, striped bass, and other species evaluated for PCB trends. The American eel is a catadromous species in which males remain in saltwater estuaries while females ascend to freshwater during the elver stage (ca. 65 mm total length) and remain there to grow and mature over several years (Hardy 1978a). Thus, the eel samples reported in Tables 66, 70, and 71 presumably were males. Figure 83 depicts the trend in PCB levels for eels in the Haverstraw Bay area (including the Tappan Zee Bridge). The data are somewhat limited on eels because in 1979 it was decided to examine other species in the Hudson River. Because 1978 PCB concentrations in eels were high (above 70 ppm on a wet-weight basis) and the significant decline in PCB contamination was unexpected, no further collections were made in the area of the Tappan Zee Bridge until 1981. However, a small sample of six eels analyzed

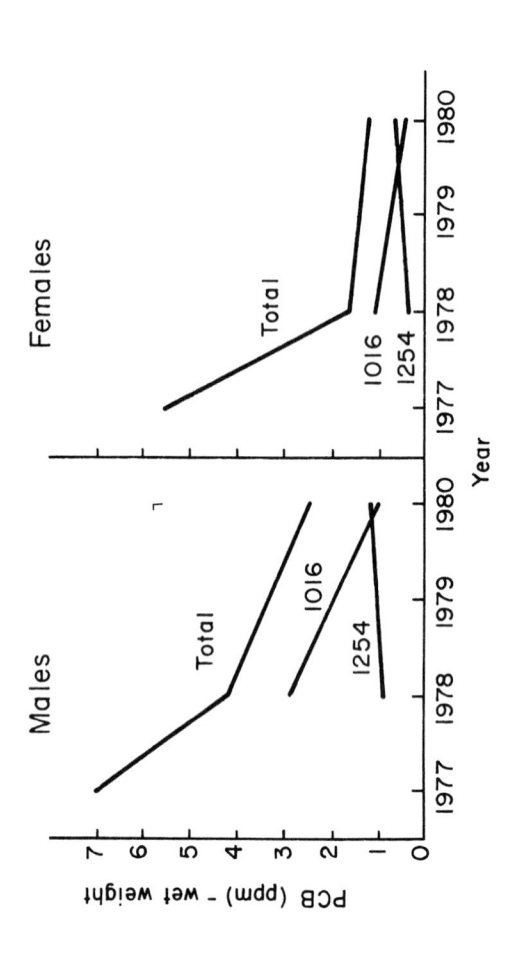

Figure 81. *PCB trends in American shad collected from the Poughkeepsie area of the Hudson River. Sexes considered separately; see text and Table 2 for explanation.*

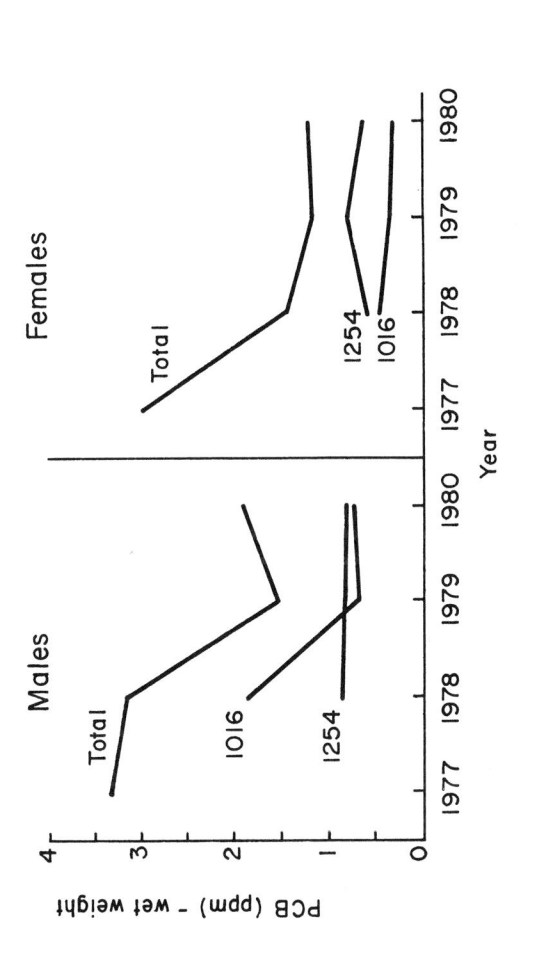

Figure 82. *PCB trends in American shad collected from the Tappan Zee Bridge area of the Hudson River. Sexes considered separately; see text and Table 2 for explanation.*

Table 71. *Annual percent changes of PCB (total, Aroclors 1016 and 1254) in Hudson River migrant/marine fish.*

		PERCENT PCB CHANGE BY YEAR				AVERAGE ANNUAL PERCENT CHANGE
SPECIES/LOCATION	PCB TYPE[a]	77–78	78–79	79–80	80–81	
BC—below Newburgh[a]	Total-W	—	—	—	—	−12.7
BbH—Albany/Troy[b]	Total-L	—	—	—	—	−20.0
	1016-L	—	—	—	—	−24.8
	1254-L	—	—	—	—	−10.0
Alw—Albany/Troy	Total-L	—	−54.1	—	—	−54.1
	1016-L	—	−69.4	—	—	−69.4
	1254-L	—	−22.2	—	—	−22.2
—Kingston	Total-L	—	—	+3.5	—	+3.5
	1016-L	—	—	−10.5	—	−10.5
	1254-L	—	—	+14.5	—	+14.5
AmS—Poughkeepsie[c]						
Male	Total-W	−40.2	—	—	—	−29.5
	1016-W	—	—	—	—	−50.0
	1254-W	—	—	—	—	+13.5
Female	Total-W	−70.4	—	—	—	−39.9
	1016-W	—	—	—	—	−41.7
	1254-W	—	—	—	—	+33.3
—Peekskill[b]						
Female	Total-W	—	—	—	—	−26.0
	1016-W	—	—	—	—	−57.0
	1254-W	—	—	—	—	−5.9
AmS—Tappan Zee Br.[e]						
Male	Total-W	−5.1	−51.6	+25.3	—	−16.8
	1016-W	—	−62.4	+5.6	—	−37.0
	1254-W	—	−4.5	−1.2	—	−2.9
Female	Total-W	−51.0	−19.9	+4.3	—	−25.7
	1016-W	—	−24.5	−10.8	—	−17.9
	1254-W	—	+33.3	−21.2	—	+2.5
AmE—Indian Pt.[f]	Total-L	—	—	—	−42.6	−43.7
	1016-L	—	—	—	−63.3	−75.6
	1254-L	—	—	—	−40.7	−28.8
RS—Newburgh	Total-L	—	—	−13.1	—	−13.1
	1016-L	—	—	−14.8	—	−14.8
	1254-L	—	—	−11.0	—	−11.0
AT—Poughkeepsie	Total-L	—	—	−51.6	—	−51.6
	1016-L	—	—	−60.7	—	−60.7
	1254-L	—	—	−21.9	—	−21.9
—Indian Pt.[g]	Total-L	—	—	—	—	−19.4
	1016-L	—	—	—	—	−30.2
	1254-L	—	—	—	—	−3.6

Table 71. *Annual percent changes of PCB (total, Aroclors 1016 and 1254) in Hudson River migrant/marine fish. (Continued)*

SPECIES/LOCATION	PCB TYPE[d]	PERCENT PCB CHANGE BY YEAR				AVERAGE ANNUAL PERCENT CHANGE
		77–78	78–79	79–80	80–81	
StB—riverwide	Total-W	—	−61.1	−12.9	−21.5	−35.7
	1016-W	—	−73.7	−33.9	−39.3	−52.7
	1254-W	—	−42.9	−2.9	−18.2	−23.1

[a] Data available only from 1976 and 1979; only total PCB quantified in 1976.
[b] Data available only from 1978 and 1980.
[c] Data available only from 1977, 1978 and 1980; only total PCB quantified in 1977.
[d] W = wet weight based calculations; L = lipid based.
[e] Only total PCB quantified in 1977.
[f] Data available only from 1978, 1980 and 1981.
[g] Data available only from 1977 and 1980.

in 1980 showed a considerable decline in total PCB on both a wet-weight and a lipid basis. Nearly all remaining PCB contamination of eels is by the Aroclor 1254 mixture. Since 1974, there has been a marked decrease in PCB concentration in striped bass (Figure 84). This species has not had a consistent PCB-lipid relationship and therefore trend data are presented on a wet-weight basis. As in other species, Aroclor 1254 has become the predominant PCB mixture. One of the striking features is that the system has developed a marked stability and there have not been further rapid declines since 1979 (Figure 84).

Wilson and Forester (1978) noted that in the years following the abatement of a PCB (Aroclor 1254) leak into Escambia Bay, Florida, concentrations in the oysters declined rapidly from over 20 ppm to 5 ppm within four years. For the next four years there was little discernible decrease and concentrations oscillated around the 5 ppm level. Another perplexing relationship developed in the 1981 striped bass data in that there was a highly significant ($P < 0.05$) positive correlation between PCB and fish length. This was not true of the data from previous years (Table 66).

Striped bass have complex life history and movement patterns (Hardy 1978b). Mixtures of coastal stocks and resident, migrant, or overwintering subpopulations preclude development of a single, simple accumulation pattern. Therefore, in presenting yearly comparisons on striped bass, the data on the spring collections from Poughkeepsie to the George Washington Bridge were combined.

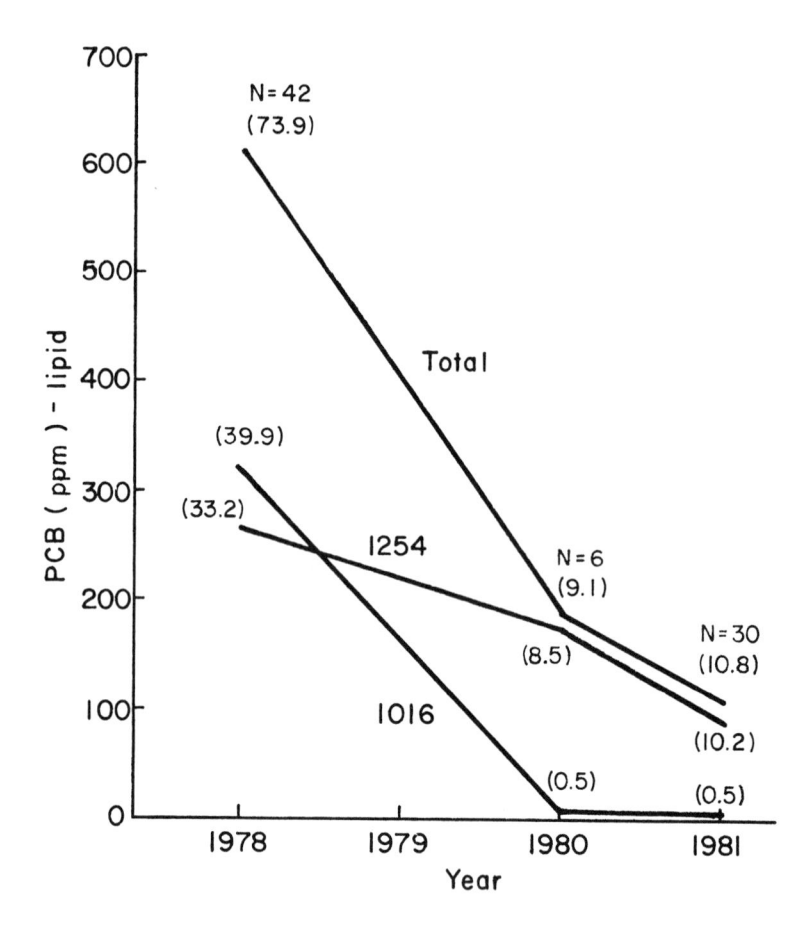

Figure 83. *PCB trends in American eel collected from the Haverstraw Bay area (Indian Point to the Tappan Zee Bridge) of the Hudson River. Average wet concentrations in ppm are shown in parentheses.*

In 1981, significant correlations emerged between PCB level and fish length and between PCB and lipid content which may indicate moderation of recent exposures which are now closer to background conditions, even though the levels are still relatively high. In 1978, PCB concentrations in striped bass from Long Island Sound and the New York Bight averaged about 3.2 ppm (NYDEC 1981c).

Most PCB decline has been due to the loss of Aroclor 1016, which is evident in Tables 70 and 71. When 1977 or 1978 data are compared to 1979 or 1980 data, there is a shift in the ratio of Aroclor 1016 to Aroclor 1254, usually from much greater than one to near or less

PCB PATTERNS: MARINE

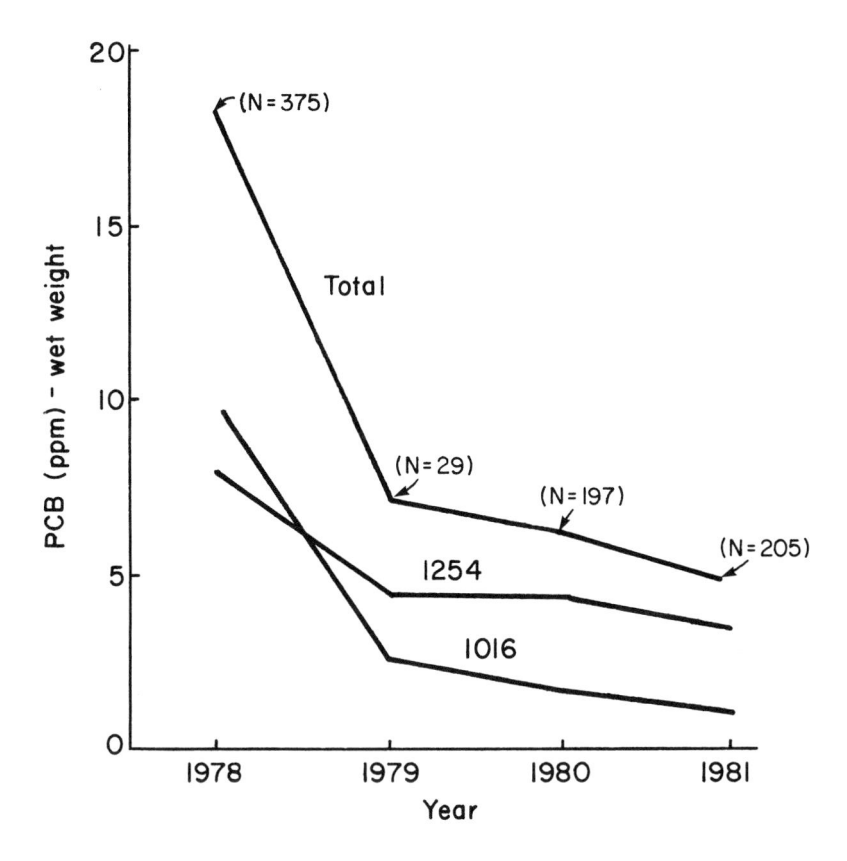

Figure 84. *PCB trends in striped bass collected from riverwide locations (Poughkeepsie to the George Washington Bridge) in the Hudson River.*

than one (Table 70). The annual percent decline in Aroclor 1016 averaged 42%, compared to 5% for Aroclor 1254 (Table 71).Tofflemire et al. (1979) analyzed types of Aroclors in 1977 sediment samples from the upper Hudson River. The ratio of Aroclor 1016 to Aroclor 1254 was 6.5 to 1. That was the only year such analyses were performed, and the present ratio is unknown. Much of the PCB may be buried under relatively clean sediment, and the present Aroclor 1254 contamination—at least in the lower Hudson River—could be due to undetermined secondary sources. Analyses of PCB levels in macroinvertebrates support the trends noted in fish (Simpson, personal communication 1981; O'Connor, personal communication 1979), but the unpublished invertebrate analyses of O'Connor indicate that these relatively sedentary organisms were collected in areas in the lower Hudson River where continuing inflow or contaminated sediments may be contributing to PCB loads in the biota. Sloan (1981) showed from analysis of PCB in American eels from Hudson River tributaries that certain streams were likely secondary PCB sources. In most of the tributaries examined, PCB concentrations were low (less than 2 ppm wet-weight basis), but some waterways—such as the Sawmill River—were producing eels with much higher PCB concentrations.

Mowrer et al. (1977) presented data on PCB content in fish (several species of cottids), mussel (*Mytilus edulis*), and superficial layers of the sediment at eighteen stations. Although they did not calculate the correlations, they did present data which yielded highly significant ($P < 0.01$) correlations between PCB levels in the sediment and in the organisms (0.85 for cottids and 0.89 for mussels). Therefore, contaminated sediment can act as a reservoir, which, in turn, may result in biotic contamination. Sewage treatment plants may also be a secondary source of contamination (Bergh and Peoples 1977).

Throughout the course of this study, we were alert to the possibility of seasonal variation in PCB contamination. Olsson et al. (1978) indicated a seasonal increase in PCB in roach from Lake Roxen, Sweden, that coincided with spawning and spring run-off. Wilson and Forester (1978) also noted a seasonal variation in oyster PCB concentrations in Escambia Bay, Florida. Newsome and Leduc (1975) noted seasonal changes in fat content of yellow perch from two lakes in Ontario, Canada.

In the Hudson River, there is a possibility that seasonal effects are damped by the uniform PCB flux from upriver sources. As noted above, expressing PCB concentrations on a lipid basis also tends to minimize short-term variations. Nevertheless, a study on seasonality of PCB contamination is desirable.

Trend Results: Comparison with Earlier Data

In addition to the extensive data discussed in this report, covering the period 1977 to 1981, a variety of PCB data pertaining to Hudson River fish analyzed between 1973 and 1976 are available in the files of the New York Department of Environmental Conservation. These earlier data, largely gathered in NYDEC's own laboratories, were generated using less rigorous sampling protocols and somewhat different preparation, analytical, and reporting procedures, and were not subject to any appreciable quality-control measures. Although these earlier results are not detailed here, a number of the collections show average lipid-based PCB concentrations substantially higher than current values, thus supporting the general declining trend observed in the late 1970s. Some early collections, however, are characterized by average PCB concentrations comparable to those observed for some species and locations from recent collections. Despite ambiguities in the trends in absolute PCB levels, all of the early data clearly show substantially greater proportions of Aroclor 1016 than are now present in comparable fish samples.

Spatial Trends

In earlier reports on the Hudson River PCB problem, much of the emphasis was focused on the geographic distribution of fish contamination (Sloan and Sheppard 1978; Sloan 1978, 1979). Since the fish monitoring project began, there have been well-defined gradients of contamination for resident species starting from the dominant PCB sources at Fort Edward and Hudson Falls (Armstrong and Sloan, this volume). As the most sensitive test of the spatial gradient in migrant and marine species, seventeen pairs of locations were selected from Table 66, each representing a downstream collection compared to the next available upstream collection for the same species in the same year. PCB concentrations compared between each site were either lipid or wet-weight bases, depending on the correlation between lipid and PCB values generally shown by the various species. A Chi-square analysis (2 × 2 contingency table) yielded random results in which the upstream areas would be just as apt to produce lower PCB levels as higher ones ($P < 0.05$). However, if the most distant points of the twelve available species/location combinations are used (Table 72), a downstream gradient is evident ($\chi^2 = 6.0; P < 0.05$). This gradient is subtle, since, in most of the comparisons, the magnitude of the difference is small. As mentioned above, the possibility of secondary PCB sources may mask some of the spatial aspects resulting from the dominant input near Fort Edward. In addition, the generally lower PCB concentrations in the estuary, along with the migrant or mobile characteristics of the mix of species, tidal flow, and perhaps other (unknown) variables may preclude detection of spatial patterns.

Current PCB Concentrations

Table 73 summarizes current wet-weight bases PCB concentrations for the eleven species of migrant and marine fish analyzed as part of the Hudson River PCB monitoring project. Although there have been major declines in PCB levels in the fish, the American eel

Table 72. *Spatial gradient in PCB concentrations indicated by upstream versus downstream collections of migrant/marine species in the Hudson River.*

SPECIES	YEAR OR DATE	LOCATION UPSTREAM	LOCATION DOWNSTREAM	AVE. TOTAL PCB (ppm) UPSTREAM	AVE. TOTAL PCB (ppm) DOWNSTREAM	CONCENTRATIONS BASED ON:
Alw	1979	Albany/Troy	Newburgh	50.1	33.8	Lipid
AmS	4/78	Poughkeepsie	Tappan Zee Br.			
—males				3.98	3.28	Wet
—females				1.66	2.73	Wet
	5/78	Poughkeepsie				
—males				4.21	3.18	Wet
—females				1.63	1.46	Wet
	1980	Catskill	Tappan Zee Br.			
—males				2.38	1.93	Wet
—females				0.92	1.22	Wet
AmE	1980	Indian Pt.	Verrazano Br.	190	98	Lipid
	1981	Poughkeepsie	Indian Pt.	129	109	Lipid
RS	1979	Kingston	Newburgh	184	213	Lipid
	1980	Newburgh	Indian Pt.	185	121	Lipid
AT	1980	Poughkeepsie	Indian Pt.	119	87	Lipid

Table 73. *Approximate average total PCB concentrations in Hudson River migrant/marine fish (wet basis) encountered below Troy.*

SPECIES	YEAR ANALYZED	APPROXIMATE AVERAGE PCB (ppm) VALUE
BC—muscle	1979	<1
—hepatopancreas		>5
AS—immature	1980	2–5
—adult	1981	~5[a]
SNS	1980	<2[b]
BbH	1980	2–5
Alw	1980	2–5
AmS	1980	1–3
AmE	1981	~10
RS	1980	3–5
AT	1980	<1
StB	1981	~5
BlF—immature ("snappers")	1979	~3

[a] Only one analyzed.
[b] Endangered species; possession is prohibited.

is still above 5 ppm on a wet-weight basis. Possession of American eels from the Hudson River has been prohibited since 1976 under Section 11.2 NYCRR, Title 6.

Since many people consider the hepatopancreas (liver or tomalley) of the blue crab a special delicacy, it should be noted that a New York State Department of Health advisory against its consumption is in effect. Although the advisory warns specifically against high cadmium concentrations, it also mentions PCB concentrations above 5 ppm.

Species which characteristically have PCB levels between 2 ppm and 5 ppm include alewife, blueback herring, American shad, bluefish ("snapper" or immature form), rainbow smelt, and, possibly, shortnose sturgeon. It is significant that 1981 analyses of striped bass indicated that for the first time since 1978 there is a positive relationship between PCB concentration and size (i.e., length). At present, this species (including a fair number of sublegal size fish) averages about

5 ppm. Data on shortnose sturgeon are limited because it is an endangered species.

Atlantic tomcod and the muscle of blue crab average less than 1 ppm PCB. Due to subtle variations now known to characterize PCB levels in these species, the tabulated information represents approximate average values of total PCB concentrations in the flesh on a wet-weight basis. As discussed earlier, the levels are variable, and actual values depend on fat content and the life history of the individual species. For detailed information see Table 66.

Even though most of these species average well below 5 ppm, many still exceed the current tolerance level of 2 ppm. In proposing the reduction of the tolerance level from 5 ppm to 2 ppm, the U.S. Food and Drug Administration (1977) noted that these chemicals are such undesirable contaminants that levels as low as 1 ppm have been considered.

In Retrospect

In 1972, Maugh wrote in *Science* "the presence and danger of polychlorinated biphenyls (PCBs) in the environment is not yet a dead issue, but it is one that appears to be dying quite rapidly." It is both gratifying and disturbing that man continues to be optimistic with regard to the abatement of environmental degradation, because attention was not focused on the PCB problem in the Hudson River until 1976. At that time, a major effort was mounted to seek solutions to the dilemma (Horn et al. 1979, Hetling et al. 1979). The recent dramatic declines in PCB levels do not provide the justification to relax the vigil on chemical contamination. Aroclor 1254 is still relatively high and overall PCB contamination of fish from the upper Hudson River is well above federal standards, and both of these conditions may persist for years. Acting on some of the cautions given in this report, we cannot project safely beyond the current data. In addition, there are other chemicals present that may continue to threaten this major waterway. Cadmium (Sloan and Karcher, 1984), polychlorinated dibenzofurans, and dioxins (Danzo 1981) are only a few. We are optimistic that the PCB problem eventually will be resolved, but complacency and apathy should not be allowed to develop as a consequence of a few encouraging trends.

CONCLUSIONS

There have been substantial declines in Hudson River PCB contamination in fish since 1977, but concentrations in the lipid are still an order of magnitude higher than what could be considered background conditions (i.e., above the major PCB source on the Hudson River, or in fish like the American shad, which have not had sufficient exposure in the river to accumulate appreciable PCB loads).

Most of the decline in PCB concentrations of migrant and marine species has been due primarily to the reduction of Aroclor 1016. The more highly chlorinated mixture, Aroclor 1254, has not declined as rapidly, although there has been some decrease of the Aroclor 1254 PCB-type. The average annual percent declines for total PCB, Aroclor 1016 and Aroclor 1254, have been 28%, 42%, and 5%, respectively. Further reductions may yet take place because of: (1) the relatively low levels of Aroclor 1016, as compared to Aroclor 1254 (which is a more persistent PCB mixture; (2) possible secondary PCB sources in the river; and (3) the lack of high-flow conditions since 1978 (which may be contributing to an artificial decrease in PCB concentrations, since there has been minimal resuspension and mobilization of contaminated sediment).

PCB concentration patterns in migrant and marine species usually were similar to those of resident and freshwater forms. Modifications of the normal accumulation pattern were evident in four migrant species that briefly enter the Hudson River to spawn and presumably feed only minimally while they are in the river. PCB concentrations in these species were inversely related to body size (surface area to volume ratio). The order of PCB concentration was: rainbow smelt $>$, blueback herring $\simeq$ alewife, $>$ male American shad, $>$ female shad.

PCB concentrations in a given species generally were slightly higher in upstream locations.

Limited data indicated little variation between body organs when PCB concentration was expressed on a lipid basis. The only marine or migrant species presently with concentrations (wet-weight basis) well above 5 ppm is the American eel. Striped bass average about 5 ppm, and other species—including blue crab (muscle), Atlantic sturgeon, shortnose sturgeon, blueback herring, alewife, American shad, rainbow smelt, Atlantic tomcod, and juvenile bluefish—have less than 5 ppm. Blue crab hepatopancreas levels are above 5 ppm.

Appendix A. *Summary of quality control data for Hudson River fish analyzed at Raltech Scientific Services, Madison, Wisconsin*

YEAR	BLANKS (ND = AROCLOR DETECTION LIMIT <0.1 N	1221	1016	1254
1978	58	all ND	all ND	all ND
1979	36	all ND	all ND	all ND
1980	23	all ND	all ND	all ND
1981	15	all ND	all ND	all ND

Duplicate analyses—Aroclors not differentiated
(expression of precision as: 100 - % difference between highest and lowest values)

YEAR	N	AVERAGE (100 -% DIFFERENCE) ± STANDARD DEVIATION
1978	174	83.7 ± 17.8
1979	108	89.4 ± 16.8
1980	117	82.4 ± 20.5
1981	45	89.3 ± 14.5

Recovery—percent of known spiked amount in fish flesh recovered during analytical procedures

YEAR	N	AROCLOR 1221	N	AROCLOR 1016	N	AROCLOR 1254
1978	18	78.9 ± 9.5	20	83.6 ± 10.5	20	82.8 ± 8.9
1979	12	74.9 ± 12.8	13	73.9 ± 6.0	11	82.6 ± 4.9
1980	44	64.8 ± 12.5	44	78.02 ± 11.9	44	96.8 ± 11.3
1981	15	79.7 ± 11.7	15	90.9 ± 9.2	15	98.0 ± 9.3

ACKNOWLEDGEMENTS

We thank Robert Brandt, Robert Mitchell, Michael Storonsky, Douglas Singleman, Andrew Kahnle, Michael Gann, Wayne Elliott, Walt Keller, Eugene Lane, Les Saltsman, and other personnel of the Bureau of Fisheries; John Kerzan, Richard Johnson, and Donald Hughes of Raltech Scientific Services; James Daley, Italo Carich, James DeZolt and members of the Hudson River PCB Advisory Committee; Robert Gabrielson, John Mylod, James Blakely, Ronald Ingold, and many other helpful Hudson River commercial fishermen; Denise Polsinelli, Diana Merchant, Laura Mandelsohn, Jacob Warnken, Katherine Day, Marian O'Sullivan, Dolores Long, and Kathy Harter for manuscript preparation assistance; and Joseph O'Connor, Jack Skea, and Lawrence Skinner for their reviews of the manuscript.

Part VIII
Management

14. Management Recommendations for a Hudson River Atlantic Sturgeon Fishery Based on an Age-Structured Population Model

John R. Young, Thomas B. Hoff, William P. Dey, and James G. Hoff

INTRODUCTION

Atlantic sturgeon (*Acipenser oxyrhynchus*) populations are highly vulnerable to overexploitation. Between 1880 and 1910, the Atlantic sturgeon fisheries in every Atlantic coast state from New York to South Carolina went from valuable commercial resources to near or total collapse (Murawski and Pacheco 1977). The extreme vulnerability of this species to overfishing is due to its long life span, delayed maturity, and high catchability while on its spawning migration. The long life span and delayed maturity make the species very slow to recover from adverse conditions. Based on reported landings (Murawski and Pacheco 1977), most coastal populations did not begin to recover from the collapse until the late 1930s or 1940s. Even then landings were one to two orders of magnitude below those of peak years. This slow response makes moderate to high exploitation extremely risky in that the fishable stock can decline to nearly zero before severely reduced recruitment is observed. When the fishery finally collapses, very few adults are left to reproduce.

Recently, reported commercial landings of Atlantic sturgeon in New York have increased to almost 6.5 metric tons in 1980 (Hoff et al. this volume). Actual landings are probably several times higher as many sturgeon either are dumped overboard or used by the fishermen themselves (Hoff 1980). This increase in reported landings and heightened interest in Hudson River sturgeon as a source of caviar (Muldoon 1981) should caution managers that this resource should be closely examined to prevent another collapse of the stock.

Preliminary management recommendations for a Hudson River fishery have been given by Dovel (1979a). In his report, Dovel recommends (1) a 48-inch minimum length for sturgeon caught in the Hudson River; (2) a 72-inch minimum size for sturgeon caught in New York coastal waters; (3) that fishing be prohibited in the navigation channel of the Hudson River, and, if necessary, that the stock be protected; (4) that commercial fishing for Atlantic sturgeon in the Hudson River be prohibited prior to 1 June; and (5) that commercial fishing activities in low-salinity estuarine areas be prohibited to protect juveniles. Hoff (1980) concurred with recommendations (1) and (2), and recommended that allowable catch be determined on a biological basis for each stock.

The purpose of this study was to examine the merits of these recommendations, particularly the 48-inch minimum length, through the use of an age-structured population model. Application of models for purposes of evaluating management alternatives, rather than for precise predictions of population sizes or yields, is the approach espoused by Holling (1978). Holling explains that models are the best way to objectively evaluate complex relationships in which interactions among the variables may produce results that are not at all intuitive. Often, a comparison of outcomes under different alternatives reveals that the population is not particularly sensitive to the exact value of the model parameters.

We first integrated the available life history data on Atlantic sturgeon into a population model that could be used to evaluate management decisions; then, we employed the model to evaluate the consequences of a 48-inch minimum size limit for a Hudson River fishery. Although many other management alternatives could also be examined with this type of a model, further analysis should consider the opinion of those who must make the management decisions.

METHODS

This exercise was based on an age-structured population model, similar in function to a Leslie matrix (Leslie 1945). One-year time steps were used to move each cohort of sturgeon from eggs through the juvenile, subadult, and adult life stages. A total of forty-one age classes (ages zero through forty) were used for this analysis (Figure 85). At the beginning of each spawning season, adults present in the Hudson River were exposed to fishing mortality, those escaping this mortality were then allowed to spawn, and the entire population was subjected to natural mortality before repeating the cycle.

MANAGEMENT RECOMMENDATIONS

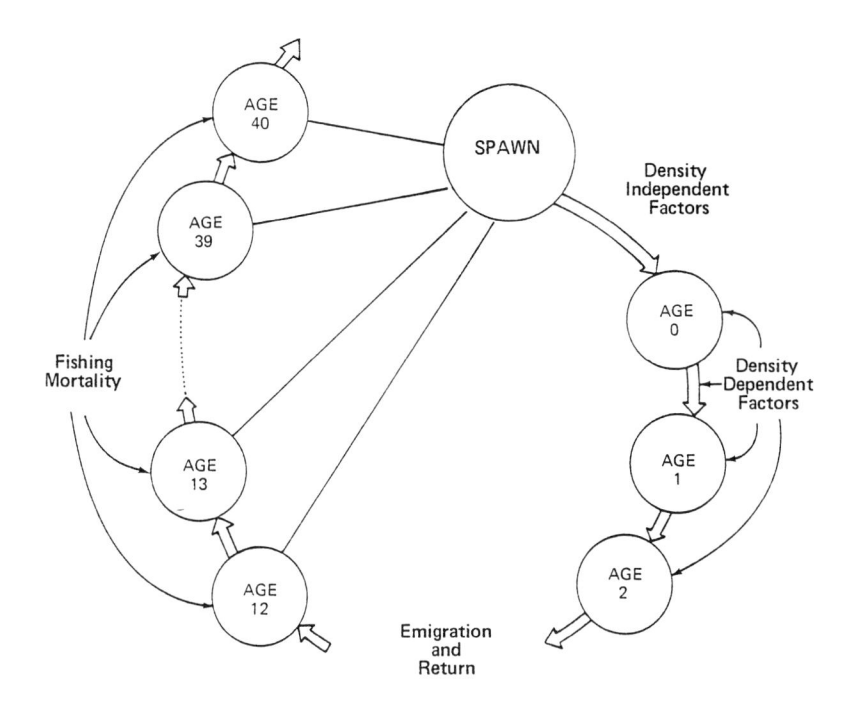

Figure 85. *Conceptual life history used as basis for population dynamics model.*

First-year survival was partitioned into density-independent and density dependent components. The density-independent term was constant for each simulation and the density-dependent term was variable, regulated by the total number of age 0, I, and II sturgeon, which all live in the estuary and share the same benthic food sources (Dovel 1979b). The density-dependent term was considered to be a linear function of juvenile population size (Figure 86). The product of the density-independent and density-dependent survival terms at equilibrium was determined by the method of Vaughan and Saila (1976).

Model parameters for the older stages (Table 74) were derived from Hudson River data or other published sturgeon data whenever possible. Natural mortality rates of age I and older fish were not available from the Hudson population and other published estimates cited in Murawski and Pacheco (1977) did not fit well with the probable longevity of the Hudson stock. Therefore, a natural mortality

Table 74. *Life history data used in population model.*

AGE	% FEMALE[a]	% SPAWNING[b]	MEAN[c] FECUNDITY (In Thousands)	WEIGHT (kg)[d]
0	0.5	0	0	0
1	0.5	0	0	0
2	0.5	0	0	0
3	0.5	0	0	0
4	0.5	0	0	0
5	0.5	0	0	0
6	0.5	0	0	0
7	0.5	0	0	0
8	0.5	0	0	0
9	0.5	0	0	0
10	0.5	0	0	0
11	0.5	0.03	205	4.8
12	0.5	0.060	319	9.3
13	0.5	0.090	432	13.7
14	0.5	0.120	546	18.2
15	0.5	0.150	660	22.7
16	0.5	0.180	773	27.2
17	0.5	0.210	887	31.7
18	0.5	0.240	1,000	36.1
19	0.5	0.270	1,114	40.6
20	0.5	0.300	1,228	45.1
21	0.5	0.333	1,341	49.6
22	0.5	0.333	1,455	54.1
23	0.5	0.333	1,568	58.5
24	0.5	0.333	1,682	63.0
25	0.5	0.333	1,796	67.5
26	0.5	0.333	1,909	72.0
27	0.5	0.333	2,023	76.5
28	0.5	0.333	2,137	80.9
29	0.5	0.333	2,250	85.4
30	0.5	0.333	2,364	89.9
31	0.5	0.333	2,477	94.4
32	0.5	0.333	2,591	98.9
33	0.5	0.333	2,705	103.3
34	0.5	0.333	2,818	107.8
35	0.5	0.333	2,931	112.3
36	0.5	0.333	3,045	116.8
37	0.5	0.333	3,159	121.3
38	0.5	0.333	3,273	125.7
39	0.5	0.333	3,386	130.2
40	0.5	0.333	3,500	134.7

[a] Huff 1975.
[b] Murawski and Pacheco 1977.
[c] Vladykov and Greeley 1963.
[d] Dovel 1979b.

MANAGEMENT RECOMMENDATIONS

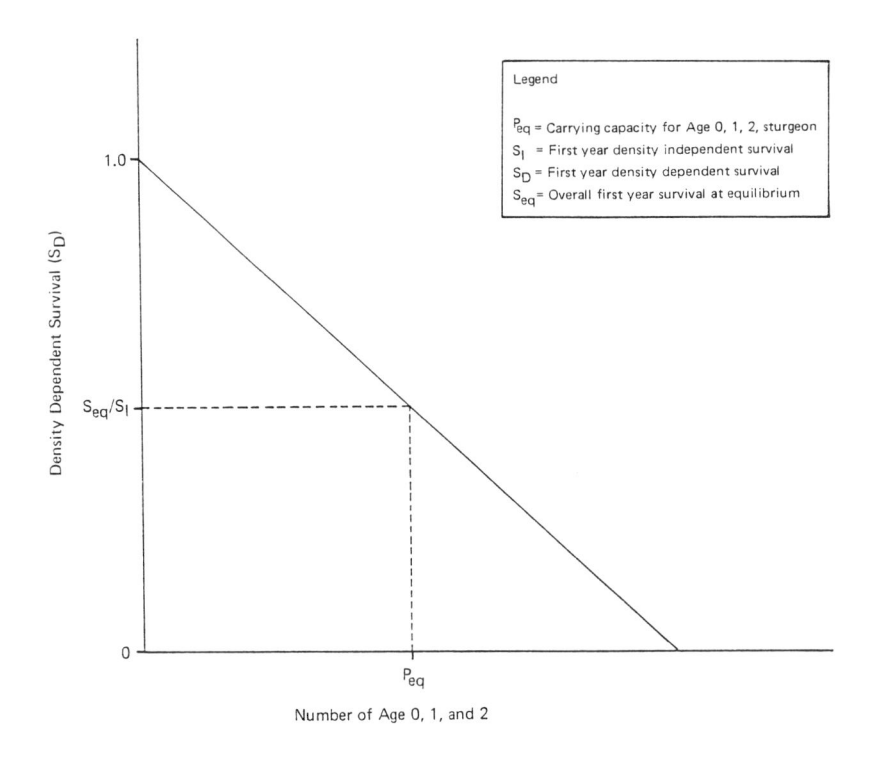

Figure 86. *Linear relationship of density-dependent survival with number of age 0, 1, and 2 Atlantic sturgeon assumed for population model.*

rate of 0.25 was determined by assuming a maximum age of forty years and a constant mortality rate which would decay approximately two hundred thousand fish in ages I–IV (Dovel 1979b) to one fish at age forty. (We assumed that Dovel's estimated one hundred thousand juvenile sturgeon in the Hudson River represented one-half the juvenile sturgeon in the Atlantic coast population).

Maturity begins at about age eleven for females and increases gradually for the next ten years until all females are mature (Murawski and Pacheco 1977). However, since sturgeon are believed to spawn only every two to four years (Murawski and Pacheco 1977), the percentage of fish actually mature for each age class was divided by three to obtain the fraction of each age participating in spawning.

Mean weight at age was estimated by a linear regression of weight as a function of lengths (Figure 87), using Hudson River data (Dovel 1979b). Fecundity was estimated as 23,500 eggs per kilogram, based on the limited data of Vladykov and Greeley (1963).

The density-independent survival rate and equilibrium population sizes were adjusted to achieve two different population growth rates without exploitation, a fivefold increase in stock size within two hundred years, and a twentyfold increase in stock size in two hundred years (Table 75). These growth rates were chosen since (1) the actual sturgeon populations are believed to be increasing from near complete depletion at the turn of this century (Scott and Crossman 1973), although the rate of increase is unknown; and (2) these values provide a range of population growth rates which permit evaluation of the sensitivity of model results to population growth rate and density dependence.

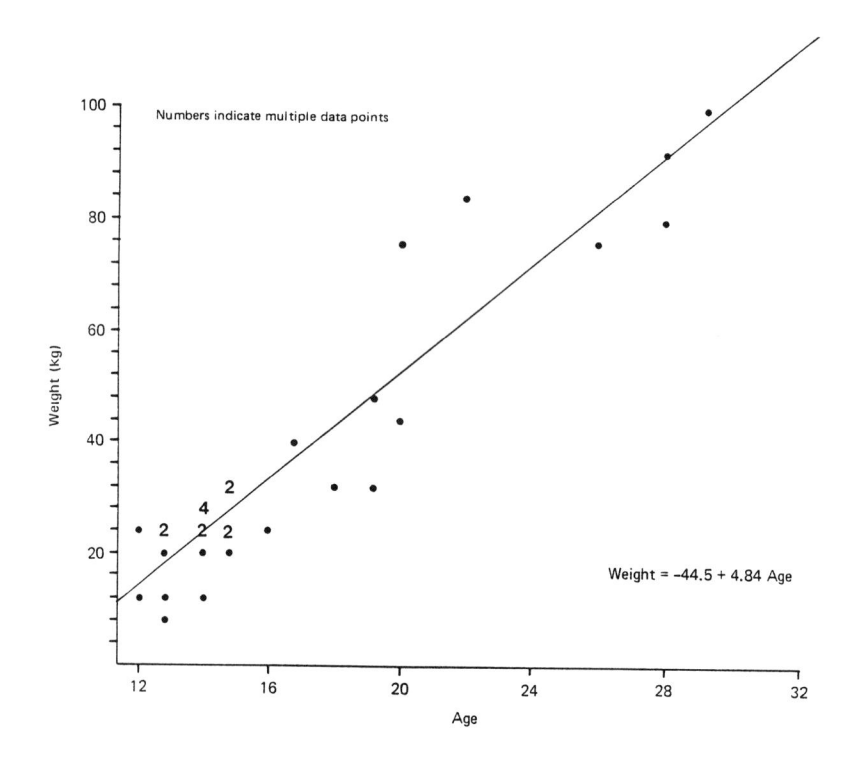

Figure 87. *Relationship between weight and age for Hudson River Atlantic sturgeon. Data from Dovel (1979b).*

Table 75. *Model parameters used in determining first year survival for population simulations.*

	200 YEAR INCREASE IN POPULATION SIZE	
	5-FOLD	20-FOLD
Equilibrium Survival S_{eq}	0.000093	0.000093
Density Independent Survival S_1	0.00015	0.00017
Equilibrium Juvenile Stock P_{eq}	1254828	5340525

After survival and equilibrium population sizes were adjusted to achieve the desired model behavior without exploitation (Figure 88), age-specific fishing mortality rates were added. All simulations in this analysis were done with uniform fishing rates applied to ages twelve–forty, which roughly correspond to Dovel's suggestions of a 48-inch minimum size, and again for ages twenty-five–forty, which would result from an 84-inch minimum size. Since only a Hudson River fishery is being evaluated, fishing mortality was applied only to the spawning migration. The exploitation rates thus represent the fraction of the spawning fish that are harvested, not the fraction of the entire population.

Although these choices of parameters and length of simulation were somewhat arbitrary and reflect the authors' opinion that management of long-lived species must use a time scale much longer than is normally employed, other choices would also be useful. These selections, however, provide a good starting point for further analyses and allow some general conclusions about sturgeon population dynamics and response to exploitation. The model was implemented in APL-11 on PDP-11/70 computing facility.

RESULTS AND DISCUSSION

Mean harvest (over two hundred years) exhibited a dome-shaped curve with increasing exploitation (Figure 89), similar to other exploitation models commonly employed in fisheries management. For the early fishing strategy (recruitment to the fishery at age twelve), the peak in mean harvest occurred near the 0.2 exploitation rate.

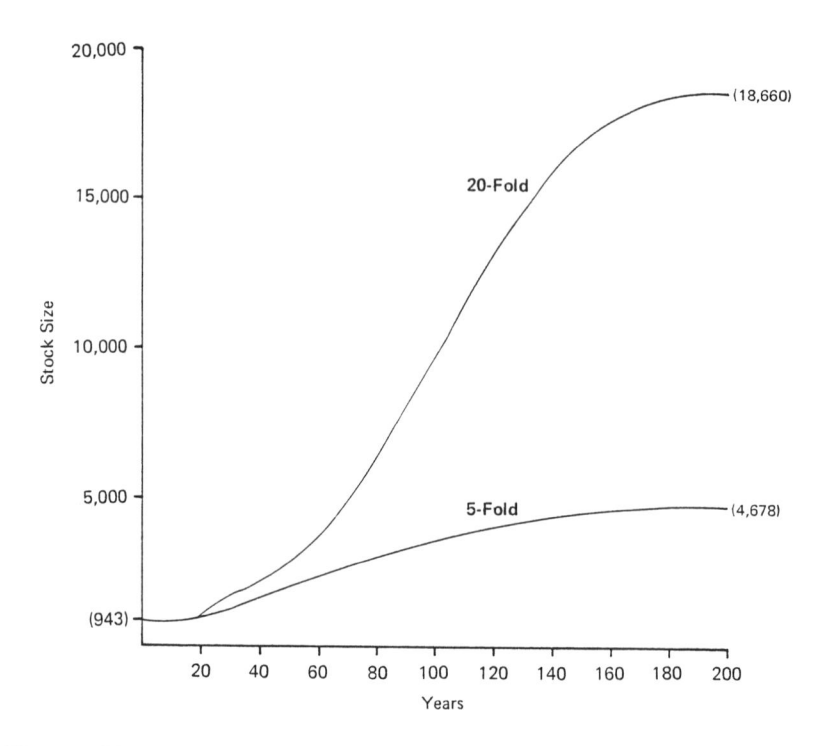

Figure 88. *Model behavior under assumptions to produce 5- and 20-fold increases in population size without fishing mortality.*

Under the late strategy (recruitment at age twenty-five), the yield curve had not yet peaked at 1.0 exploitation. This result is due to the density-dependent function and the fact that even 100% exploitation of spawning fish over twenty-four years of age will allow each fish the opportunity to spawn several times before being recruited to the fishable population. The highest mean harvest under the stated assumptions about population growth rate and density-dependence was 44 metric tons annually. Although the validity of these assumptions cannot be proven and is not crucial to evaluation of the 48-inch minimum size, the peak mean harvest is nevertheless far below the peak reported landing of 195 metric tons (428,000 lbs) given by Murawski and Pacheco (1977).

The differences in mean stock size for the two fishing strategies are even more striking than the differences in mean harvest. For the early fishing strategy, the stock size (defined as the number of fish less than twenty years of age) declines rapidly with increasing exploitation (Figure 89). This occurs because exploitation beginning at age

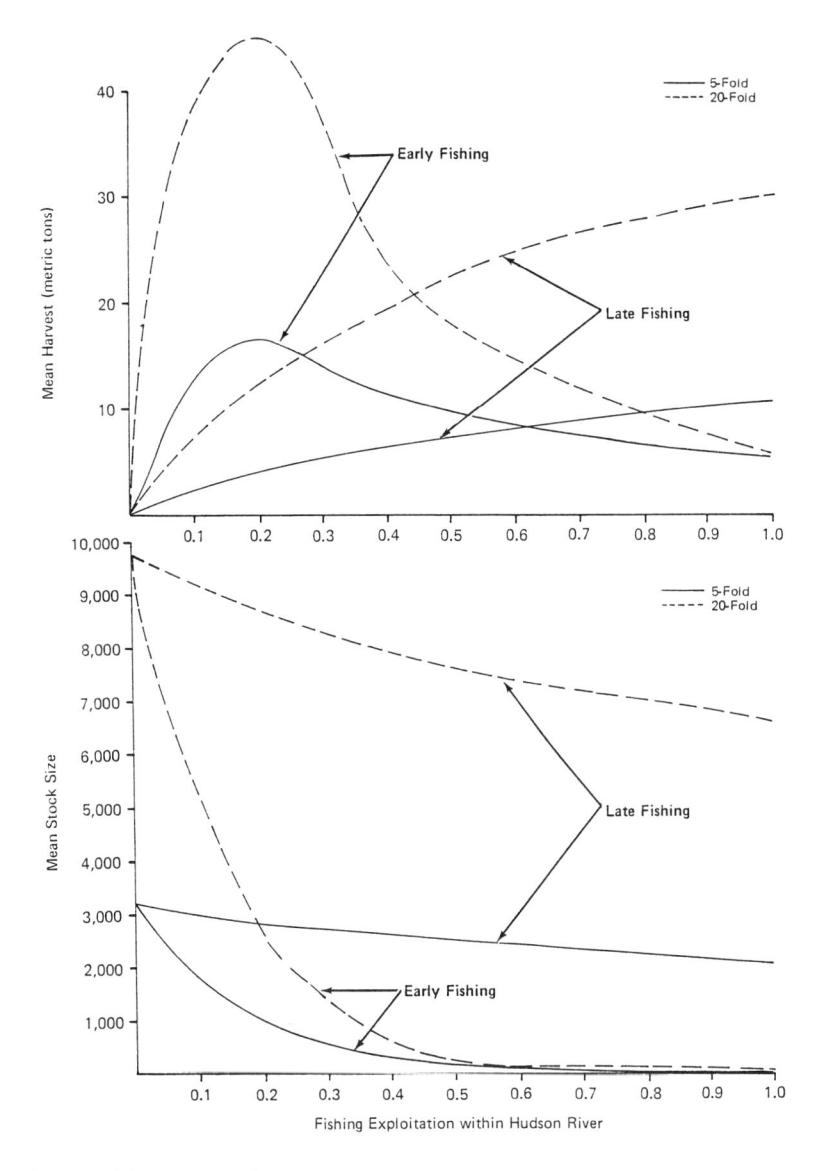

Figure 89. *Mean harvest and stock size as a function of exploitation and fishing strategy under 5-fold and 20-fold growth assumptions.*

twelve would allow few fish to live long enough to reach age twenty. Thus, the number of eggs spawned annually under high exploitation with the early fishing strategy is extremely small; however, the density dependence in age zero survival in the model allows sufficient recruitment to keep the population from going extinct. At an exploitation rate of 0.5, the predicted mean stock sizes are small under either assumption about asymptotic stock size, even though the mean annual harvest under the twentyfold increase assumption is still moderately high, approximately 18 metric tons, compared to only 9 metric tons for the fivefold increase.

The late fishing strategy depresses mean stock size much more slowly as exploitation rate is increased. This occurs because only fish over age twenty-five are exploited, which allows accumulation in prerecruit age groups. This accumulation of young adult fish gives the population a cushion which would also provide protection from unpredictable natural catastrophies. Although the predicted mean stock size is highly dependent on the assumptions controlling population growth at all exploitation rates, the late fishing strategy is far more protective of the stock than the early strategy, regardless of population growth rate.

The temporal pattern of annual harvest and stock size was also quite different for the early and late fishing strategies. For comparison, the exploitation rate was set to 0.45, a value that would give approximately the same two hundred year mean harvest for both fishing strategies when the twentyfold increase assumptions were used (Figure 89). Under the early fishing strategy, the harvest varied from 15 to 25 metric tons, with a mean of 21 metric tons (Figure 90). The harvest dropped rapidly in the first few years; then, after some oscillation, as the age structure of the model population stabilized, began a gradual but steady increase. However, even after two hundred years, harvest was still below the initial harvest. For the late strategy, harvest the first year was about 4 metric tons, dropped to 2 metric tons while the population age structure reached a new steady state, then increased in a logistic fashion. At two hundred years, the harvest had reached 43 metric tons and had not yet peaked. Mean harvest for the late fishing strategy was also 21 metric tons.

Stock size under the early strategy declined from the initial value of 943 to less than 300 within twenty years (Figure 90). Although it began to increase once the new stable age structure was achieved, it was still below five hundred individuals after two hundred years. Conversely, the late fishing strategy allowed stock size to increase,

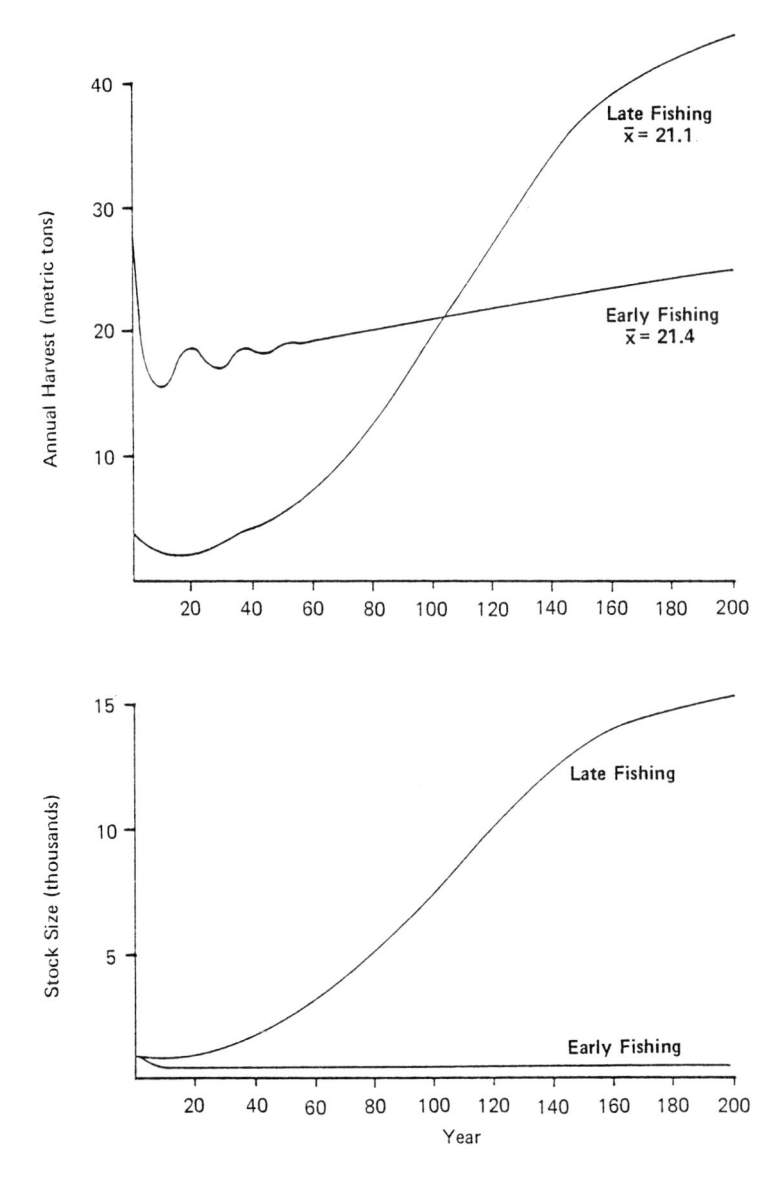

Figure 90. *Predicted annual harvest and stock size for early and late fishing strategies under 0.45 fishing exploitation within Hudson River.*

after a slight initial drop, to over fifteen thousand at the end of one hundred years. This is only about 3,000 less than the predicted stock size for an unexploited population (Figure 88).

CONCLUSIONS

The long life span and late maturity of sturgeon combine to make response of the population to exploitation very slow. In this regard, from ten to fifteen years would not be sufficient to determine the ultimate success of a management plan. Although no good records are available for comparison, the Hudson population is likely still in the process of recovering from the severe depletion that occurred at the turn of the century. If the effects of mismanagement of this resource must be endured for nearly a century, or longer, then managers must be very careful to weigh long-term population density goals against short-term economic benefits. Policies which would allow stock depletion much below current estimates should be carefully reviewed, particularly since two other factors which could have significant effects on stock size—environmental variability and exploitation in coastal waters—were not even considered in this analysis.

Of the two fishing strategies examined, the late strategy (exploiting only fish over twenty-four years of age) clearly gives much more protection to the stock. The severe decline predicted for stock size under exploitation as low as 0.2 with early fishing suggests that a 48-inch minimum size is too small and could result in another severe depletion of the population in a few years if fishing pressure is heavy. It would be much safer to restrict fishing to age groups which have been fully recruited to the adult stock. Although equal yields, in terms of weight of sturgeon landed, could be achieved with either fishing strategy, the early strategy will rely increasingly on young fish to achieve the yield. These conclusions hold true whether the actual population growth rate is relatively low (fivefold increase) or high (twentyfold increase).

RECOMMENDATIONS

The recent increase in sturgeon landings and interest in Atlantic sturgeon flesh and roe make it imperative that a sound management plan be formulated and implemented. A sound management plan can only be developed by integrating the existing information about sturgeon population dynamics with appropriate regulations. The model we have developed is an attempt at integrating the information so that the relative merits and risks of various management decisions could be evaluated. Although we examined only two management aspects, age at recruitment and exploitation rate within the Hudson River, the model could be easily adapted to examine other types of management problems.

The fishery itself could be used to collect the data to replace many of the assumptions we were forced to use in developing the model. If fishermen were required by law to register each fish landed, as in the Atlantic tuna fishery, and minimal funds were committed to data collection from landings, a better data base could be quickly established.

In summary, our specific recommendations for the sturgeon fishery are the following:

1. Minimum size should be set in excess of 48 inches to allow fish the opportunity to mature and spawn before being harvested. A better minimum size would be at least 72 inches, as recommended by Dovel (1979a) and Hoff (1980) for the coastal waters. Although the larger minimum size would probably limit initial harvests, it would ensure that the spawning stock would not be rapidly depleted.

2. As recommended by Hoff (1980), we are proposing that a maximum allowable catch be established for the Hudson River. A conservative initial allowable catch would be 10 metric tons until more information on stock size and age structure can be obtained.

3. A special license should be required to fish for sturgeon, and reporting requirements for landings should be established and strictly enforced.

4. A program should be developed to collect data from the fishery. This would be the most cost-effective way to get the data, since fishermen themselves, who obtain the economic benefits of the resource, would supply the fishing effort.

5. A team approach could develop and implement a long-term management plan. This would give the fishermen a voice in formulating policy and encourage them to cooperate in data collection. The team should reevaluate the success of the management plan on a periodic basis and make necessary changes in regulations.

ACKNOWLEDGEMENTS

The authors thank EA for the use of computing facilities, secretarial, and graphics assistance. In addition, Doug Carlson, John Gladden, and Byron Young provided many useful suggestions for improvement of the manuscript.

Literature Cited

Abood, K. A. 1974. Circulation in the Hudson estuary. Ann. N.Y. Acad. Sci. 250: 29–111.

Abood, K. A. 1977. Evaluation of circulation in partially stratified estuaries as typified by the Hudson River. Ph.D. thesis. Rutgers Univ., New Brunswick, N.J.

Abood, K. A., K. A. Konrad, J. Shirk, and P. McGroddy. 1973. The effects of power plants on physical and chemical water quality with specific attention to temperature, dissolved oxygen and chlorine. In: Proc. Third Symposium on Hudson River Ecology. March 1973. Hudson River Environmental Society.

Adams, P. B. 1980. Life history patterns in marine fishes and their consequences for fisheries management. Fish. Bull. 78(1): 1–12.

Albrecht, A. B. 1964. Some observations on factors associated with survival of striped bass eggs and larvae. Calif. Fish Game 50(2): 100–113.

Aleveras, R. A. 1973. Occurrence of a lookdown in the Hudson River. N.Y. Fish Game J. 20: 1.

American Fisheries Society Committee on Names of Fishes. 1980. A list of common and scientific names of fishes from the United States and Canada (Fourth Edition). Spec. Publ. No. 12. Amer. Fish. Soc. 174 pp.

Anderson, R. S. and L. G. Raasveldt. 1974. *Gammarus* predation and the possible effects of *Gammarus* and *Chaoborus* feeding on the zooplankton composition in some small lakes in Western Canada. Can. Wildl. Serv. Occas. Pap. 18 (CW69-1-18).

Anonymous. 1980. Action plan. Emergency striped base study. U.S. Fish Wildl. Serv. and Nat. Mar. Fish. Serv. 22 pp.

Armstrong, R. W. and R. J. Sloan. 1980. Trends in levels of several known chemical contaminants in fish from New York State waters. DEC Bureau of Environmental Protection and Wildlife. Tech. Rept. 80-2, iii + 77 pp.

Bache, C. A., J. W. Serum, W. D. Youngs, and D. J. Lisk. 1972. Polychlorinated biphenyl residues: Accumulation in Cayuga Lake lake trout with age. Sci. 177: 1191–1192.

Bagenal, T. G. and E. Braum. 1971. Eggs and early life history, pp. 166–198 in: Methods for assessment of fish production in fresh waters. (W. E. Ricker, ed.) IBP Handbook No. 3, 2nd Ed. Blackwell Sci. Pub. Oxford and Edinburgh.

Barlocher, F. and B. Kendrick. 1973. Fungi in the diet of *Gammarus pseudolimnaeus* (Amphipoda). Oikos 24: 295–300.

Barlocher, F. and B. Kendrick. 1975. Assimilation efficiency of *Gammarus pseudolimnaeus* (Amphipoda) feeding on fungal mycelium or autumn-shed leaves. Oikos 26: 55–59.

Barnthouse, L. W., R. J. Klauda, and R. L. Klauda. 1987. Assessing ecological impacts of power plants: Lessons from the Hudson River case. Amer. Fish. Soc. Monogr. 4. Bethesda, Md.

Bath, D. W. 1973. A limnological investigation of Sterling Lake, Orange County, New York. M.S. Thesis, New York University. 148 pp.

Bath, D. W. 1974. Synopsis of a meeting held on February 13, 1974, to discuss the fine points of identification of fish eggs and larvae of the Hudson River. New York University Medical Center. Unpublished.

Bath, D. W., C. A. Beebe, R. H. Ryder, and J. H. Hecht. 1976. A list of common and scientific names of fishes in the Hudson River. In: Proc. 4th Symposium on Hudson River Ecology. Hudson River Environmental Society, paper 33, 6 pp.

367

LITERATURE CITED

Baumann, P. C. and J. F. Kitchell. 1974. Diel patterns of distribution and feeding of bluegill (*Lepomis macrochirus*) in Lake Wingra, Wisconsin. Trans. Amer. Fish. Soc. 103(2): 255–260.

Bayless, J. D. 1972. Artificial propagation and hybridization of striped bass, *Morone saxatilis* (Walbaum). South Carolina Mar. Res. Dept. 135 pp.

Beaven, M. and J. Mihursky. 1980. Food and feeding habits of larval striped bass: an analysis of larval striped bass stomachs from 1976 Potomac Estuary collections. UMCEES Reference No. 79-45-CBL. Chesapeake Biological Laboratory.

Berggren, T. J. and J. T. Lieberman. 1978. Relative contribution of Hudson, Chesapeake, and Roanoke striped bass, *Morone saxatilis*, stocks to the Atlantic coast fishery. Fish. Bull. 76: 335–345.

Bergh, A. K. and R. S. Peoples. 1977. Distribution of polychlorinated biphenyls in a municipal wastewater treatment plant and environs. Sci. Total Environ. 8: 197–204.

Bigelow, H. B. and W. C. Schroeder. 1953. Fishes of the Gulf of Maine. U.S. Fish Wildl. Serv., Fish. Bull. 74 vol. 53: vii + 1–577.

Bonn, E. W., W. M. Bailey, J. D. Bayless, K. E. Erickson, and R. E. Stevens (eds.). 1976. Guidelines for striped bass culture. South. Div. Amer. Fish. Soc. 103 pp.

Bonomo, M. and M. Daly. 1981. Trawl comparison in the northern Newburgh Bay area of the Hudson River. Paper presented at the annual meeting of the New York Chapter, American Fisheries Society, Marcy, New York.

Booth, R. A. 1967. A description of the larval stages of tomcod, *Microgadus tomcod*, with comments on its spawning ecology. Ph.D. Thesis, Univ. Connecticut. Storrs. 53 pp.

Boreman, J. and H. M. Austin. 1985. Production and harvest of anadromous striped bass stocks along the Atlantic Coast. Trans. Amer. Fish. Soc. 114(1): 3–7.

Boreman, J., C. P. Goodyear and S. W. Christensen. 1978. An empirical transport model for evaluating entrainment of aquatic organisms by power plants. U.S. Fish and Wildl. Serv., FWS/OBS-78/90. Ann Arbor, U.S.A.

Bousfield, E. L. 1958. Fresh-water amphipod crustaceans of glaciated North America. Can. Field-Naturalist 72: 55–113.

Bousfield, E. L. 1969. New records of *Gammarus* (Crustacea: Amphipoda) from the Middle Atlantic States Region. Chesapeake Sci. 10(1): 1–17.

Bousfield, E. L. 1973. Shallow-water Gammaridean Amphipoda of New England. Cornell University Press, Ithaca, 312 pp.

Brawn, V. M. 1961. Aggressive behavior in the cod (*Gadus callarius* L.). Behaviour 18: 107–147.

Breder, C. M., Jr. 1948. Field book of marine fishes of the Atlantic coast from Labrador to Texas. G. P. Putnam's Sons, New York and London, pp. xxxvii + 1–332.

Burbidge, R. G. 1974. Distribution, growth, selective feeding, and energy transformations of young-of-the-year blueback herring, *Alosa aestivalis* (Mitchill), in the James River, Virginia. Trans. Amer. Fish. Soc. 103(2): 297–311.

Burdick, G. E. 1954. An analysis of factors, including pollution, having possible influence on the abundance of shad in the Hudson River. N.Y. Fish Game J. 1: 188–205.

LITERATURE CITED

Busby, W. 1966. Flow, quality, and salinity in the Hudson River estuary. pp. 135–145 in Hudson River Ecology. Hudson River Valley Commission.
California Department of Fish and Game. 1974. Interagency ecological study program for the Sacramento-San Joaquin estuary. Cooperative study by Calif. Dept. Fish and Game, Calif. Dept. Water Resources, and U.S. Bureau of Reclamation. Third Ann. Rept. (1973). 81 pp.
Cannon, T. C., S. M. Jinks, L. R. King, and G. J. Lauer. 1978. Survival of entrained ichthyoplankton and macroinvertebrates at Hudson River power plants. In: Fourth National Workshop on Entrainment and Impingement. (L. D. Jensen, ed.).
Carlson, F. T. and J. A. McCann. 1969. Evaluation of a proposed pumped-storage project at Cornwall, New York, in relation to fish in the Hudson River: Hudson River fisheries investigations, 1965–1968. Hudson River Policy Committee, DEC.
Carriker, M. R. 1967. Ecology of estuarine benthic invertebrates: a perspective. pp. 442–487 in: Estuaries (G. H. Lauff, ed.). American Association for the Advancement of Science, Washington, D.C.
Carscadden, J. E. 1975. Studies on the American shad (*Alosa sapidissima* Wilson) in the St. John River and Miramichi River, New Brunswick with special reference to homing and r-K selection. Ph.D. dissertation, McGill University, Montreal.
Chadwick, H. K. 1968. Mortality rates in the California striped bass population. Calif. Fish and Game. 54(4): 228–246.
Chambers, J. R., J. A. Musick, and J. Davis. 1976. Methods of distinguishing larval alewife from blueback herring. Chesapeake Sci. 17(2): 93–100.
Cheng, C. 1942. On the fecundity of some Gammarids. J. Mar. Biol. Ass. U.K. 25: 467–487.
Chittenden, M. E., Jr. 1969. Life history and ecology of the American shad, *Alosa sapidissima*, in the Delaware River. Ph.D. dissertation. Rutgers University, New Brunswick, New Jersey.
Chittenden, M. E., Jr. 1972. Responses of young American shad, *Alosa sapidissima*, to low temperatures. Trans. Amer. Fish. Soc. 101(4): 680–685.
CHGE Corp. 1977. Roseton Generating Station. Near-field effects of once-through cooling system operation on Hudson River Biota. Prepared by Lawler, Matusky and Skelly Engineers and Ecological Analysts, Inc.
CHGE Corp. 1978. Roseton and Danskammer Point generating stations habitat selection studies. Prepared by Lawler, Matusky and Skelly Engineers.
Christensen, S. W., W. Van Winkle, L. W. Barnthouse and D. S. Vaughan. 1981. Science and the law: Confluence and conflict on the Hudson River. EIA Review 2/1: 63–90.
Clemens, H. P. 1950. Life cycle and ecology of *Gammarus fasciatus* Say. Contr. Franz Theodore Stone Inst. Hydrobiol. 12: 1–61.
Clock, J. A. and T. G. Huggins. 1982. Preliminary results of biological studies of a modified frontwash vertical traveling screen. In: Proc. of the Workshop on Advanced Intake Technology.
Cole, L. C. 1954. The population consequences of life history phenomena. Quart. Rev. Biol. 29(2): 103–137.

LITERATURE CITED

Cooper, J. C., C. E. Newton, and F. R. Cantelmo. 1979. A physical/chemical overview of the Hudson River estuary. (Unpublished).
Cox, P. 1921. Histories of new food fishes. V. The tomcod. Bull. Biol. Board Canada. No. 5, 16 pp.
Cronin, E. L., J. C. Daiber, and E. M. Hulbert. 1962. Quantitative seasonal aspects of zooplankton in the Delaware River estuary. Chesapeake Sci. 3: 63–93.
Curran, H. W. and D. T. Ries. 1937. Fisheries investigations in the lower Hudson River. pp. 124–125 in: A biological survey of the Lower Hudson watershed. Biol. Survey, N.Y. State Conserv. Dept. 11.
Cushing, D. H. 1972. The production cycle and the numbers of marine fish. Symp. Zool. Soc. Lond. 29: 213–232.
Dadswell, M. J. 1975. Biology of the shortnose sturgeon (*Acipenser brevirostrum*) in the Saint John estuary, New Brunswick, Canada. In: Baseline survey and living resources potential study of the Saint John estuary. Vol. III. Fish and Fisheries. Huntsman Marine Laboratory, St. Andrews, N. B. 75 pp.
Dadswell, M. J. 1976. Biology of the shortnose sturgeon (*Acipenser brevirostrum*) in the Saint John River estuary, New Brunswick, Canada. Trans. Atlantic Chapter Canadian Soc. Environ. Biol. Ann. Meet. 1975: 20–72.
Dadswell, M. J. 1979. Testimony on shortnose sturgeon and Atlantic sturgeon on behalf of National Marine Fisheries Service. Submitted to the U.S. EPA, Region II. 19 pp.
Dadswell, M. J. 1981. Synopsis of biological data on the shortnose sturgeon *Acipenser brevirostrum* LeSueur, 1818. In: Shortnose Sturgeon Recovery Plan. 75 pp. manuscript.
Danzo, A. 1981. New toxic chemical found in Hudson River fish. The Knickerbocker News, Albany, N.Y., 46(1): 1a and 10a. Thurs., August 6, 1981.
Darmer, K. I. 1969. Hydrologic characteristics of the Hudson River estuary. pp. 40–55 in: Hudson River Ecology, G. P. Howell and G. L. Lauer, eds. DEC.
Davis, J. R. and R. P. Cheek. 1966. Distribution, food habits, and growth of young clupeids, Cape Fear River system, North Carolina. Proc. 20th Ann. Conf. Southeastern Assoc. Game Fish Comm., pp. 250–260.
Davis, J. R. and R. P. Cheek. 1973. Spawning sites and nurseries of fishes of the genus *Alosa* in Virginia. pp. 140–141 in: Proceedings of a workshop on egg, larval, and juvenile stages of fish in Atlantic coast estuaries. (A. L. Pacheco, ed.). Tech. Publ. No. 1, Mid. Atlantic Coast Fish. Center. NMFS.
Day, J. H. 1967. The biology of Knysha estuary, South Africa. pp. 397–407 in: Estuaries. (G. H. Lauff, ed.). American Association for the Advancement of Science, Washington, D.C.
DeAngelis, D. L. and C. C. Coutant. 1982. Genesis of bimodal size distributions in species cohorts. Trans. Amer. Fish. Soc. 111(3): 384–388.
Dennert, H. G., A. L. Dennert, and J. H. Stock. 1968. Range extension in 1967 of the alien amphipod, *Gammarus tigrinus* Sexton, 1939, in the Netherlands. Bull. Zool. Mus. Univ. Amsterdam 1(7): 79–81.
de Sylva, D. P., F. A. Kalber, Jr., and C. N. Shuster, Jr. 1962. Fishes and ecological conditions in the shore zone of the Delaware River estuary, with notes on other species collected in deeper water. Univ. Delaware Marine Lab. Information Series, Pub. No. 5. 164 pp.

LITERATURE CITED

Dew, C. B. 1973. Comments on the recent incidence of gizzard shad (*Dorosoma cepedianum*) in the lower Hudson River. In: Proc. Third Symposium on Hudson River Ecology. March 1973.

Dew, C. B. 1981. Biological characteristics of commercially caught Hudson River striped bass. pp. 82–102 in: Papers of the Fifth Symposium on Hudson River Ecology, 1980. (J. W. Rachlin and G. Tauber, eds.) HRES.

Dew, C. B. and J. Hecht. 1976. Ecology and population dynamics of Atlantic tomcod (*Microgadus tomcod*) in the Hudson River estuary. In: Proc. Fourth Symp. on Hudson River Ecology. March 1976.

Dey, W. P. 1981. Mortality and growth of young-of-the-year striped bass in the Hudson River estuary. Trans. Amer. Fish. Soc. 110(1): 151–157.

Dey, W. P. and P. C. Baumann. 1978. Community dynamics of shore zone fish populations in the Hudson River estuary. Poster paper presented at the 108th Annual American Fisheries Society Meeting, Kingston, Rhode Island.

Domermuth, R. B. and R. J. Reed. 1980. Food of juvenile American shad, *Alosa sapidissima*, juvenile blueback herring, *Alosa aestivalis*, and pumpkinseed, *Lepomis gibbosus*, in the Connecticut River below Holyoke Dam, Massachusetts. Estuaries, 3(1): 65–68.

Dorgelo, J. 1974. Comparative ecophysiology of gammarids (Crustacea: Amphipoda) from marine, brackish and freshwater habitats exposed to the influence of salinity-temperature combinations. I. effects on survival. Hydrobiol. Bull. (Amsterdam), 8: 90–108.

Doroshev, S. I. 1970. Biological features of the eggs, larvae, and young of the striped bass [*Roccus saxatilis* (Walbaum)] in connection with the problem of its acclimatization in the USSR. J. Ichthyol. 10(2): 235–247.

Dovel, W. L. 1976. Sturgeons of the Hudson River. Performance Report (October 1, 1975 to March 31, 1976) for DEC. 12 pp.

Dovel, W. L. 1977. Performance report for biology and management of shortnose and Atlantic sturgeons of the Hudson River. 1 April 1976–31 March 1977. DEC. 130 pp.

Dovel, W. L. 1978a. Performance report for biology and management of shortnose and Atlantic sturgeons of the Hudson River. 1 April 1977–31 March 1978. DEC. 154 pp.

Dovel, W. L. 1978b. Sturgeons of the Hudson River. Draft of Final Performance report for DEC. 181 pp.

Dovel, W. L. 1979a. Atlantic and shortnose sturgeon in the Hudson River estuary. Testimony rept. for U.S. EPA, The Oceanic Society, Conn. 26 pp.

Dovel, W. L. 1979b. The endangered shortnose sturgeon of the Hudson River: Its life history and vulnerability to the activities of man. Periodic Progress Report No. 1, April 1–30. Prepared for FERC. 10 pp.

Dovel, W. L. 1979c. Same title. Periodic Progress Report No. 2, 1–31 May. Prepared for FERC. 5 pp.

Dovel, W. L. 1979d. Same title. Periodic Progress Report No. 3, 1–30 June. Prepared for FERC. 6 pp.

Dovel, W. L. 1979e. Same title. Periodic Progress Report No. 4, 1 July–30 September. Prepared for FERC. 6 pp.

Dovel, W. L. 1979f. Performance report for biology and management of shortnose

and Atlantic sturgeons of the Hudson River. Period Extension 1 April 1978–30 September 1978. 169 pp.
Dovel, W. L. 1979g. The biology and management of shortnose and Atlantic sturgeon of the Hudson River. DEC, Project AFS9-R 54 pp. plus 14 p. appendix.
Dovel, W. L. 1980a. The endangered shortnose sturgeon of the Hudson River: Its life history and vulnerability to the activities of man. Periodic Progress Report No. 5. Prepared for FERC. 5 pp. plus data tables.
Dovel, W. L. 1980b. Same title. Periodic Progress Report No. 6 (Third quarterly). Prepared for FERC. 19 pp.
Dovel, W. L. 1980c. Same title. Periodic Progress Report No. 7. (Fourth quarterly report). Prepared for FERC. 32 pp.
Dovel, W. L. 1981a. Same title. Final Report. Prepared for FERC. 139 pp.
Dovel, W. L. 1981b. Ichthyoplankton of the lower Hudson estuary, New York. N.Y. Fish Game J. 28: 21–39.
Dovel, W. L., E. H. Buckley, M. E. Crandall, J. Fortier, S. M. Fredericks, W. L. Hopkins, S. S. Ristich, and D. L. Sirois. 1977. An atlas of biological resources of the Hudson estuary. Boyce Thompson Institute for Plant Research, Yonkers, New York. 104 pp.
Dowd, C. F. and E. D. Houde. 1980. Combined effects of prey concentration and photoperiod on survival and growth of larval sea bream, *Archosargus rhomboidalis* (Sparidae). Marine Ecology. 3: 181–185.
Dustman, E. H., L. F. Stickel, L. J. Blus, W. L. Reichel, and S. N. Wiemeyer. 1971. The occurrence and significance of polychlorinated biphenyls in the environment. pp. 118–133 in: Trans. 36th N. Amer. Wildl. Natur. Res. Conf. Wildl. Mgt. Inst., Washington, D.C.
EA. 1976a. Danskammer Point Generating Station impingement and entrainment survival studies. 1975 annual report. Prepared for CHGE. xv + 210 pp.
EA. 1976b. Roseton Generating Station impingement and entrainment survival studies. 1975 Annual Report. Prepared for CHGE. xv + 279 pp.
EA. 1976c. Lovett Generating Station entrainment survival and abundance studies. Vol. 1. 1975 annual interpretative report. Prepared for ORU. Part I. vii + 84 pp. Part II. iv + 46 pp.
EA. 1976d. Bowline Point Generating Station entrainment survival and abundance studies. Vol. 1. 1975 annual interpretative report. Prepared for ORU. Part I xvii + 213 pp. Part II. vii + 78 pp.
EA. 1977a. Danskammer Point Generating Station impingement survival studies. 1976 annual report. Prepared for CHGE. iii + 46 pp.
EA. 1977b. Impingement survival studies at Roseton and Danskammer Point Generating Stations. December 1977 progress report. Prepared for CHGE. vi + 86 pp.
EA. 1977c. Lovett Generating Station entrainment and impingement studies. 1976 annual report. Prepared for ORU. vi + 117 pp.
EA. 1977d. Bowline Point Generating Station entrainment and impingement studies. 1976 annual report. Prepared for ORU. ix + 225 pp.
EA. 1978a. Hudson River thermal effects studies for representative species. Final report. Prepared for CHGE, Con. Ed., and ORU. iii + 913 pp.
EA. 1978b. Thermal effects literature review for Hudson River representative important species. Prepared for CHGE, Con. Ed., and ORU. vi + 127 pp.

LITERATURE CITED

EA. 1978c. Field studies of the effects of thermal discharges from Roseton and Danskammer Point Generating Stations on Hudson River fishes. Prepared for CHGE. v + 177 pp.

EA. 1978d. Indian Point Generating Station entrainment survival and related studies. 1977 annual report. Prepared for Con. Ed. vi + 71 pp.

EA. 1978e. Impingement survival studies at the Roseton and Danskammer Point Generating Stations. August 1978 progress report. Prepared for CHGE. ii + 22 pp.

EA. 1978f. Roseton Generating Station entrainment survival studies. 1976 annual report. Prepared for CHGE. iii + 174 pp.

EA. 1978g. Roseton Generating Station entrainment survival studies. 1977 annual report. Prepared for CHGE. iv + 56 pp.

EA. 1978h. Entrainment abundance studies at the Roseton and Danskammer Point Generating Stations. December 1978 progress report. Prepared for CHGE. i + 67 pp.

EA. 1978i. Lovett Generating Station entrainment abundance studies. 1977 annual report. Prepared for ORU. iii + 23 pp.

EA. 1978j. Bowline Point Generating Station entrainment abundance studies. 1977 annual report. Prepared for ORU. iv + 68 pp.

EA. 1978k. Lovett Generating Station entrainment survival studies. 1977 annual report. Prepared for ORU. iv + 47 pp.

EA. 1978l. Bowline Point Generating Station entrainment survival studies. 1977 annual report. Prepared for ORU. v + 68 pp.

EA. 1979a. Effects of heat shock on predation of striped bass larvae by yearling white perch. iii + 49 pp.

EA. 1979b. Evaluation of the effectiveness of a continuously operating fine mesh traveling screen for reducing ichthyoplankton entrainment at the Indian Point Generating Station. Prepared for CHGE, Con. Ed., ORU, PASNY. iii + 28 pp.

EA. 1979c. Indian Point Generating Station entrainment survival and related studies. 1978 annual report. Prepared for Consolidated Edison and PASNY. viii + 194 pp.

EA. 1979d. Bowline Point impingement studies 1975–1978. Overview Report. Prepared for ORU. v + 69 pp.

EA. 1979e. Bowline Point Generating Station entrainment abundance and survival studies. 1978 annual report. Prepared for ORU. x + 200 pp.

EA. 1980a. Roseton Generating Station entrainment studies. 1978 annual report. Prepared for CHGE. iv + 44 pp.

EA. 1980b. Indian Point Generating Station entrainment and near-field river studies. 1977 annual report. Prepared for Con. Ed. and PASNY. iv + 232 pp.

EA. 1980c. Indian Point Generating Station entrainment and near-field river studies. 1978 annual report. Prepared for Con. Ed. and PASNY. ix + 360 pp.

EA. 1981a. Indian Point Generating Station entrainment survival and related studies. 1979 annual report. Prepared for Con. Ed. and PASNY. xi + 140 pp.

EA. 1981b. Bowline Point Generating Station entrainment abundance and survival studies. 1979 annual report with an overview of 1975–1979 studies. Prepared for ORU. xiv + 329 pp.

EA. 1981c. Indian Point Generating Station entrainment and near-field river studies. 1979 annual report. Prepared for Con. Ed. and PASNY.

EA. 1981d. Bowline Point Generating Station entrainment abundance studies. 1980 annual report. Prepared for ORU. i + 109 pp.

LITERATURE CITED

Edwards, S. J. and J. B. Hutchinson, Jr. 1980. Effectiveness of a barrier net in reducing white perch (*Morone americana*) and striped bass (*Morone saxatilis*) impingement. Environ. Sci. Tech. 14(2): 210–213.

Elder, J. W. 1959. The dispersion of marked fluid in turbulent sheer flow. J. Fluid Mech. 5: 544.

Eldridge, J. 1976. An illustrated key to the adult copepods of the Hudson River, Roseton/Danskammer Point vicinity. LMS, Unpublished.

Eldridge, M. B., J. King, D. Eng, and M. J. Bowers. 1977. Role of the oil globule in survival and growth of striped bass (*Morone saxatilis*) larvae. Proc. 57th Ann. Conf. Western Assoc. State. Game and Fish Commissioners, pp. 303–313.

Eldridge, M. B., J. A. Whipple, D. Eng, M. J. Bowers, and B. M. Jarvis. 1981. Effects of food and feeding factors on laboratory-reared striped bass larvae. Trans. Amer. Fish. Soc. 110(1): 111–120.

Ellis, R. H., S. J. Koepp, J. Weis, E. D. Santoro, and S. L. Cheng. 1981. A comprehensive monitoring and assessment program for selected heavy metals in New Jersey aquatic fauna. NJMSC.

Emlen, J. M. 1970. Age specificity and ecological theory. Ecology 51: 588–601.

Englert, T. L. and F. N. Aydin. 1975. An intra-tidal population model of young-of-the-year striped bass in the Hudson River. Advances in Computer Methods for Partial Differential Equations. Proc. AICA International Symposium, p. 273.

Eraslan, A. H., W. Van Winkle, R. D. Sharp, S. W. Christensen, C. P. Goodyear, R. M. Rush, and W. Fulkerson. 1976. A computer simulation model for the striped bass young-of-the-year population in the Hudson River. Oak Ridge National Laboratory, Environmental Sciences Division Publication 766, ORNL/NUREG-8, Oak Ridge, Tennessee.

Eustace, I. J. 1974. Zinc, cadmium, copper, and manganese in species of finfish and shellfish caught in the Derwent Estuary, Tasmania. Aust. J. Mar. Freshwater Res. 25: 209–217.

Fabricius, E. 1954. Aquarium observations on the spawning behavior of the burbot, *Lota vulgaris*. Ann. Rept. for 1953, and short papers, Inst. Freshwater Res., Drottningholm, 35: 51–57.

Fenneman, N. M. 1938. Physiography of eastern United States. vol. XIV. McGraw Hill, N.Y. 714 pp.

Flemming, J., W. Anderson, J. Motyer, and M. J. Dadswell. ms. Utilization of an intertidal mudflat by tomcod (*Microgadus tomcod*) and smooth flounder (*Liopsetta putnami*) and their impact on benthic prey organisms, Peck's Cove, Cumberland Basin, New Brunswick.

Fowler, C. W. 1980. Density dependence as related to life history strategy. Ecology 62(3): 602–610.

Fried, S. M. and J. D. McCleave. 1973. Occurrence of the shortnose sturgeon (*Acipenser brevirostrum*), an endangered species in Montsweag Bay, Maine. J. Fish. Res. Bd. Canada. 30: 563–564.

Friedmann, B. R. and C. T. Hamilton. 1980. The fish life of Upper New York Bay. Underwater Natur. 12(2): 18–21.

Gable, M. F. and R. A. Croker. 1977. The salt marsh amphipod, *Gammarus palustris* Bousfield, 1969 at the northern limit of its distribution. I. Ecology and life cycle. Estuarine and Coastal Marine Sci. 5: 123–134.

LITERATURE CITED

Gadgil, M. and W. H. Bossert. 1970. Life historical consequences of natural selection. Amer. Natur. 104: 1–24.
Gardner, M. B. 1981. Mechanisms of size selectivity by planktivorous fish: a test of hypotheses. Ecology 62(3): 571–578.
Garrod, D. J. and B. J. Knights. 1979. Fish stocks: their life history characteristics and response to exploitation. in: Fish Phenology: anabolic adaptiveness in teleosts. P. J. Miller (ed.). Symp. Zool. Soc. Lond. 44: pp. 307–326.
Giese, G. L. and J. W. Barr. 1967. The Hudson River estuary, a preliminary investigation of flow and water characteristics. N.Y. State Water Comm. Bull. vol. 61, 39 pp.
Giesel, J. T. 1974. Fitness and polymorphism for net fecundity distribution in iteroparous populations. Am. Natur. 108: 321–331.
Giesel, J. T. 1976. Reproductive strategies as adaptations to life in temporally heterogeneous environments. Ann. Rev. Ecol. Syst. 7: 57–79.
Ginn, T. C. 1977. An ecological investigation of Hudson River macrozooplankton in the vicinity of a nuclear power plant. Ph.D. Thesis, New York University. Dept. Biol.
Girisch, H. B., J. C. Dieleman, G. W. Petersen, and S. Pinkster. 1974. The migration of two sympatric gammarid species in a French estuary. Bijdr. Dierk. 44: 239–273.
Gorham, S. W. and D. E. McAllister. 1974. The shortnose sturgeon (*Acipenser brevirostrum*), in the Saint John River, New Brunswick, Canada, a rare and possibly endangered species. Syllogeus No. 5, National Museum of Canada, Ottawa.
Grabe, S. A. 1978. Food and feeding habits of juvenile Atlantic tomcod, *Microgadus tomcod*, from Haverstraw Bay, Hudson River. Fish. Bull. 76(1): 89–94.
Grabe, S. A. 1980. Food of age 1 and 2 Atlantic tomcod, *Microgadus tomcod*, from Haverstraw Bay, Hudson River. Fish. Bull. 77(4): 1003–1006.
Grabe, S. A. and R. E. Schmidt. 1978. Overlap and diet variations in feeding habits of *Alosa* spp. juveniles from the lower Hudson estuary. Paper presented at Annual Meeting, New York Chapter, American Fisheries Society, Marcy, New York.
Grant, G. C. 1974. The age composition of striped bass catches in Virginia rivers, 1967–1971, and a description of the fishery. Fish. Bull. 72: 193–199.
Greeley, J. R. 1935. Fishes of the watershed with annotated list. pp. 63–101 in: A biological survey of the Mohawk-Hudson watershed. Suppl. 24th Ann. Rept. N.Y. State Conserv. Dept.
Greeley, J. R. 1937. Fishes of the area with annotated list. pp. 45–103. in: A biological survey of the lower Hudson watershed. Suppl. 26th Ann. Rept. N.Y. Conserv. Dept.
Greig, R. A. and R. McGrath. 1977. Trace metals in sediments of Raritan Bay. Mar. Pollut. Bull. 8(8): 188–192.
Greze, I. I. 1968. Feeding habits and food requirements of some amphipods in the Black Sea. Mar. Biol. 1: 316–321.
Griffiths, D. 1975. Prey availability and the food of predators. Ecology. 56(5): 1209–1214.
Grove, T. L., T. J. Berggren, and D. A. Powers. 1976. The use of innate tags to segregate spawning stocks of striped bass, *Morone saxatilis*. pp. 166–176 in: Estuarine Processes (M. Wiley, ed.), vol. 1.

LITERATURE CITED

Gunderson, D. R. 1980. Using r-K selection theory to predict natural mortality. Can J. Fish. Aquat. Sci. 37: 2266–2271.
Hamley, J. M. 1975. Review of gill net selectivity. J. Fish. Res. Bd. Canada 32: 1943–1969.
Hamer, R. C., H. A. Hensel, and R. E. Tiller. 1948. Maryland commercial fisheries statistics, 1944–1945. Chesapeake Biol. Lab. Publ. No. 69.
Hansen, L. G. 1979. Selective accumulation and depletion of polychlorinated biphenyl components: Food animal implications. Ann. N.Y. Acad. Sci. 320: 238–246.
Hansen, M. J. and D. H. Wahl. 1981. Selection of small *Daphnia pulex* by yellow perch fry in Oneida Lake, New York. Trans. Amer. Fish. Soc. 110(1): 64–71.
Hardy, J. D., Jr. 1978a. Development of Fishes of the Mid-Atlantic bight. Vol. II. Anguillidae through Syngnathidae. Center Environ. Estuarine Studies. Univ. Maryland. Contrib. No. 784. 458 pp.
Hardy, J. D., Jr. 1978b. Development of Fishes of the Mid-Atlantic bight. Vol. III. Aphredoderidae through Rachycentridae. Center Environ. Estuarine Studies. Univ. Maryland. Contrib. No. 785. 394 pp.
Hardy, J. D., Jr. and L. L. Hudson. 1975. Description of the eggs and juveniles of the Atlantic tomcod, *Microgadus tomcod*. Ref 75–11, Chesapeake Biol. Lab., Solomons, Md. 14 pp.
Hatch, W. R. and W. C. Ott. 1968. Determination of sub-microgram quantities of mercury by atomic absorption spectrophotometry. Anal. Chem. 40(14): 2085–2087.
Hazen and Sawyer, Inc. 1979. The city of New York section 208 areawide waste treatment management planning program. Prepared for the City of New York Department of Environmental Protection.
Heidt, A. R. and R. J. Gilbert. 1978. The shortnose sturgeon in the Altamaha River drainage, Georgia. MS Rept. Contract 03-7-043-35-165. NMFS. 16 pp.
Helwig, J. T. and K. S. Council. 1979. SAS user's guide. SAS Institute, Inc. Cary, North Carolina. 494 pp.
Hetling, L. J., E. Horn, and T. J. Tofflemire. 1978. Summary of Hudson River PCB study results. New York State Dept. Environ. Conserv. Tech. Paper 52, 88 pp.
Hetling, L. J., T. J. Tofflemire, E. G. Horn, R. Thomas, and R. Mt. Pleasant. 1979. The Hudson River PCB problem: Management alternatives. Ann. N.Y. Acad. Sci. 320: 630–650.
Hildebrand, S. F. 1963. Fishes of the western North Atlantic. Mem. Sears Found. Mar. Res. 1(3): 324–332.
Hildebrand, S. F. and W. C. Schroeder. 1928. Fishes of Chesapeake Bay. Bull. U.S. Bur. Fish. 43, part 1, 366 pp.
Hirschfield, H. I., J. W. Rachlin, and E. Leff. 1966. A survey of invertebrates from selected sites of the lower Hudson River. pp. 220–257 in: Hudson River Ecology, DEC.
Hjort, J. 1926. Fluctuations in the year class of important food fishes. J. Cons., Cons. Int. Explor. Mer. 1: 5–39.
Hoff, J. G. 1980. Review of the present status of the stocks of the Atlantic sturgeon, *Acipenser oxyrhynchus* Mitchill. Prepared for NMFS. 136 pp.
Hoff, J. G. 1979. Annotated bibliography and subject index on the shortnose sturgeon, *Acipenser brevirostrum*. NOAA Tech. Rept. NMFS SSRF-731. 16 pp.

LITERATURE CITED

Hoff, T. B., R. J. Klauda, and B. S. Belding. 1977a. Incidental catch and distribution of shortnose sturgeon and Atlantic sturgeon in the Hudson River Estuary, 1969 through 1977. Prepared for workshop of Shortnose Sturgeon Recovery Team, Philadelphia, Pennsylvania.
Hoff, T. B., R. J. Klauda, and B. S. Belding. 1977b. Data on distribution and incidental catch of shortnose sturgeon (*Acipenser brevirostrum*) in the Hudson River estuary 1969 to present. TI MS Rept. 21 pp.
Hoff, T. B. and R. J. Klauda. 1979a. Data on shortnose sturgeon (*Acipenser brevirostrum*) collected incidentally from 1969 through 1979 in sampling programs conducted for the Hudson River Ecological Study. Prepared for workshop of Shortnose Sturgeon Recovery Team, Danvers, Massachusetts.
Hoff, T. B. and R. J. Klauda. 1979b. Distribution and some life history aspects of shortnose sturgeon in the lower Hudson estuary. Prepared for and presented at the Shortnose Sturgeon Recovery Team Workshop, December 1979, Danvers, Mass. 31 pp.
Hoff, T. B., R. J. Klauda, and J. R. Young. 1986. Contributions to the biology of the shortnose sturgeon in the Hudson River Estuary. This volume.
Holland, D. G. 1976. The inland distribution of brackish-water *Gammarus* species in the area of the Mersey and Weaver River Authority. Freshwater Biol. 6: 277–285.
Holling, C. S. (ed.). 1978. Adaptive environmental assessment and management. John Wiley and Sons. New York. 377 pp.
Hogan, T. M. and B. S. Williams. 1976. Occurrence of the gill parasite, *Ergasilus labracis* on striped bass, white perch, and tomcod in the Hudson River. N.Y. Fish Game J. 23: 97.
Holsapple, J. G. and L. E. Foster. 1975. Reproduction of white perch in the lower Hudson River. N.Y. Fish Game J. 22: 122–127.
Horn, E. G., L. J. Hetling, and T. J. Tofflemire. 1979. The problem of PCBs in the Hudson River system. Ann. N.Y. Acad. Sci. 320: 591–609.
Houde, E. D. 1978. Critical food concentrations of three species of subtropical marine fishes. Bull. Mar. Sci. 28(3): 395–411.
Howe, A. B. 1971. Biological investigations of the Atlantic tomcod, *Microgadus tomcod* Walbaum, in the Wewantic River estuary, Massachusetts, 1967. M.S. thesis, Univ. Massachusetts, Amherst. 89 pp.
Howells, G. P. 1972. The estuary of the Hudson River, USA. Proc. Royal Soc. London, Bull. 180: 521–534.
Hudson River Research Council (HRRC). 1980. Results of Hudson River field weeks: April 1977 and August 1978, xxii + 50 pp.
Hudson River Valley Commission. 1966. The Hudson fish and wildlife: a report on fish and wildlife resources in the Hudson River Valley. NYDEC, Albany.
Huff, J. A. 1975. Life history of Gulf of Mexico sturgeon, *Acipenser oxyrhynchus desoti* in Suwanee River, Florida. Florida Mar. Res. Publ. 16, 32 pp.
Humphries, E. T. and K. B. Cummings. 1973. An evaluation of striped bass fingerling culture. Trans. Amer. Fish. Soc. 102(1): 13–20.
Hunter, J. R. 1976. Report of a colloquium on larval fish mortality studies and their relation to fishery research. NOAA Tech. Rept. NMFS Circ. 395.
Hutchinson, G. E. 1967. A treatise on limnology. Vol. 2. John Wiley and Sons, Inc. New York.

LITERATURE CITED

Hynes, H. B. N. 1955. The reproductive cycle of some British freshwater Gammaridae. J. Animal Ecol. 24; 352–387.
IA. 1977. Impingement and entrainment at the Werner Generating Station and a study of the fishes of the Raritan River and Bay near the station. Report submitted to Jersey Central Power and Light Company.
Interstate Sanitation Commission (ISC). 1978. Annual report on the water pollution activities and the interstate air pollution program.
Interstate Sanitation Commission (ISC). 1980. Annual report on the water pollution activities and the interstate air pollution program.
Ivlev, V. S. 1961. Experimental ecology of the feeding of fishes. Yale Univ. Press. New Haven.
Jacobs, R. P. and V. A. Crecco. 1980. Age and growth of larval and juvenile American shad (*Alosa sapidissima*) utilizing daily growth increments of otoliths. Paper presented at the 36th Annual Northeast Fish and Wildlife Conference, Ellenville, New York.
Jinks, S. M., M. E. Loftus, and G. J. Lauer. 1981. Advances in techniques for assessment of ichthyoplankton entrainment survival. In: Issues Associated with Impact Assessment (L. D. Jensen, ed.). EA Communications.
Johnson, H. B., H. W. Stevens, and W. W. Hassler. 1979. Cooperative management program for Albemarle Sound-Roanoke River striped bass. North Carolina Dept. Nat. Res. and Comm. Develop., Div. Mar. Fish., Morehead City. Annual progress report for project AFS-14-2. 63 pp.
Jones, P. W., F. D. Martin, and J. D. Hardy, Jr. 1978. Development of fishes of the mid-Atlantic bight. An Atlas of egg, larval and juvenile stages. Vol. 1. Acipenseridae through Ictaluridae. U.S. Dept. Int. U.S. Fish Wildl. Serv. 366 pp.
Katz, H. M. 1978. Circadian rhythms in juvenile American shad, *Alosa sapidissima*. J. Fish Biol. 12: 609–614.
Kawasaki, T. 1980. Fundamental relations among the selection of life history in the marine teleosts. Bull. Japanese Soc. Sci. Fish. 46(3): 289–293.
Keast, A. 1965. Resource subdivision amongst co-habiting fish species in a bay, Lake Opinicon, Ontario. Proc. 8th Conf. Great Lakes Res. pp. 106–132.
Keast, A. 1970. Food specializations and bioenergetic interrelationships in the fish faunas of some small Ontario waterways. pp. 377–411 in Marine Food Chains, (J. H. Steele, ed.). Univ. Calif. Press.
Keck, G. and J. Raffenot. 1979. Chemical contamination by PCBs in the fishes of a French river: the Furans (Jura). Bull. Environ. Contam. Toxicol. 21: 689–696.
Kellog, R. L., J. J. Salermo, and D. L. Latimer. 1978. Effects of acute and chronic thermal exposure on the eggs of three Hudson River anadromous fishes. In: Energy and Environmental Stress in Aquatic Systems, (J. H. Thorpe and J. W. Gibbons, eds.). Tech. Info. Serv., U.S. Dept. Energy.
Kernehan, R. J., M. R. Headrick, and R. F. Smith. 1981. Early life history of striped bass in the Chesapeake and Delaware Canal and vicinity. Trans. Amer. Fish. Soc. 110(1): 137–150.
King, L. R., J. B. Hutchison, Jr., and T. G. Huggins. 1978. Impingement survival studies on white perch, striped bass, and Atlantic tomcod at three Hudson River power plants. In: Fourth National Workshop on Entrainment and Impingement, (L. D. Jensen, ed.). EA Communications.

LITERATURE CITED

King, L. R., B. A. Smith, R. L. Kellog, and E. S. Perry. 1981. Comparisons of ichthyoplankton collected with a pump and stationary nets in a power plant discharge canal. In: Issues Associated with Impact Assessment, (L. D. Jensen, ed.). EA Communications.

Kinne, O. 1960. Growth, moulting frequency, heartbeat, number of eggs and incubation time in *Gammarus zaddachi* exposed to different environments. Crustaceana 2(10): 26–36.

Kinne, O. 1971. Marine Ecology. Vol. 1. John Wiley and Sons, New York.

Kissil, G. W. 1974. Spawning of the anadromous alewife, *Alosa pseudoharengus*, in Bride Lake, Connecticut. Trans. Amer. Fish. Soc. 103(2): 312–317.

Klauda, R. J., M. Nittel, and K. P. Campbell. 1976. The commercial fishery for American shad in the Hudson River: fishing effort and stock abundance trends. pp. 107–134 In: Proc. American shad workshop, University of Massachusetts, Amherst. U.S. Fish and Wildl. Serv. and NMFS.

Klauda, R. J., W. P. Dey, T. B. Hoff, J. B. McLaren, and Q. E. Ross. 1980. Biology of juvenile Hudson River striped bass. pp. 101–123 In: Proc. 5th Ann. Mar. Rec. Fish. Symp. (H. Clepper, ed.). Sport Fishing Institute, Washington, D.C.

Klauda, R. J., P. H. Muessig, and J. A. Matousek. 1987. Fisheries data sets compiled by utility-sponsored research in the Hudson River Estuary. (This volume).

Klauda, R. J., T. H. Peck, and G. K. Rice. 1981. Accumulation of polychlorinated biphenyls in Atlantic tomcod (*Microgadus tomcod*) collected from the Hudson River Estuary, New York. Bull. Environ. Contam. Toxicol. 27: 829–835.

Kneip, T. J. and J. M. O'Connor. 1980. Cadmium in Foundry Cove crabs: health hazard assessment. Final Report to Health Research Council. New York State Health Planning Commission, Albany, N.Y.

Koo, T. S. Y. 1970. The striped bass fishery in the Atlantic states. Chesapeake Sci. 11: 73–93.

Koski, R. T. 1973. Life history and ecology of the hogchoker, *Trinectes maculatus*, in its northern range. Ph.D. dissertation. University of Connecticut, Storrs, Connecticut.

Koski, R. T., E. C. Kelley, and B. E. Turnbough. 1971. A record-sized shortnose sturgeon from the Hudson River. N.Y. Fish Game J. 18: 75.

Kostalos, M. and R. L. Seymour. 1976. Role of microbial enriched detritus in the nutrition of *Gammarus minus* (Amphipoda). Oikos 27: 512–516.

Kretser, W. A. 1973. Aspects of striped bass larvae, *Morone saxatilis*, captured in the Hudson River near Cornwall, New York, April–August, 1968. Prepared for Con. Ed.

Lawler, J. P. 1972. Effect of entrainment and impingement at Indian Point on the population of the Hudson River striped bass. Docket No. 50–247, October 30.

Legendre, V. and R. Lagueux. 1948. The tomcod (*Microgadus tomcod*) as a permanent fresh-water resident of Lake St. John, Province of Quebec. Canad. Field-Natur. 65: 157.

Leggett, W. C. 1976. The American shad (*Alosa sapidissima*), with special reference to its migration and population dynamics in the Connecticut River. pp. 169–225 in: The Connecticut River ecology study. (D. Merriman and L. M. Thorpe, eds.). Amer. Fish. Soc. Monogr. No. 1.

Leggett, W. C. and J. E. Carscadden. 1978. Latitudinal variation in reproductive characteristics of American shad (*Alosa sapidissima*): evidence for population specific life history strategies in fish. J. Fish. Res. Bd. Canada 35: 1469–1478.

Leggett, W. C. and R. R. Whitney. 1972. Water temperature and the migration of the American shad. U.S. Fish Wildl. Serv. Bull. 70(3): 659–670.

Leland, H. V., D. J. Wilkes, and E. D. Copenhauer. 1976. Heavy metals and related trace elements. J. Water Pollut. Control Fed. 48(6): 1459–1486.

Leslie, P. H. 1945. On the use of matrices in certain population mathematics. Biometrika 33: 183–212.

LeSueur, C. A. 1818. Description of several species of Chondropterygious fishes of North America, with their varieties. Trans. Amer. Philos. Soc. 1: 383–395.

Levesque, R. C. and R. J. Reed. 1972. Food availability and consumption by young Connecticut River shad, *Alosa sapidissima*). J. Fish. Res. Bd. Canada. 29: 1495–1499.

Lieberman, J. T. and P. H. Muessig. 1978. Evaluation of an air bubbler to mitigate fish impingement at an electric generating plant. Estuaries 1: 129–132.

Limburg, K. E., M. A. Moran, and W. H. McDowell. 1986. The Hudson River Ecosystem. Springer-Verlag. New York. 344 pp.

Lippson, A. J., M. S. Haire, A. F. Holland, F. Jacobs, J. Jensen, R. L. Moran-Johnson, T., T. Polgar, and W. A. Richkus. 1980. Environmental Atlas of the Potomac Estuary. Martin Marietta Corp. Baltimore, MD. 279 pp.

LMS. 1974a. Cornwall gear evaluation study. Prepared for Con. Ed.

LMS. 1974b. Danskammer Point fish impingement study. Prepared for CHGE.

LMS. 1975a. 1973–1974 ecological studies of the Hudson River in the vicinity of Lloyd, New York. Prepared for New York State Atomic and Space Development Authority. x + 195 pp. Appendices 227 pp.

LMS. 1975b. 1974 Hudson River aquatic ecology studies—Bowline Point and Lovett Generating Stations. Prepared for ORU. xiii + 260 pp. Appendices xxii + 516 pp.

LMS. 1975c. 1973 Hudson River aquatic ecology studies at Roseton/Danskammer Point. Vols. I–VI. Prepared for CHGE. xxix + 1720 pp.

LMS. 1975d. Aquatic ecology studies—Kingston, N.Y.—1973. Prepared for CHGE. iii + 280 pp.

LMS. 1975e. Thermal survey at the Roseton/Danskammer Point Generating Stations. Prepared for CHGE. vi + 41 pp.

LMS. 1975f. Albany Steam Electric Generating Station impingement survey, April 1974–March 1975. Prepared for NMPC. vi + 192 pp.

LMS. 1975g. Report on development of a real-time, two dimensional model of the Hudson River striped bass population. Prepared for Con. Ed. xi + 122 pp.

LMS. 1976a. 1975 Hudson River aquatic ecology studies—Bowline Point and Lovett Generating Stations. Prepared for ORU. xx + 361 pp.

LMS. 1976b. Danskammer Point reduced flow study evaluation of the influence of reduced flow on fish impingement. Prepared for CHGE.

LMS. 1976c. Danskammer Point thermal analysis—report evaluation of thermal influence on fish populations. Prepared for CHGE.

LMS. 1976d. Preliminary evaluation of the effectiveness of a chain barrier and of reduction in cooling water flow in reducing impingement at Danskammer Point Generating Station. Prepared for CHGE.

LITERATURE CITED

LMS. 1976e. 1976 Roseton intake avoidance study—preliminary report on an evaluation of intake avoidance as a function of size for white perch. Prepared for CHGE.

LMS. 1976f. Environmental impact assessment: Water quality analysis: Hudson River. Prepared for the National Commission on Water Quality. NTIS, Springfield, Virginia. PB-251-099.

LMS. 1977a. Influence of Indian Point Unit 2 and other steam electric generating plants on the Hudson River estuary with emphasis on striped bass and other fish populations—Supplement 1. Prepared for Con. Ed.

LMS. 1977b. 1976 Hudson River aquatic ecology studies at Bowline Point Generating Station. Prepared for ORU. xi + 298 pp.

LMS. 1977c. Hudson river aquatic ecology studies at the Lovett Generating Station. Prepared for ORU.

LMS. 1977d. Roseton/Danskammer—viability studies of impinged Atlantic Tomcod—1974–1975. Prepared for CHGE.

LMS. 1977e. Dissolved oxygen loss due to once-through cooling at the Roseton Generating Station. Prepared for CHGE.

LMS. 1977f. Roseton—fish migration route study evaluation of river cross-sectional distribution patterns of fish larvae and fish eggs. Prepared for CHGE.

LMS. 1977g. Evaluation of the influence of a chain barrier and of reduced flow on size distribution of impinged fish. Prepared for CHGE.

LMS. 1978a. Mitigation of impingement impact at Hudson River power plants. Prepared for Con. Ed.

LMS. 1978b. 1977 impingement studies at the Lovett Generating Station. Prepared for ORU.

LMS. 1978c. 1977 Hudson River aquatic ecology studies at the Bowline Point Generating Station. Prepared for ORU. xv + 320 pp.

LMS. 1978d. Ichthyoplankton gear comparison study. Prepared for ORU. i + 23 pp.

LMS. 1978e. 1974 annual progress report. Prepared for CHGE.

LMS. 1978f. 1975 annual progress report. Prepared for CHGE. iii + 315 pp.

LMS. 1978g. Roseton/Danskammer Generating Stations—Habitat selection studies. Prepared for CHGE.

LMS. 1979a. Mitigation of entrainment and impingement impact at Hudson River power plants. Prepared for Con. Ed.

LMS. 1979b. Application of the niche analysis to four biotic communities at the Roseton-Danskammer Point Generating Stations, Hudson River, in 1973. (Research Grant No. 1200-400-010) Vols. I and II. Prepared for CHGE.

LMS. 1979c. 1978 impingement studies at the Lovett Generating Station. Prepared for ORU.

LMS. 1979d. 1978 Hudson River aquatic ecology studies at the Bowline Point Generating Station. Prepared for ORU. viii + 359 pp.

LMS. 1979e. 1976 annual progress report. Vols. I–IV. Prepared for CHGE. xviii + 410 pp.

LMS. 1979f. 1977 annual progress report. Vols. I–V. Prepared for CHGE. xxvii + 500 pp.

LMS. 1979g. Fish impingement monitoring screen carry-over study—Roseton Electric Generating Station, June 1978–February 1979. Prepared for CHGE.

LITERATURE CITED

LMS. 1980a. TI/LMS ichthyoplankton (1975–1979) and river fish (1977–1979) comparison in the Croton/Haverstraw Bay region of the Hudson River. Prepared for ORU.
LMS. 1980b. Evaluation of lower trophic level aquatic communities in the vicinity of the Bowline Point Generating Station—1971–1977. Prepared for ORU.
LMS. 1980c. 1979 impingement studies at the Lovett Generating Station. Prepared for ORU.
LMS. 1980d. 1978 annual progress report. Vols. I–III. Prepared for CHGE. xxiii + 432 pp.
LMS. 1980e. 1979 annual progress report. Vols. I–III. Prepared for CHGE. xvi + 276 pp.
LMS. 1980f. Biological and water quality data collected in the Hudson River near the Proposed Westway Project during 1979–1980. Vol. I. Prepared for DOT and Parsons, Brinckerhoff, Quade and Douglas, Inc.
LMS. 1980g. Trawl comparisons in the northern Newburgh Bay area of the Hudson River, Yankee trawl vs. otter trawl. Prepared for CHGE.
LMS. 1980h. Ichthyoplankton entrainment survival study at Roseton Generating Station. Prepared for CHGE. iv + 393 pp.
LMS. 1981a. 1979 Hudson River aquatic ecology studies at Bowline Point Generating Station. Prepared for ORU. x + 376 pp.
LMS. 1981b. 1980 impingement studies at the Lovett Generating Station. Prepared for ORU. iii + 269 pp.
LMS. 1981c. 1980 Bowline Point aquatic ecology studies. Prepared for ORU.
LMS. 1981d. 1980 annual progress report. Prepared for CHGE. (Revised 1982). xii + 209 pp.
LMS. 1981e. Trawl methodology study: Constant distance vs. constant time. Prepared for CHGE.
LMS. 1981f. Report on 1976 and 1977 data analyses and application of life cycle models of the Hudson River striped bass population. Prepared for Con. Ed.
Lockwood, A. P. M. 1962. The osmoregulation of crustacea. Biol. Rev. 37: 512–516.
Loesch, J. G. 1969. A study of the blueback herring, *Alosa aestivalis* (Mitchill) in Connecticut waters. Ph.D. Thesis. University of Connecticut. 78 pp.
Loesch, J. G., W. H. Kriete, Jr., and E. J. Foell. 1982. Effects of light intensity on the catchability of juvenile anadromous *Alosa* species. Trans. Amer. Fish. Soc. 111(1): 41–44.
Loesch, J. G. and W. A. Lund, Jr. 1977. A contribution to the life history of the blueback herring. *Alosa aestivalis.* Trans. Amer. Fish. Soc. 106(6): 583–589.
Lyles, C. H. 1965. Fishery statistics of the United States. 1964. U.S. Fish Wildl. Serv. Stat. Dig. No. 57. 541 pp.
Malcolm Pirnie, Inc. 1976. Preliminary appraisal sediment transport relations upper Hudson River. Engineering Rep. Malcolm Pirnie, Inc., White Plains, N.Y.
Mansueti, R. J. 1958. Eggs, larvae and young of the striped bass, *Roccus saxatilis.* Contrib. 112. Chesapeake Biol. Lab. Solomons.
Mansueti, R. J. 1961. Age, growth, and movements of the striped bass, *Roccus saxatilis,* taken is size selective fishing gear in Maryland. Chesapeake Sci. 2: 9–36.

LITERATURE CITED

Mansueti, R. J. 1964. Eggs, larvae, and young of white perch, *Roccus americanus*, with comments on its ecology in the estuary. Chesapeake Sci. 5(12): 3–45.

Mansueti, A. J. and J. D. Hardy. 1967. Development of fishes of the Chesapeake Bay region, an atlas of egg, larval, and juvenile stage. Part 1. Nat. Res. Inst. Univ. Maryland.

Marcy, B. C., Jr. 1969. Age determinations from scales of *Alosa pseudoharengus* (Wilson) and *Alosa aestivalis* (Mitchill) in Connecticut waters. Trans. Amer. Fish. Soc. 98(4): 622–630.

Marcy, B. C., Jr. 1972. Spawning of the American shad, *Alosa sapidissima*, in the lower Connecticut River. Ches. Sci. 13: 116–119.

Marcy, B. C., Jr. 1976a. Fishes of the lower Connecticut River and the effects of the Connecticut Yankee Plant. pp. 61–113. in: The Connecticut River ecological study. (D. Merriman and L. M. Thorpe, eds.). Amer. Fish. Soc. Monogr. No. 1.

Marcy, B. C., Jr. 1976b. Planktonic fish eggs and larvae of the lower Connecticut River and the effects of the Connecticut Yankee plant including entrainment. pp. 115–139. ibid.

Marcy, B. C., Jr. 1976c. Early life history studies of American shad in the lower Connecticut River and the effects of the Connecticut Yankee plant. ibid. pp. 141–178.

Massman, W. H. 1962. Water temperatures, salinities, and fishes collected during trawl surveys of Chesapeake Bay and York and Pamunkey Rivers, 1956–1959. Virginia Inst. Mar. Sci. Spec. Sci. Rept. No. 27, 51 pp.

Massman, W. H. 1963. Summer food of juvenile American shad in Virginia Waters. Ches. Sci. 4: 167–171.

Mather, F. 1886. Work at Cold Spring Harbor, Long Island, during 1883 and 1884. Rept. U.S. Comm. Fish., 1884, Appendix pp. 129–142.

Mather, F. 1887. Report of operations at Cold Spring Harbor, New York, during the season of 1885. ibid. 1885, Appendix pp. 109–116.

Mather, F. 1889. Report of operations at Cold Spring Harbor, New York, during the season of 1886. ibid. 1886, Appendix pp. 721–728.

Mather, F. 1900. Modern fish culture in fresh and salt water. Forest and Stream Publishing Co., New York. 333 pp.

Matousek, J. A. 1973. Gear selectivity in fish eggs and larvae collections. In: Third Symposium on Hudson River Ecology, HRES March 1973.

Matousek, J. A., L. G. Arvidson, and R. L. Wyman. 1981. Diel hydroacoustic assessment of fish distributions. In: 38th Northeastern Fish and Wildlife Conf., April 1981. (Manuscript available upon request for Lawler, Matusky, and Skelly, Engineers).

Maugh, T. H., II. 1972. Polychlorinated biphenyls: Still prevalent but less of a problem. Science 178: 388.

McCleave, J. D., S. M. Fried, and A. K. Towt. 1977. Daily movements of shortnose sturgeon, *Acipenser brevirostrum*, in a Maine estuary. Copeia (1): 149–157.

McCormick, J. M. and S. Koepp. 1979. Distribution, diversity and toxicologic response of resident species, as correlated with changes in the physiochemical environment of Newark Bay. New Jersey Seagrant Biennial Report, 1976–1978. New Jersey Mar. Sci. Consortium.

LITERATURE CITED

McEachran, J. D., D. F. Boesch, and J. A. Musick. 1976. Food division within two sympatric species-pairs of skates (Pisces: Rajidae). Mar. Biol. 35: 301–317.
McFadden, J. T. (ed.). 1977. Influence of Indian Point Unit 2 and other steam electric generating plants on the Hudson River estuary, with emphasis on striped bass and other fish populations. Prepared for Con. Ed. 1+970 pp.
McFadden, J. T., TI and LMS. 1978. Influence of the proposed Cornwall pumped storage project and steam electric generating plants on the Hudson River estuary with emphasis on striped bass and other fish populations. Revised. Prepared for Con. Ed. lxii + 1159 pp.
McFadden, J. T. and J. P. Lawler (eds.). 1977. Supplement I to Influence of Indian Point Unit 2 and other stream electric generating plants on the Hudson River estuary, with emphasis on striped bass and other fish populations. Prepared for Con. Ed. ii + 366 pp.
McGie, A. M. and R. E. Muller. 1979. Age, growth, and population trends of striped bass, *Morone saxatilis*, in Oregon. Oregon Dept. Fish and Wildlife Inform. Rept. Series, Fish. Number 79–8. 57 pp.
McGroddy, P. M. and R. L. Wyman. 1977. Efficiency of nets and a new device for sampling living fish larvae. J. Fish. Res. Bd. Canada. 34(4): 571–574.
McLaren, J. B., J. C. Cooper, T. B. Hoff, and V. Lander. 1981. Movements of Hudson River striped Bass. Trans. Amer. Fish Soc. 110(1): 158–167.
McLeese, D. W., C. D. Metcalfe, and D. S. Pezzack. 1980. Uptake of PCBs from sediment by *Nereis virens* and *Crangon septemspinosa*. Arch. Environ. Contam. Toxicol. 9: 507–518.
Medeiros, W. H. 1974. The Hudson River shad fishery: Background, management problems and recommendations. N.Y. Sea Grant. Inst. 53 pp.
Mehrle, P. M., T. A. Haines, S. Hamilton, J. L. Ludke, F. L. Mayer, and M. A. Ribick. 1982. Relationship between contaminants and bone development in East-Coast striped bass. Trans. Amer. Fish. Soc. 111(2): 231–241.
Menzie, C. A. 1980. The Chironomid (Insecta: Diptera) and other fauna of a *Myriphyllum spicatum* L. plant bed in the lower Hudson River. Estuaries 3(1): 38–54.
Meshaw, J. C. 1969. A study of the feeding selectivity of striped bass fry and fingerlings in relation to zooplankton availability. M.S. thesis, North Carolina State Univ.
Messieh, S. N. 1977. Population structure and biology of alewives (*Alosa pseudoharengus*) and blueback herring (*A. aestivalis*) in the Saint John River, New Brunswick. Env. Biol. Fish. 2: 195–210.
Miller, R. R. 1972. Threatened freshwater fishes of the United States. Trans. Amer. Fish. Soc. 101(2): 239–252.
Milstein, C. B. 1981. Abundance and distribution of juvenile *Alosa* species off southern New Jersey. Trans. Amer. Fish. Soc. 110(2): 306–309.
Monk, D. C. 1976. The distribution of cellulose in freshwater invertebrates of different feeding habits. Freshwater Biol. 6: 471–475.
Montgomery, J. R. 1974. Individual variation of trace metal content in fish. ERDA NTIS CONF. 741023-5. Radioecology Div., Puerto Rico Nuclear Center, Mayaguez.
Mowrer, J., J. Calambokidis, N. Musgrove, B. Drager, M. V. Beug, and S. G. Herman. 1977. Polychlorinated biphenyls in cottids, mussels and sediment

LITERATURE CITED

in southern Puget Sound, Washington. Bull. Environ. Contam. Toxicol. 18(5): 588–594.
Mueller, J. A., J. S. Jeris, A. R. Anderson, and C. Hughes. 1975. Contaminant inputs to the New York Bight. NOAA Tech. Memo. ERL MESA-6. Rep. MESA Program.
Muldoon, T. 1981. Sturgeon come back to Hudson, fostering hope in caviar trade. National Fisherman, April 1981, p. 64.
Murawski, S. A. and A. L. Pacheco. 1977. Biological and fisheries data on Atlantic sturgeon, *Acipenser oxyrhynchus* (Mitchill). NMFS Tech. Ser. Rept. No. 10. Sandy Hook, N.J. 69 pp.
Murdock, W., S. Avery, and M. E. B. Smyth. 1979. Switching in predatory fish. Ecology 56: 1094–1105.
Murphy, G. I. 1968. Pattern in life history and the environment. Amer. Natur. 102: 390–404.
National Research Council. 1979. Polychlorinated biphenyls. National Academy of Sciences, Washington, D.C. 185 pp.
Newsome, G. E. and G. Leduc. 1975. Seasonal changes of fat content in the yellow perch (*Perca flavescens*) of two Laurentian lakes. J. Fish. Res. Bd. Canada. 32: 2213–2221.
NYDEC. 1976a. Hearing testimony on PCB discharges. Legal File #2833.
NYDEC. 1976b. Agreement between General Electric and NYDEC on PCB discharges. Legal File #2833.
NYDEC. 1979. Toxic substances in fish and wildlife. Vol. 3, no. 1. Division of Fish and Wildlife, DEC. Albany. 21 pp.
NYDEC. 1980. Hudson estuary fisheries development program. Report to State Legislature. 12 pp.
NYDEC. 1981a. New York state fishing, small game hunting, trapping regulations guide. Division of Fish and Wildlife DEC. Albany. 96 pp.
NYDEC. 1981b. Reducing toxics: Fish filleting guide. DEC pamphlet. 2 pp.
NYDEC. 1981c. Toxic substances in fish and wildlife. Vol. 4, No. 1. Division of Fish and Wildlife DEC, Albany. 138 pp.
NY Department of Health. 1981. Untitled press release on cadmium in Hudson River blue crab. 2 pp.
New York University Medical Center. 1975. Effects of entrainment by the Indian Point power plant on Hudson River biota: A progress report for 1974 to Con. Ed.
Niagara Mohawk Power Company. 1976. Albany steam electric generating station, 316(a) demonstration submission NPDES permit NY 0005959. Prepared by LMS.
Nichols, J. T. and C. M. Breder, Jr. 1926. The marine fishes of New York and southern New England. Zoologica 9: 1–192.
Nilsson, L. M. and C. Otto. 1977. Effects of population density and the presence of *Gammarus pulex* L. (Amphipoda) on the growth of larvae of *Potamophylax cingulatus* Steph. (Trichoptera). Hydrobiol. 54(2): 109–112.
Nittel, M. 1976. Food habits of Atlantic tomcod (*Microgadus tomcod*) in the Hudson River. In: Proc. 4th Symposium on Hudson River Ecology. HRES.
Occhiogrosso, T. J., L. R. King, B. Muchmore, D. P. Ostrye, D. Belcher, B. A. Smith, and S. M. Jinks. 1981. A portable automated plankton abundance

LITERATURE CITED

sampling device. In: Issues Associated with Impact Assessment (L. D. Jensen, ed.). EA Communications.

O'Connor, J. M. and J. W. Rachlin. 1982. Perspectives on metals in New York Bight organisms: I. Factors controlling accumulation and body burdens. pp. 655–673 in: Ecological Stress and the New York Bight: Science and Management. (G. F. Mayer, ed.). Estuarine Research Foundation, Columbia, S. C.

O'Connor, J. S. and H. M. Stanford. 1979. Chemical pollutants of the New York Bight; priorities for research. Special report.

Olsson, M., S. Jensen, and L. Reutergard. 1978. Seasonal variation of PCB levels in fish—An important factor in planning aquatic monitoring programs. Ambio. 7(2): 66–69.

Ontario Ministry of the Environment. 1978. Health implications of contaminants in fish. Ontario M. O. E. Toronto. 132 pp.

ORU. 1977. Bowline Point Generating Station. Near-field effects of once-through cooling system operation on Hudson River Biota. Prepared by LMS and EA.

Pacheco, A. L. 1973. Alewife, blueback herring, and American shad. p. 266 in: Proceedings of a workshop on egg, larval, and juvenile stages of fish in Atlantic coast estuaries. (A. L. Pacheco, ed.). Tech. Publ. No. 1. NMFS. Mid. Atlantic Coast. Fish. Center.

Pearcy, W. G. and S. W. Richards. 1962. Distribution and ecology of the fishes of the Mystic River estuary. Ecol. 43(2): 248–259.

Pecovitch, A. W. 1979. Distribution and some life history aspects of the shortnose sturgeon (*Acipenser brevirostrum*) in the upper Hudson River estuary. Hazleton Environmental Science Corporation, Northbrook, Ill. 67 pp.

Pennak, R. W. 1953. Freshwater invertebrates of the United States. The Ronald Press Company, New York.

Perlmutter, A., E. Leff, E. E. Schmidt, R. Heller, and M. Siciliana. 1966. Distribution and abundance of fishes along the shore of the lower Hudson River during the summer of 1966. In: Hudson River Ecology. Hudson River Valley Commission. pp. 147–200.

Perlmutter, A., E. E. Schmidt, and E. Leff. 1967. Distribution and abundance of fish along the shores of the lower Hudson River during the summer of 1965. N.Y. Fish Game J. 14: 47–75.

Perlmutter, A., E. E. Schmidt, R. Heller, F. C. Ford, and S. Sininsky. 1968. Distribution and abundance of fish along the shores of the lower Hudson during the summer of 1967. Ecological Survey of the Hudson River, Progress Report 3: 2–1, 2–42.

Peterson, R. E. and P. D. Guiney. 1979. Disposition of polychlorinated biphenyls in fish. pp. 21–36 in: Pesticide and xenobiotic metabolism in aquatic organisms. (M. A. Q. Khan, J. J. Lech, and J. J. Menn., eds.) ACS Symposium Series. No. 99. American Chemical Society.

Peterson, R. H., P. H. Johansen, and J. L. Metcalfe. 1980. Observations on early life stages of Atlantic tomcod, *Microgadus tomcod*. Fish. Bull. 78(1): 147–158.

Pianka, E. R. and W. S. Parker. 1975. Age-specific reproductive tactics. Amer. Natur. 109:453–464.

Pinkster, S. 1975. The introduction of the alien amphipod *Gammarus tigrinus* Sexton, 1939 (Crustacea, Amphipoda) in the Netherlands and its competition with indigenous species. Hydro. Biol. Bull. 9(3): 131–138.

LITERATURE CITED

Poje, G. V. 1977. The growth and reproduction of *Gammarus tigrinus* Sexton (Crustacea, Amphipoda). M.S. thesis, New York University, Dept. Biology.
Polgar T. T., J. A. Mihursky, R. E. Ulanowicz, R. P. Morgan, II, and J. S. Wilson. 1976. An analysis of 1974 striped bass spawning success in the Potomac Estuary. pp. 151–165 In: Estuarine Process, vol. 1. Academic press, New York.
Power, E. A. 1962. Fishery statistics of the United States. 1960. U.S. Fish Wild. Serv. Stat. Digest No. 53. 460 pp.
Pyke, G. H. 1979. Optimal foraging in fish in predator-prey systems in fisheries management. In: Int. Symp. of predator-prey systems in fish communities and their role in fisheries management Atlanta, Georgia, 24–27 July 1978. (H. Clepper, ed.). Sport Fishery Institute, Washington.
QLM. 1966. Algae sampling of the Hudson River—Summary Report. Prepared for Con. Ed.
QLM and Oceanographic Analysts, Inc. 1969a. Effect of Roseton plant cooling water discharge on Hudson River temperature distribution and ecology. Prepared for CHGE. i + 566 pp.
QLM. 1969b. Effect of Lovett plant cooling water discharge on Hudson River temperature distribution and ecology. Prepared for ORU.
QLM. 1969c. Effect of Indian Point cooling water discharges on Hudson River temperature distribution. Prepared for Con. Ed.
QLM. 1969d. Effect of Bowline cooling water discharge on Hudson River temperature distribution and ecology. Prepared for ORU.
QLM. 1971a. Environmental effects on Hudson River—Albany Steam Station discharge. Prepared for NMPC.
QLM. 1971b. Environmental effects of Bowline Generating Station on Hudson River. Prepared for ORU.
QLM. 1971c. Effects of Cornwall Pumped Storage Plant on Hudson River saltwater intrusion. Prepared for Con. Ed.
QLM. 1973a. Interim report on larval fish studies in the vicinity of Bowline Generating Station, 1971–1972. Prepared for ORU.
QLM. 1973b. Interim report on plankton studies in the vicinity of Bowline Generating Station, 1971–1972. Prepared for ORU.
QLM. 1973c. Interim report on impingement studies in the vicinity of Bowline Generating Station, 1973. Prepared for ORU.
QLM. 1973d. Interim report on Hudson River water quality—Bowline Generating Station, 1971–1972. Prepared for ORU.
QLM. 1973e. Interim report on benthos in the vicinity of Bowline Generating Station, 1971–1972. Prepared for ORU.
QLM. 1973f. Cornwall gear evaluation study. Prepared for Con. Ed.
QLM. 1973g. Roseton/Danskammer Point Generating Stations aquatic ecology studies 1971–1972. Prepared for CHGE. xi + 369 pp. Appendices 96 pp.
QLM. 1973h. Aquatic ecology studies, Kingston, New York 1971–1972. Prepared for CHGE. xii + 358 pp.
QLM. 1973i. Statistical analysis of fish impingement data at Indian Point Generating Station. Prepared for Con. Ed.
QLM. 1973j. The biological effects of entrainment at Indian Point. Prepared for Con. Ed.

LITERATURE CITED

QLM. 1974a. 1973 Hudson River aquatic ecology study at Bowline and Lovett Generating Stations. Vols. 1–5. Prepared for ORU.
QLM. 1974b. Danskammer Point Generating Station 1972 fish impingement study. Prepared for CHGE. iv + 184 pp.
Raytheon, Inc. 1971. Indian Point ecology survey. Final Report. Submitted to Con. Ed.
Reese, C. P. 1975. Life cycle of the amphipod *Gammarus palustris* Bousfield. Estuarine Coastal Mar. Sci. 3: 413–419.
Remaine, A. and C. Schlieper. 1971. Biology of Brackish water. John Wiley and Sons, Inc. New York. 372 pp.
Reish, D., G. Gesey, T. J. Kawling, F. G. Wilkes, A. J. Mearns, P. S. Oshida, and S. S. Rossi. 1980. pp. 1533–1575 in: Marine and estuarine water pollution. J. Water Pollu. Control Fed. Suppl. Review Volume.
Richards, S. W. 1959. Pelagic fish eggs and larvae of Long Island Sound. Bull. Bingham Oceanogr. Coll. 17: 95–124.
Richkus, W. A. 1975. Migrating behavior and growth of juvenile anadromous alewives, *Alosa pseudoharengus*, in a Rhode Island drainage. Trans. Amer. Fish. Soc. 104(3): 483–493.
Richkus, W. A., J. K. Summers, T. T. Polgar, A. F. Holland, R. Ross, G. F. Johnson, and P. Souza. 1980. Applicability of fisheries stock models in management. Prepared for Martin Marietta Corporation Environmental Center for Coastal Resources Division, Tidewater Administration, Maryland Department of Natural Resources, Annapolis, Maryland.
Ricker, W. E. 1975. Computation and interpretation of biological statistics of fish populations. Bull. Fish. Res. Bd. Canada. 191, 382 pp.
Roy, J. M., G. Beaulieu, and G. Labrecque. 1975. Observations sur le poulamon, *Microgadus tomcod* (Walbaum), de l'estuaire du Saint-Laurent et de la Baie de Chaleurs. Ministere de Industrie et du Commerce. Direction des peches maritimes, Direction de la recherche. Cahiers d'information 70: 1–56.
Ryder, J. A. 1887. On the development of osseus fishes, including marine and freshwater forms. Rept. U.S. Comm. Fish., Appendix pp. 489–604.
Ryder, J. A. 1888. On the development of the common sturgeon *Acipenser sturio*. Amer. Nat. 22 (259): 659–660.
Sandler, R. and D. Schoenbrod. 1981. The Hudson River power plant settlement. Materials prepared for a conference sponsored by New York University School of Law and Natural Resources Defense Council, Inc. 353 pp.
Sazaki, M., W. Heubach, and J. E. Skinner. 1973. Some preliminary results on the swimming ability and impingement tolerance of young-of-the-year steelhead trout, king salmon, and striped bass. Final Rept. for Anadromous Fisheries Act Proj. Calif. AFS13. Development of fish screen design criteria.
Schaefer, M. B. 1951. Estimation of the size of animal populations by marking experiments. Fish. Bull. 69: 191–203.
Schaffer, W. M. 1974a. Selection for optimal life histories: effects of age structure. Ecol. 55:291–303.
Schaffer, W. M. 1974b. Optimal reproductive effort in fluctuating environments. Am. Natur. 108:783–790.
Schaffer, W. M. 1979. The theory of life history evolution and its application

to Atlantic salmon. In: Fish Phenology: anabolic adaptiveness in teleosts. Symp. Zool. Soc. Lond. 44: 307–326.

Schaffer, W. M. and P. F. Elson. 1975. The adaptive significance of variations in life history among local populations of Atlantic salmon in North America. Ecology 56: 577–590.

Scott, W. B. and E. J. Crossman. 1973. Freshwater Fishes of Canada. Fish. Res. Bd. Canada Bull. 184: 1–966.

Seber, G. A. F. 1973. The estimation of animal abundance and related parameters. Hafner Press, N.Y. 506 pp.

Sexton, E. W. 1928. On the rearing and breeding of *Gammarus* in laboratory conditions. J. Mar. Biol. Assoc. 15: 33–55.

Sheader, M. and F. Chia. 1970. Development, fecundity and brooding behavior of the amphipod, *Marinogammarus obtusatus*. J. Mar. Biol. Assoc. U.K. 50: 1079–1099.

Shelton, W. L., W. D. Davies, T. A. King, and T. J. Timmons. 1979. Variation in the growth of the initial year class of largemouth bass in West Point Reservoir, Alabama and Georgia. Trans. Amer. Fish. Soc. 108: 142–149.

Shortnose Sturgeon Recovery Team. 1981. Shortnose sturgeon *Acipenser brevirostrum* LeSueur, 1818, Recovery Plan. Prepared for NMFS. Draft. January 1981, 120 pp.

Shuberth, C. J. 1968. The geology of New York City and environs. Natural History Press, Garden City, New York.

Shyamasundari, K. 1976. Effects of salinity and temperature on development of eggs in the tube building amphipod *Corophium triaenonyx* Stebbing. Biol. Bull. 150: 286–293.

Simpson, H. J., R. Bopp, and D. Thurber. 1973. Salt movement patterns in the Hudson. In: Proc. Third Symp. on Hudson River Ecology. HRES.

Skea, J. 1979. Effects of trimming and cooking on levels of mirex, PCB, and DDE in Lake Ontario smallmouth bass, brown trout, and chinook salmon. Fisheries Abstr., 35th N.E. Fish Wildl. Conf. Providence, R.I. pp. 7–8.

Skea, J. C., H. A. Simonin, H. J. Dean, J. R. Colquhoun, J. J. Spagnoli, and G. D. Veith. 1979. Bioaccumulation of Aroclor 1016 in Hudson River fish. Bull. Environ. Contam. Toxicol. 22: 332–336.

Sloan, R. J. 1978. PCBs in Hudson River fish. Presented at "Technical Report to Hudson River PCB Advisory Committee." Jan 19–20, Albany, N.Y.

Sloan, R. J. 1979. Monitoring PCB contamination levels in Hudson River fish. Presented at "PCBs in the Hudson River: 2-1/2 years of research. Technical report to the Hudson River PCB Advisory Committee." June 11–12, Albany, New York.

Sloan, R. J. 1981. Using PCB contamination to understand American eel life history. Abstr., Ann. mtg. N.Y. Chapter Amer. Fish. Soc., Marcy, N.Y. pp. 6–7.

Sloan, R. J. and R. Karcher. in press. On the origin of high cadmium concentrations in Hudson River blue crabs (*Callinectes sapidus* Rathbun). Northeast Environmental Science.

Sloan, R. J. and J. D. Sheppard. 1978. Spatial and temporal distributions of PCB concentrations in Hudson River fish: a preliminary assessment. Fisheries Wildl. Abstr., 34th N.E. Fish Wildl. Conf., White Sulphur Springs, W. Va. p. 32.

Smit, H. 1974. Extension de l'aire de repartition de *Gammarus tigrinus* Sexton

LITERATURE CITED

en 1973 aux pays-bas, et quelques remarques sur la concurrence avec les gammares indigenes (Crustacea, Amphipoda). Bull. Zool. Mus. Univ. Amsterdam. 4(5): 33: 323–330.

Smith, C. E., T. H. Peck, R. J. Klauda, and J. B. McLaren. 1979. Hepatomas in Atlantic tomcod, *Microgadus tomcod* (Walbaum) collected in the Hudson River estuary in New York. J. Fish Diseases 2: 313–319.

Smith, C. L. 1976. The Hudson River fish fauna. Paper no. 32 In: Proc. Fourth Symp. on Hudson River Ecology. HRES. 11 pp.

Snedecor, G. W. and W. G. Cochran. 1967. Statistical methods (Sixth Edition). The Iowa State University Press. Ames. 593 pp.

Sokal, R. R. and F. J. Rohlf. 1969. Biometry. The principles and practice of statistics in biological research. W. H. Freeman and Co. San Francisco. 776 pp.

Spagnoli, J. J. and L. C. Skinner. 1977. PCBs in fish from selected waters of New York State. Pest. Monit. J. 11(2): 69–87.

Spooner, G. M. 1947. The distribution of *Gammarus* in estuaries, Part. I. J. Mar. Biol. Assoc. U.K. 27: 1–52.

Statistical Abstracts of the United States. 1978. 99th Annual edition. U.S. Dept. Commerce, Bureau of Census, Washington, D.C.

Stearns, S. C. 1976. Life-history tactics: a review of the ideas. Quart. Rev. Biol. 51: 3–47.

Steel, R. G. D. and J. H. Torrie. 1960. Principles and procedures of statistics with special reference to the biological sciences. McGraw-Hill Book Co. New York. 481 pp.

Steele, D. H. and V. J. Steele. 1973. The biology of *Gammarus* (Crustacea, Amphipoda) in the northwestern Atlantic. VII. The duration of embryonic development in five species at various temperatures. Can. J. Zool. 51: 995–999.

Steele, D. H. and V. J. Steele. 1975. The biology of *Gammarus* (Crustacea, Amphipoda) in the northwestern Atlantic. XI. Comparison and discussion. Can. J. Zool. 53: 1116–1126.

Steimle, F., J. Caracciolo, and J. Pearce. 1982. Impacts of dumping on New York Bight Apex Benthos. pp. 213–223 in: Ecological Stress and the New York Bight: Science and Management (G. F. Mayer, ed.). Estuarine Research Federation, Columbia, S.C.

Stira, R. J. and B. A. Smith. 1976. The distribution of early life history stages of American shad (*Alosa sapidissima*) in the Hudson River estuary. pp. 179–189 in: Proceedings of a workshop on American Shad, Univ. of Massachusetts, Amherst. U.S. Fish and Wildl. Serv. and NMFS.

Strand, I. E., V. J. Norton, and J. G. Adriance. 1980. Economic aspects of commercial striped bass harvest. In: Proc. Fifth Annual Mar. Rec. Fish. Symp. (H. Clepper, ed.). Sport Fishing Institute, Washington, D.C. pp. 51–62.

Strauss, R. E. 1979. Reliability estimates for Ivlev's electivity index, the forage ratio, and a proposed linear index of food selection. Trans. Amer. Fish. Soc. 108(4): 344–352.

Tabery, M. A., A. Ricciardi, and T. J. Chambers. 1978. Occurrence of larval inshore lizard fish in the Hudson River estuary. N.Y. Fish Game J. 25: 87–89.

Talbot, G. B. 1954. Factors associated with fluctuations in abundance of Hudson River Shad. U.S. Fish and Wildl. Serv. Bull. 101(56): 373–413.

LITERATURE CITED

Taubert, B. D. 1980. Reproduction of shortnose sturgeon (*Acipenser brevirostrum*) in Holyoke Pool of the Connecticut River, Massachusetts. Copeia (1): 114–117.

Taubert, B. D. and M. J. Dadswell. 1980. Description of some larval shortnose sturgeon (*Acipenser brevirostrum*) from the Holyoke Pool, Connecticut River, Massachusetts, U.S.A., and the Saint John River, New Brunswick, Canada. Can. J. Zool. 58: 1125–1128.

Taubert, B. and J. Tranquilli. 1980. Use of daily rings in otoliths of largemouth bass and threadfin shad to determine age, growth, and time of spawning. Paper presented at the 110th Annual Meeting, American Fisheries Society, Louisville, Kentucky.

TI. 1972a. Hudson River ecological study in the area of Indian Point. First semi-annual report. Vol. 1. Biological Sampling. Prepared for Con. Ed. xii + 216 pp.

TI. 1972b. Hudson River ecological study in the area of Indian Point. First semi-annual report. Vol. 2. Standard procedures. Prepared for Con. Ed. vii + 97 pp.

TI. 1973a. Hudson River ecological study in the area of Indian Point. First annual report. Prepared for Con. Ed. xlii + 414 pp.

TI. 1973b. Hudson River environmental studies in the area of Ossining. Prepared for Con. Ed.

TI. 1973c. An atlas of chemical, physical and meteorological parameters in the area of Indian Point, Hudson River, New York. Prepared for Con. Ed.

TI. 1974a. Feasibility of culturing and stocking Hudson River striped bass. 1973 annual report. Prepared for Con. Ed. viii + 63 pp.

TI. 1974b. Indian Point impingement study report for the period 15 June 1972 through 31 December 1973. Prepared for Con. Ed. xvi + 166 pp.

TI. 1974c. Hudson River ecological study in the area of Indian Point. 1973 annual report. Prepared for Con. Ed. xxix + 246 pp.

TI. 1974d. Acute and chronic effects of evaporative cooling tower blowdown and power plant chemical discharges on white perch (*Morone americana*) and striped bass (*Morone saxatilis*). Prepared for Con. Ed. vii + 39 pp.

TI. 1974e. Second semi-annual report related to the feasibility study for spawning, hatching and stocking striped bass in the Hudson River. Prepared for Con. Ed. iii + 11 pp.

TI. 1974f. First annual progress report on fisheries investigations of the Hudson River as related to the Cornwall pumped-storage hydroelectric plant, Cornwall, New York. Prepared for Con. Ed. vii + 58 pp.

TI. 1975a. First annual report for the multiplant impact study of the Hudson River estuary. Vol. I. Text. Prepared for Con. Ed. xx + 318 pp.

TI. 1975b. First annual report for the multiplant impact study of the Hudson River estuary. Vol. II. Appendices. Prepared for Con. Ed. iii + 496 pp.

TI. 1975c. Hudson River ecological study in the area of Indian Point. 1974 annual report. Prepared for Con. Ed. xv + 160 pp.

TI. 1975d. Final report of the synoptic subpopulation analysis, phase I: Report on the feasibility of using innate tags to identify striped bass (*Morone saxatilis*) from various spawning rivers. Prepared for Con. Ed.

TI. 1975e. Indian Point impingement study report for the period 1 January 1974 through 31 December 1974. Prepared for Con. Ed. xviii + 182 pp.

TI. 1975f. Feasibility of culturing and stocking Hudson River striped bass. 1974 annual report. Prepared for Con. Ed. ix + 88 pp.
TI. 1976a. Fisheries survey of the Hudson River. March–December 1973. Revised edition. Vol. IV. Prepared for Con. Ed. xxii + 564 pp.
TI. 1976b. Indian Point impingement studies for the period 1 January 1975 through 31 December 1975. Prepared for Con. Ed. vii + 109.
TI. 1976c. A synthesis of available data pertaining to major physicochemical variables within the Hudson River estuary emphasizing the period from 1972 through 1975. Prepared for Con. Ed. ix + 152 pp.
TI. 1976d. Report on relative contribution of Hudson River striped bass to the Atlantic coastal fishery. Prepared for Con. Ed. v + 101 pp.
TI. 1976e. Predation by bluefish in the lower Hudson River. Prepared for Con. Ed. ii + 23 pp.
TI. 1976f. Hudson River ecological study in the area of Indian Point. 1975 annual report. Prepared for Con. Ed. vi + 329 pp.
TI. 1976g. Hudson River ecological study in the area of Indian Point. Thermal effects report. Prepared for Con. Ed. xvii + 231 pp.
TI. 1976h. Liberty State Park ecological study: Final report. Prepared for the Port Authority of New York and New Jersey.
TI. 1977a. 1974 and 1975 gear evaluation studies. Prepared for Con. Ed. vi + 83 pp.
TI. 1977b. Production of striped bass for experimental purposes. 1976 hatchery report. Prepared for Con. Ed. vi + 21 pp.
TI. 1977c. Feasibility of culturing and stocking Hudson River striped bass. 1975 annual report. Prepared for Con. Ed. xii + 170 pp.
TI. 1977d. Feasibility of culturing and stocking Hudson River striped bass an overview, 1973–1975. Prepared for Con. Ed. viii + 99 pp.
TI. 1977e. 1974 year-class report for the multiplant impact study of the Hudson River estuary. Vol. I. Text. Prepared for Con. Ed. xiv + 191 pp.
TI. 1977f. 1974 year-class report for the multiplant impact study of the Hudson River estuary. Vol. II. Appendices. Prepared for Con. Ed. 305 pp.
TI. 1977g. 1974 year-class report for the multiplant impact study of the Hudson River estuary study. Vol. III. Lower estuary study. Prepared for Con. Ed. viii + 97 pp.
TI. 1977h. Hudson River ecological study in the area of Indian Point. 1976 annual report. Prepared for Con. Ed. x + 106 pp.
TI. 1977i. Production of striped bass for power plant entrainment studies. 1977 hatchery report. Prepared for Con. Ed. iv + 23 pp.
TI. 1978a. 1975 year-class report for the multiplant impact study of the Hudson River estuary. Prepared for Con. Ed. xiii + 378 pp.
TI. 1978b. Initial and extended survival of fish collected from a fine mesh continuously operating traveling screen at the Indian Point generating station, for the period 15 June–22 December 1977. Prepared for Con. Ed. iv + 26 pp.
TI. 1978c. Catch efficiency of 100-ft (30-m) beach seines for estimating density of young-of-the-year striped bass and white perch in the shore zone of the Hudson River estuary. Prepared for Con. Ed. iv + 64 pp.
TI. 1978d. Evaluation of a submerged weir to reduce fish impingement at Indian Point for the period 25 May–29 July 1977. Prepared for Con. Ed. ii + 20 pp.
TI. 1979a. 1976 year-class report for the multiplant impact study of the Hudson River estuary. Prepared for Con. Ed. xxiv + 645 pp.

LITERATURE CITED

TI. 1979b. Efficiency of a 100-ft beach seine for estimating shore zone densities at night of juvenile striped bass, juvenile white perch, and yearling and older (150mm) white perch. Prepared for Con. Ed. vii + 104 pp.

TI. 1979c. Hudson River ecological study in the area of Indian Point. 1977 annual report. Prepared for Con. Ed. xi + 265 pp.

TI. 1979d. Production of striped bass for experimental purposes. 1978 hatchery report. Prepared for Con. Ed. viii + 94 pp.

TI. 1979e. Hatchery letter report. September 1979. Prepared for Con. Ed.

TI. 1979f. Collection efficiency and survival estimates of fish impinged on a fine mesh continuously operating traveling screen at the Indian Point generating station for the period 8 August to 10 November 1978. Prepared for Con. Ed.

TI. 1980a. 1977 year class report for the multiplant impact study of the Hudson River Estuary. Prepared for Con. Ed. xx + 564 pp.

TI. 1980b. Hudson River ecological study in the area of Indian Point. 1978 annual report. Prepared for Con. Ed.

TI. 1980c. Report on 1978–1979 studies to evaluate the catch efficiency of the 1.0-m^2 epibenthic sled. Prepared for Con. Ed. ii + 31 pp.

TI. 1980d. Hudson River ecological study in the area of Indian Point. 1979 annual report. Prepared for Con. Ed. x + 172 pp.

TI. 1980e. 1978 year class report for the multiplant impact study of the Hudson River estuary. Prepared for Con. Ed. viii + 846 pp.

TI. 1981a. 1979 bottom trawl comparability study for the interregional trawl survey. Prepared for Con. Ed. iii + 40 pp.

TI. 1981b. 1979 year class report for the multiplant impact study of the Hudson River estuary. Prepared for Con. Ed. xxxv + 509 pp. Appendices xxii + 176 pp.

Thoits, C. F. 1958. A compendium of the life history and ecology of the white perch, *Morone americana*. Mass. Div. Fish and Game Fish Bull. 24. 16 pp.

Timmons, T. J., W. L. Shelton, and W. D. Davies. 1980. Differential growth of largemouth bass in West Point Reservoir, Alabama-Georgia. Trans. Amer. Fish. Soc. 109(2): 176–186.

Tofflemire, T. J., L. J. Hetling, and S. O. Quinn. 1979. PCB in the upper Hudson River: Sediment distributions, water interactions and dredging. Tech. Rept. 55, Bur. Water Research, DEC. 68 pp.

Tranter, D. J. and P. E. Smith. 1968. Filtration performance. pp. 27–56. in: Reviews on zooplankton sampling methods (D. J. Tranter, ed.). Monograph of Oceanographic Methodology, UNESCO.

Trent, L. and W. W. Hassler. 1968. Gill net selection, migration, size and age composition, sex ratio, harvest efficiency, and management of striped bass in the Roanoke River, North Carolina. Chesapeake Sci. 9: 217–232.

Ulanowicz, R. E., and T. T. Polgar. 1980. Influence of anadromous spawning behavior and optimal environmental conditions upon striped bass (*Morone saxatilis*) year-class success. Canadian J. Fish. Aquatic Sci. 37(2): 143–154.

U.S. Food and Drug Administration. 1977. Polychlorinated biphenyls (PCBs). Fed. Reg. 42(63): 17487–17494.

U.S. Geological Survey. 1977. Water resources data for New York water year 1977. Vol. 1. U.S. Geol. Surv. U.S. Dept. Interior, Albany, NY. USGS-WDR-NY-1. 566 pp.

LITERATURE CITED

U.S.E.P.A. 1978. National pollution discharge elimination system. Region 1, Boston, Mass.

Van Dolah, R. F., L. E. Shapiro, and C. P. Rees. 1975. Analysis of an intertidal population of the amphipod *Gammarus palustris* using a modified version of the egg ratio method. Mar. Biol. 33: 323–330.

Van Dolah, R. F. and E. Bird. 1980. A comparison of reproductive patterns in epifaunal and infaunal gammaridean amphipods. Estuar. Coastal Mar. Sci. 11: 593–604.

Van Winkle, W., B. L. Kirk, and B. W. Rust. 1979. Periodicities in Atlantic Coast striped bass (*Morone saxatilis*) commercial fisheries data. J. Fish. Res. Bd. Canada. 36: 54–62.

Vaughan, D. S. and S. B. Saila. 1976. A method of determining mortality rates using the Leslie matrix. Trans. Amer. Fish. Soc. 105(3): 380–383.

Vernberg, W. B. and F. J. Vernberg. 1976. Physiological adaptations of estuarine animals. Oceanus. 19(5): 48–54.

Vladykov, V. D. and J. R. Greeley. 1963. Order Acipenseroidei. In: Fishes of the western North Atlantic. Mem. Sears Found. Mar. Res. 1, 630 pp.

Wallace, D. N. 1978. Two anomalies of fish larval transport and their importance in environmental assessment. N.Y. Fish Game J. 37(2): 59–71.

Ward and Whipple. 1959. Fresh Water Biology. (W. T. Edmondson, ed.). John Wiley and Sons, Inc., New York.

Watson, D. A., G. A. Roth, and L. M. Pristash. 1981. A fisheries data base using SAS. pp. 359–362. in: Proc. 6th Ann. SAS User's Group International Conference. pp. 359–362.

Watson, L. C. 1978. Tomcod eggs: Sticky or slippery. Texas Instruments Science Services News 4(2).

Watson, L. C. MS. Methods for spawning and hatching Atlantic tomcod (*Microgadus tomcod*).

Weaver, J. E. 1975. Food selectivity, feeding chronology, and energy transformations of juvenile alewife, (*Alosa pseudoharengus*) in the James River near Hopewell, Virginia. Ph.D. dissertation. University of Virginia, Charlottesville.

Werner, E. E., and D. J. Hall. 1976. Niche shifts in sunfishes: Experimental evidence and significance. Science 191: 404–406.

Werntz, H. O. 1963. Osmotic regulation in marine and freshwater gammarids. Biol. Bull. 124:225–239.

Wetzel, R. G. 1975. Limnology. W. B. Saunders Company, Philadelphia.

Whittle, K. J., R. Hardy, A. V. Holden, R. Johnston, and R. J. Pentreath. 1977. Occurrence and fate of organic and inorganic contaminants in marine animals. Ann. N.Y. Acad. Sci. 298: 47–79.

Williams, A. B. and K. H. Bynum. 1972. A ten-year study of meroplankton in North Carolina estuaries: Amphipods. Chesapeake Sci. 13(3): 251–266.

Williams, G. C. 1966. Adaptation and natural selection. Princeton Univ. Press. Princeton N.J. 307 pp.

Wilson, A. J. and J. Forester. 1978. Persistence of Aroclor 1254 in a contaminated estuary. Bull. Environ. Contam. Toxicol. 19: 637–640.

Youngs, W. D., W. H. Gutenmann, and D.J. Lisk. 1972. Residues of DDT in lake trout as a function of age. Environ. Sci. Technol. 6(5): 451–452.

LITERATURE CITED

Zimmerman, R., R. Gibson, and J. Harrington. 1979. Herbivory and detritivory among gammaridean amphipods from a Florida seagrass community. Mar. Biol. 54: 41–48.

List of Contributors

Dr Karim A. Abood, LMS One Blue Hill Plaza, Pearl River, NY 10965

Dr. Roger W. Armstrong, Chemistry Department, Russel Sage College, Troy, NY 12180

John M. Bartels, RD5 Wappingers Falls, NY 12590

Dr. William P. Dey, EA Science and Technology, RD 2 Goshen Turnpike, Middletown, NY 10940

Gerard DiNardo, Montclair State College, Upper Montclair, NJ 07043

Dr. Thomas L. Englert, LMS One Blue Hill Plaza, Pearl River, NY 10965

Marcia N. Gardinier, IBM, Sterling Forest, NY 10979

Douglas A. Hjorth, Chas. T. Main, Inc., Prudential Center, Boston, MA 02199

Dr. James G. Hoff, Southeastern Massachusetts University, North Dartmouth, MA 02747

Thomas B. Hoff, Mid-Atlantic Fishery Management Council, Room 2115 Federal Building, 300 South New Street, Dover, DE 19901

Pamela J. Keeser, LMS One Blue Hill Plaza, Pearl River, NY 10965

Dr. Ronald J. Klauda, Johns Hopkins University, Applied Physicas Laboratory, Shady Side, MD 20764

Dr. Stephen H. Koepp, Biology Department, Montclair State College, Upper Montclair, NJ 07043

Jeffrey A. Leslie, 3 Greenway, Sloatsburg, NY

Dr. James B. McLaren, Beak Consultants, Inc., 12072 Main Road, Akron, NY 14001

Dr. Richard E. Moos, CH2M Hill, Rocky Mountain Regional Office, PO Box 22508, Denver, CO 80222

Paul H. Muessig, EA Science and Technology, RD 2 Goshen Turnpike, Middletown, NY 10940

Dr. Joseph M. O'Connor, New York University Medical Center, Institute of Environmental Medicine, A. J. Lanza Laboratories, Long Meadow Road, Tuxedo, NY 101987

Dr. Gerald J. Poje, National Wildlife Federation, 1412 16th Street N.W., Washington DC 20036-2266

LIST OF CONTRIBUTORS

Stacy A. Riordan, New York University Medical Center, Institute of Environmental Medicine, Tuxedo, NY 10987

Edward D. Santoro, William F. Cosulich Assoc., 3000 Hadley Road, South Plainfield, NJ 07080

Dr. Robert E. Schmidt, Hudsonia, Ltd., Box 96 Bard College, Annandale, NY 12504

Dr. Ronald A. Sloan, NYDEC, 50 Wolf Road, Albany, NY 12233

Dr. C. Lavett Smith, American Museum of Natural History, New York, NY 10024

Dr. David Sugerman, ATT Bell Laboratories, 185 Monmouth Parkway, West Long Branch, NJ 07764

John R. Young, Consolidated Edison, Environmental Affairs Room 306-S, 4 Irving Place, New York, NY 10003

Index

Acipenser brevirostrum. See shortnose sturgeon
Acipenser oxyrbynchus. See Atlantic sturgeon
Acipenser spp.: comparative abundance of shortnose and Atlantic sturgeon, 187, table 42; unidentified juveniles, 186, table 41
Advective transport factor, 150
Age-structured population model, 354–365
Albany: growth of fishes near, 36; migratory species, 33; oxygen levels, 290, 292
Albany Steam Electric Generating Station: impingement of *Alosa* spp., 213; near-field surveys, 32–33
Alewife: age and growth studies, 36; in Albany area, 33; bimodal length frequency distribution, 211; biological characteristics, 33; catch of juveniles, figure 44; distribution of juveniles, 202, 205, figure 44; effects of water temperature, 205; heavy metals in, tables 55–56, 58; migration of juveniles, 211; modal lengths of juveniles, figure 45; length frequency, figure 46; reproduction 36; seasonal occurrence of juveniles, 202–205; spawning habitat, 193, 210
Alosa aestivalis. See blueback herring
Alosa mediocris. See hickory shad
Alosa pseudoharengus. See Alewife
Alosa sapidissima. See American shad
Alosa spp.: diel movements and segregation, 214; distribution of eggs, 197, figure 41; feeding differences, 214–215; movements, 193–215; niche segregation, 213–215; passive drift of larvae, 211; sampling of eggs and larvae, 196–197; sampling of juveniles, 197; study area, 194, figure 39; seasonal occurrence of eggs and larvae, 200. *See also river herrings*
American eel: in Albany area, 33; arsenic in, 281–282; heavy metals in, tables 55–56, 58

American shad: abundance of juveniles, 202, figure 42; distribution of eggs and larvae, 198, figure 40; distribution of juveniles, figure 43; effect of temperature, 202; commercial fishery lift periods, 92; PCB in, figures 81–82; distribution of eggs and larvae, 196–198; modal lengths of juveniles, 202, figure 43; time of migration of juveniles, 211; seasonal occurrence of juveniles, 200–202; spawning habitat, 193, 200, 210; spawning at suboptimal temperatures, 210
American shad roe: PCB concentration, 332, table 68
Amphipods: dominance in estuaries, 255–256
Anarchis sp.: food for *Gammarus*, 258
Anchoa mitchilli. See bay anchovy
Anguilla rostrata. See American eel
Annelid larvae, 139
Arochlor: 1016/1254 ratio, table 70; 1221, 305; 1242, 305; 1254, 305. *See also* PCBs
Arsenic: analytical methods, 276; in selected animals, 281–282, tables 55–58; in American eel, 281–282; in mummichog, 282; in striped bass, 282, in bluefish, 282; by tropic level, 67
Atlantic herring: heavy metals in, tables 55–57; occurrence in Hudson River, 210–211
Atlantic mackerel: heavy metals in, table 57
Atlantic menhaden: heavy metals in, 281, tables 55–58; presence in Hudson River, 210
Atlantic silversides: heavy metals in, tables 55–58
Atlantic sturgeon: age at maturity, 357; age specific mortality rate, 359; commercial fishery, 353–354; distribution studies, 17; fecundity, 358; first year survival, 359; collapse of fishery, 353; conceptual life history model, figure 85; life

INDEX

Atlantic sturgeon (*Cont.*)
history data, table 74; management recommendations, 353–365; maximum age, 357; mean harvest and stock size, 361–363; population model assumptions, 353–365, figure 86; relationship between weight and age, 358, figure 87; survival, 355; vulnerability to fishery, 353

Atlantic tomcod: age and growth studies, 33; biological characteristics studies, 33; boxtrap survey, 222, figure 52, table 44; catch rates, figure 53; culture of, 77; distribution studies, 17; distribution of eggs, 238–240; distribution of juveniles, 227, 245–251, figures 63–64; distribution of post yolk-sac larvae, 242–245, figure 60; distribution of yolk-sac larvae, 240–242, table 46; downstream migration of larvae, 244, figure 62; early life stages, 224–227; effects of temperature on, 250–251; geographic range, 219; mark and recapture studies, 22–23; heavy metals in, tables 55–58; juveniles, 227, figures 63–64; impingement studies, 61; life history, 219–251; maturity of gonads, 224, table 43; percentage of females in spawning condition, figure 54; physiology and behavior studies, 36; recovery of tagged fish outside Hudson estuary, 236, figure 58; regional density of life history stages, 226; reproduction studies, 36; sampling areas, 220, figures 50–51; sampling methods, 222–224; sexual maturity, 227–228; sex ratio, figure 56; spawning behavior, 236–238, figure 59; spawning location, 234, 235, figure 57; spawning rivers, 219; spawning time, 228; temporal variation in abundance, figure 57; upriver movement of juveniles, 250–251

Atlantic western trawl, 73
Atomic absorption spectroscopy, 276

Bay anchovy: heavy metals in, table 55
Bag seine, 32
BB. *See* brown bullhead
Beach seine, 15, 17, 73, 197, 274
Beam trawl, 73
Bioaccumulation: of heavy metals, 285–286
Biological characteristics: definition, 10; TI data, 30, table 8; LMS data 33, 36, table 9
Blueback herring: age and growth, 36; in Albany area, 33; biological characteristics, 36; catch, figure 47; comparative length frequency, figure 48; distribution and movements, 205–210; distribution of juveniles, 205, 208; heavy metals in, tables 55–58; mean catch per seine haul, figure 47; migration later than alewife and shad, 208; seasonal occurrence of juveniles, 205–210; size at maturity, 211; size distribution, figure 48; spawning habitat, 193, 200, 210
Blue crab (blue claw crab): mercury in, 281; copper in, 281, heavy metals in, tables 55–58; PCB advisory, 347
Bluefish: cadmium in, 281; copper in, 281; food habits study, 33; heavy metals in, tables 55–58; lead in, 281; mercury in, 281
Bosmina sp. 138–145
Bottom trawl, 73, 175, 186, 188, 202, 219, 227, 247
Bowline Point power plant: distribution studies, 25; entrainment studies, 41, 57; impingement studies, 64; trawl stations, 32
Boxtrap, 22, 73, 222–223, 229, 234, 236, 242, 244
Boxtrap survey: for Atlantic tomcod, 222, figure 52
Boyce Thompson Institute, 85
Brevoortia tyrannus. See Atlantic menhaden
Brown bullhead: heavy metals in, table 56; PCB concentrations, cor-

400

INDEX

Brown bullhead (*Cont.*) relation with length and lipid content, table 61; PCB levels, table 65

Cadmium: analytical methods, 276; in selected animals 281; by trophic level, figure 66; in different areas, figure 66
Callinectes sapidus, See blue crab
Carassius auratus, See goldfish
Carp: PCB levels, table 65
Catch composition striped bass, 105–108
Catfish: heavy metals in, table 55–56
Caviar, 353
Chain pickerel: PCB levels, table 65
Chaoborus: competition with *Gammarus*, 256
Chironomid larvae: food for juvenile *Alosa* spp., 214–215
Cirripedia larvae, 139
Cladocerans: food of *Alosa* spp., 214; food of *Morone* spp, 138–145
Clupea harengus, See Atlantic herring
Commercial fishery: regulations, 92, for shortnose sturgeon, 172; for striped bass, 89–123; incidental catch of sturgeons, 175. *See also* Striped bass commercial fishery
Computers: needed for access to data sets, 12–15
Cold vapor analysis for mercury, 276
Copepod nauplii, 138–147
Copper: analytical methods, 276; in selected animals, tables 55–58; by trophic level, figure 67
CP, *See* chain pickerel
Crab traps, 274
Critical age: striped bass, 120–122, figure 10

Culture, 10: Atlantic tomcod, 77; striped bass, 77; white perch, 77; EA data, 77–81; TI data, 77, table 22
Culture methods: *Gammarus*, 257
Cynoscion regalis. *See* Northern weakfish
Cyprinus carpio. *See* Carp

Danskammer Point power plant, 10; distribution studies, 25; entrainment studies, 57; impingement studies, 64; trawl stations, 32
Daphnia pulex: food for *Gammarus*, 259
Data sets: availability, 12–15, table 1
Density dependent survival: Atlantic sturgeon, 357
Diel activity patterns: in *Alosa* spp., 214
Diptera: larvae fed on by blueback herring, 214
Dissolved oxygen: trends in Hudson River, 287–303; predicted summer profiles, figure 69; recommendations, 303; seasonal variation in Albany area, 294, figures 71–72, table 59; seasonal variation in New York area, 296; fig 73–74; spring profiles, 292, figure 70; summer profiles, 280, figure 68
Distribution, abundance and species composition: studies, 10, 15; EA data (near-field studies), 24–25, table 6; LMS data (near-field studies) 25–33, table 7, TI data (near-field surveys) 15–17, table 3; TI data (Far-field studies) 17–24, tables 4–5; methods, 15–16; special studies, table 21
Dockside value striped bass, 97
Dorosoma cepedianum. *See* gizzard shad
Dovel, W., 85, 176, 357
Drift nets, 32

EA Science and Technology, formerly Ecological Analysts, Inc., 9
Ectinosoma cuticorne, 138
Eggs and larvae: *Alosa* spp., 196
Egg Index (EI): definition, 126–136; in striped bass, figure 14; relation to size of female striped bass, table 34; variation, 130–132, figure 14
Electric power: environmental conflicts, 2; impact on fish populations, 7

401

INDEX

Entrainment: definition, 10; EA studies, 37–41, tables 12–13; LMS data, 41, 57, table 14
Epibenthic sled: farfield surveys, 17; avoidance of, 22; sampling for *Alosa* spp., 196; sampling for tomcod, 222, 227, 247
Epiphytic animals: food of *Alosa* spp., 215
Esox lucius. See northern pike
Esox niger. See chain pickerel
Estuaries: characteristics of, 255; adaptations for life in, 255–256
Estuary wide sampling, 10
Eurytemora affinis, 140–147

Far-field sampling: definition, 10; area, 22
Federal Energy Regulatory Commission, 85
Federal Water Pollution Control Act, 288–289
Feeding habits: *Alosa* spp., 214
Feeding selectivity: analytical procedures, 137; white perch, 141–143; striped bass, 143
Fin rot: high incidence in shortnose sturgeon, 188
Fishery sampling gear: Bag seine, 32; Experimental gill nets, 32; Otter trawl, 32
Fishery strategies, 359–364
Fisheries data sets, 7–35
Fishing areas: striped bass, 92
Fishing effort: striped bass, 98, table 26, figure 4
Fishing mortality: Atlantic sturgeon, 359
Fishing season, 92
Flame atomic absorption spectrometry: for cadmium, lead, copper, and zinc, 276
Flow meters, 196
Formicidae: food of American shad, 214
Foundry Cove: cadmium contamination, 285

Freshwater Flows: annual cycle, 61, 244; effects on Atlantic tomcod postlarvae, 244–245, figure 47, 62; patterns, figure 49
Fundulus heteroclitus. See mummichog
Fundulus majalis. See striped killifish
Fyke nets, 274

Gammarids: adaptations for life in estuaries, 256–257; population density, 256
Gammarus daiberi: urine, 256
Gammarus locusta, 256
Gammarus lacustris, 256
Gammarus minus, 257
Gammarus mucronatus, 257
Gammarus pulex, 256
Gammarus pseudolimnaeus, 257
Gammarus spp.: life history parameters, table 54
Gammarus tigrinus: as fish predator, 270; effect of diet on growth and reproduction, 255–270, tables 50–53; factors affecting abundance, 269; food habits, 255–270; growth, 259–262, table 49; introduction into Europe, 257; numbers of young, 264; study methods, 258–259; range, 257; reproduction, 262–65, 267–9
Gastropod veligers, 139
Gear performance: EA studies, 73, table 19; LMS studies, 76–80, table 20; TI studies, 64, 73, table 18
George Washington Bridge, 22
Gill nets, 32, 73, 90–92, 94, 98, 102, 105, 116, 120, 122, 125, 172, 175–176, 184, 187, 188, 274, 304
Gill nets: area fished, figure 4; drift nets, 91; materials, 91; stake and anchor nets, 90
Gill net regulations: length, 92; mesh size, 92
Gizzard shad, occurrence in Hudson River, 210
Glenmont: Dissolved oxygen, 294, figures 71–72

INDEX

Gold. *See* goldfish
Goldfish: PCB levels, table 65; PCB correlated with length and lipid content, table 61
Gonad Index (GI): definition, 126–132; seasonal variation, figures 11–13
Gonad maturity: criteria for tomcod, 222–223, table 43
Grass shrimp: heavy metals in, tables 56–58
Green Island freshwater flows, table 59
Growth patterns, 36

Halicyclops fosteri, 138
Half-life of PCB: in resident fish, table 63
Hard clam: heavy metals in, Table 58
Haul seines, 22, 94, 125
Haul seines outlawed, 92
Haverstraw Bay, 36
Heavy metals: analytical techniques, 274; preparation of samples, 274; sampling location, figure 65; sampling schedule, 274; sources in Hudson River, 274
Heavy metals in selected organisms, 273–286
Hickory shad: heavy metals in, table 57; presence in river, 211
Hippolyte sp. *See* grass shrimp
Hogchoker, heavy metals in, tables 55–56
Hook and Line, 274
Hoop nets, 274
Hudson River: cross section, figure 51; freshwater flow patterns, figure 49, 61; morphometry, 174, figure 34; power plant sites, figure 1; sampling zones for *Alosa* spp, 194–5, figure 39; sampling regions for Atlantic tomcod, 221, figure 39; sampling stations for heavy metals, figure 50; as nursery for shortnose sturgeon, 188
Hudson River Ecological Study: definition, 7; goals, 8; history, 7–8; quality control, 9; sampling area, figure 1
Hudson River Environmental Society, 4
Hudson River Estuary: description, 173–175; gradient, 173, salinity, 175
Hudson River Field weeks, 288, 292
Hudson River Foundation for Science and Environmental Research, 8
Hudson River Settlement, 3

Ichthyoplankton methods, 135
Ichthyoplankton survey, 22, 76, 225, 234, 245
Ictalurus catus. See white catfish
Ictalurus nebulosus. See brown bullhead
Ictalurus sp. See catfish
Impingement: alewife and blueback herring, 213; sturgeon, 175
Impingement studies: EA data, 61–63, table 16; LMS data 64, table 17; TI data 57–61, table 15; special studies, table 21
Indian Point: biological characteristics studies, 33; distribution studies, 15, 24; entrainment studies, 41; impingement studies, 57
Indian Point Power Plant, 10
Interregional Trawl Survey, 145

Juveniles: *Alosa* spp., 197. *See also* individual species

Kennebec-Sheepscot River shortnose sturgeon, table 40
Kingston: growth studies, 36; near-field surveys, 32
Kingston-Rhinecliff Bridge Fishery Survey, 32–33

Landings of striped bass, 93
Largemouth Bass: heavy metals in, 273, PCBs in, table 65, figure 77
Lawler, Matusky and Skelly, Engineers, 9

403

INDEX

Lead: analysis, 276; in bluefish, 281; in predatory fish, 286; in selected species, table 55–58; by trophic level, figure 66
Lepomis auritus. See redbreast sunfish
Leslie Matrix, 354
Lift periods, 92
Lippson, A. J., 85
Long Island, 22
LMB. See Largemouth bass
LMS. See Lawler, Matusky and Skelly, Engineers
Lovett near-field surveys, 32
Lovett Power Plant, Distribution studies, 25; entrainment studies, 41; impingement, 64
Lower Estuary Survey, 22

Mark and Recapture studies, 22–23, table 5
Menidia menidia. See Atlantic silverside
Mercury: analytical methods, 274, 276; bioaccumulation of, 285; in selected animals, 276, 281, tables 55–58; by trophic levels, figure 66
Microgadus tomcod. See Atlantic tomcod
Micropterus salmoides. See largemouth bass
Microzooplankton, 135
Minnow trap, 274
Modiolus demissus. See ribbed mussel
Mohawk River, 200
Morone americana. See white perch
Morone saxatilis. See striped bass
Mougotis sp.: food for *Gammarus*, 258
Movements of striped bass and white perch larvae, 148–168
Mummichog, tables 56–58
Mya arenaria. See soft clam
Myriophyllum spicatum: epiphytic animals fed on by *Alosa* spp., 215; food for *Gammarus*, 257; habitat for *Gammarus*, 259
Mytilus edulis. See blue mussel

Narrows, the, 296, figure 73
National Marine Fisheries Service, 90, 98, table 26
Near-field studies: definition, 10
Newburgh Bay, 32, 36, 76–77
New York City dissolved oxygen levels, 290
New York State Department of Environmental Conservation, 85
New York State Department of Transportation, 85
New York University, 24
Niche segregation of *Alosa* spp., 213–14
Nitzchia sp.: food of *Gammarus*, 258–259
Northern weakfish: heavy metals in, tables 55, 58
Northern Pike: PCB levels, table 65
NP. See northern pike

Optimal foraging theory, 145
Ossining distribution studies, 15, 17, 24
Ossining Power Plant, 10
Otter trawl, 15, 22, 32, 73, 77, 175, 197, 227
Otter trawl efficiency study, 77

Paralichthys dentatus. See summer flounder
PCBs: analysis, 305; body size and accumulation, 332–34, table 69; collection and preparation of samples, 304–5, 325; concentrations in different organs, 330–31; table 67; current levels, 323–24, 345, tables 65, 73; data summary, 326–30, table 66; declines and river flow conditions, 322; half-life calculations, 306, table 63; historical summary, 304, table 60; lipid correlated concentration, 306–07; patterns in resident and freshwater fishes, 304–324; patterns in migrant and marine species, 325–350; perspectives and cautions, 348; quality control, 326; resuspension of pcb laden sediments, 321; spatial

404

PCB *(Cont.)*
trends, 345; table 72; temporal trends, 311–322, 334–345, tables 62, 64, 70–71, figures 78–84
Peekskill, 135
Perca flavescens. See yellow perch
Physiology and behavior studies, 9, 36–37, EA data 37, table 11; TI data, 36–37, table 10
Piscivores: accumulation of heavy metals in, 286
Pksd. *See* pumpkinseed
Plankton nets, 41, 57, 135
Pomatomus saltatrix. See bluefish
Pomoxic annularis. See white crappie
Population model: for Atlantic sturgeon, 354–365
Port Authority of New York and New Jersey, 85
Post yolk-sac larvae. See individual species entries
Poughkeepsie area: PCB in American shad, figure 81, 82
PP. *See* Walleye
Pseudopleuronectes americanus. See winter flounder
Pumpkinseed: PCB correlated with length and lipid content, 76, table 61; PCB levels, table 65
Purse seine, 73

Quirk, Lawler and Matusky. *See* LMS; Lawler, Matusky and Skelly Engineers

Raltech Scientific Services, 305, 325–326
Raytheon, 15
Rbs. *See* redbreast sunfish
Real Time Life Cycle model: defining equations, 148–149; comparisons of predictions with field data, 150, figures 21–24
Redbreast sunfish: PCB correlated with length and lipid content, table 61
Regulations: Commercial fishery, 92

Reproductive capacity of *Gammarus,* 259
Ribbed mussel, Heavy metals in, tables 57–58
River herrings: distribution of eggs and larvae, 149
River Mile index: definition, 159; as a measure of downstream movement of life stages, 160; values, 1974–80, table 36
Roseton Power Plant, 10
Roseton trawl survey, 32
Roseton power plant: distribution studies, 25; entrainment, 41, 57; impingement, 64
Rotifers, 138–142
RTLC. *See* Real Time Life Cycle model

St. John River system: shortnose sturgeon from, 185
Salt front, 161, 194, table 37, figures 29–32
Selenium, 273
Sewage treatment plants: North River, 290; Passaic Valley, 290
Sex ratio, 229
Scomber scombrus. See Atlantic mackerel
Shortnose Sturgeon Recovery Team, 172, 187–88
Shortnose sturgeon: biology, 171–189; catch, 176–78, 186–88, table 38, figure 35; comparative morphology, 182–185, tables 39–40; diet, 173; distribution study, 17, figure 36; downstream migration after spawning, 176; exploitation, 172; habitat, 171–73; identification, 171; incidence of fin rot, 188; larvae and juveniles, 184–186; length and weight, 178, 181, 184, figures 37–38; management, 186–189; nursery area, 188; PCB levels in, 188; population, 187; relative numbers of shortnose and Atlantic sturgeon, 187; sampling program, 175; spawning, 173, 175, 188

INDEX

Soft clam, heavy metals in, tables 57–58
Spatial segregation: *Alosa* spp., 214
Spawning behavior: Atlantic tomcod, 236–239
Standard operating procedures, Hudson River Ecological Study, 9
Statistical Analysis System, 12–15, table 2
Statistical file structure, 1, 85, table 2
Steam electric generating plants, 10, figure 1
Sterling Lake, 258, table 48
Stickleback, heavy metals in, table 58
Stony Point, 135
Stizostedion vitreum vitreum. See walleye (Pp)
Strauss's Linear Feeding Selectivity Index, 137
Striped bass: abundance, table 28; biological characteristics studies, 17, 25, 33; commercial fishery, 89–123; figures 2–9, tables 23–30; critical age, 120–122, figure 10; culture, 77; distribution of larval stages, figures 21–29; distribution studies, 17, 25; effects of pollution, 36; egg diameter, 126–133, figure 15; feeding of larvae, 138–146, figures 17–19, table 35; larval development, 145; egg structure, 146–147; exploitation rates, 116–120; feeding selectivity of larval stages, 134–147, figures 17–19; impingement studies, 61; influence of salt front on distribution of early stages, 160–166; mark and recapture studies, 22–23; maturity index, 125–126; movement of larval stages, 148–168, figure 33; eggs and larvae predation by *Gammarus*, 270; heavy metals in, 273, tables 55–58; influence of salt front on distribution of larval stages, figures 29–39; prey utilization, table 35; reproduction studies, 36; reproductive effort, 124–133; selectivity of fishery, 114–116, figure 9; sex ratio, 114, figure 8; tagging program, 116–120, tables 31–33; thermal physiology studies, 36; trends in relative abundance, 102–104; vertical migrations, 149; year class strength, 107–114
Striped killifish: heavy metals in, table 57
Summer flounder: heavy metals in, table 58
Surface trawl, 73

Tappan Zee, figure 82
Temperature effects studies, 36–37
Terrestrial insects, 214
Texas Instruments, 9
316(b) studies, 168
Thermal discharge effects studies, 24, table 6
TI. *See also* Texas Instruments
Trap nets, 32
Traveling screens, 64
Trinectes maculatus. See Hogchoker
Troy Dam, 10
Trophic levels, 282, figures 66–67
Try trawl, 197
Tucker trawl, 17, 73, 175, 196, 222, 225, 227, 247

United States Environmental Protection Agency, 85
Utility sponsored research, 7–85

Verplanck fish hatchery, 77

Walleye (Pp): PCB levels, table 65
Water temperature, 202, figure 55
WC. *See* white catfish
West Point, 232
Westside Highway project 300, figures 74–75
White catfish: PCB levels, table 65
White Perch: biological characteristics studies, 33; culture, 77; distribution of larval stages, figure 31; effects of pollution, 36; feeding selectivity of larval stages, 134–147, figures 17–18, 20; impingement studies, 61; influence of salt front on distribution of larval

INDEX

White Perch (*Cont.*)
stages, figures 31–33; mark and recapture studies, 22–23; mercury concentration in, 276–77m tables 55–57; morphology of larvae, 146–147; movement of larval stages, 144–168; figures 21–28, 33; PCB levels in, table 65; prey utilization, table 35; reproduction studies, 36; thermal physiology studies, 36; vertical migrations, 149
Winter flounder: heavy metals in, tables 55–58
WP. *See* white perch

Yankee trawl, 77
Yellow perch: PCB concentration correlated with length and lipid content, table 61; PCB levels in, table 65
Yolk-sac larvae. *See* individual species entries
YP. See yellow perch

Zinc: analytical methods, 276; in selected animals, tables 55–58; in blue crab, 281; in blueback herring, 281, in mummichog, 281; by trophic level, figure 67

407